Stochastic Processes
Harmonizable Theory

SERIES ON MULTIVARIATE ANALYSIS

Editor: M M Rao

ISSN: 1793-1169

Series on
Multivariate Analysis
●— **Vol. 12** —●

Stochastic Processes
Harmonizable Theory

M M Rao
University of California, Riverside, USA

World Scientific

NEW JERSEY · LONDON · SINGAPORE · BEIJING · SHANGHAI · HONG KONG · TAIPEI · CHENNAI · TOKYO

Published by

World Scientific Publishing Co. Pte. Ltd.

5 Toh Tuck Link, Singapore 596224

USA office: 27 Warren Street, Suite 401-402, Hackensack, NJ 07601

UK office: 57 Shelton Street, Covent Garden, London WC2H 9HE

Library of Congress Cataloging-in-Publication Data
Names: Rao, M. M. (Malempati Madhusudana), 1929–   author.
Title: Stochastic processes : harmonizable theory / M.M. Rao, University of California, Riverside.
Description: 1st edition. | Hackensack, NJ : World Scientific Publishing Co. Pte. Ltd., [2020] |
Series: Series on multivariate analysis,
    1793-1169 ; vol. 12 | Includes bibliographical references and index.
Identifiers: LCCN 2020012114 | ISBN 9789811213656 (hardcover) | ISBN 9789811213663 (ebook) |
    ISBN 9789811213670 (ebook other)
Subjects: LCSH: Stochastic processes.
Classification: LCC QA274 .R3725 2020 | DDC 519.2/3--dc23
LC record available at https://lccn.loc.gov/2020012114

British Library Cataloguing-in-Publication Data
A catalogue record for this book is available from the British Library.

For any available supplementary material, please visit
https://www.worldscientific.com/worldscibooks/10.1142/11648#t=suppl

Printed in Singapore

To the Memories of Friends

P. R. Krishnaiah

for changing my attempts to join the Indian Air Force Academy in
DehraDun to Minnesota's Stochastic Analysis study

and

Herbert Heyer

for the advocacy and example to use abstract methods in
Stochastic Analysis

Preface

The following work constitutes the third and final part of the original planned three-volume treatment of 'Stochastic Analysis'. The first one appeared in 1995, as the second edition of an earlier publication (the 1979 version), and the second part, termed 'Inference Theory' (of stochastic processes) appeared in 2010 (as a revised, and enlarged work of the original 2000 year publication). These two volumes are published originally by Netherlands' publishers. That publication house was acquired by Springer and then reissued under their own name (2nd edition of 2nd volume). The current volume completes the originally planned coverage of the trilogy. Many new developments are covered in these volumes. I shall now indicate the contents of this volume for a brief view of the included topics presented in six chapters. Each chapter also has its contents outlined in an early paragraph.

Chapter 1 details a characterization of the general concept of (weak) harmonizability extending the classical weak stationarity of processes, initiated by Khintchine, extended by Loéve and the general form obtained by Bochner (after some initial special cases treated separately by L. Schwartz and Yu. A. Rozanov). They showed the central part of Fourier or harmonic analysis is essential in stochastic theory as well, and the first two chapters of this volume, using this framework, are devoted to analyzing the structural aspects based on the first two moments. Most of the traditional treatments concentrate on the second order (i.e., covariance) properties, assuming the processes to be centered. But the mean values are not generally constants. First, they are characterized for harmonizable classes and then the primary focus shifts to the positive definiteness of covariances which are characterized. As a key consequence, the existence of **Lévy's Brownian Motion** and its properties are established. Thereafter the general harmonizability aspect becomes the key component of our study.

From this point on, the property of V-boundedness of process, becomes primary and Bochner's characterization of it, and consequences take on a central place in the ensuing analysis. Chapters 2 and 3 consider *extensions* of weakly harmonizable processes, as well as a detailed study in much of the *(new) classes by Karhunen and Cramér*, and their dilations to larger (or super) Hilbert spaces. They have several consequences. *Also integral representations of (Gel'fand's) **local functionals** given for*

vii

*the first time, as well as a new and notable probabilistic proof of the long open and interesting **Riemann hypothesis**, with a detailed discussion, are included here.*

In these chapters some applications of the work to the harmonizable processes and other extensions including the Cramér, Karhunen classes using generalized Fourier analysis methods as well as some summability analysis of Kampé de Feriet and Frankiel (and independently of Parzen's) on nonstationary processes are studied. They are of interest in applications involving works on signal extraction problems.

When the indexing set is higher (2 or more) dimensional, then one has to study random fields so that new methods and analyses are needed. This is considered from Chapter 4 on. Here the concept of isotropy arises and we need to use properties of Bessel and related special functions. In these cases, a weakening of Lebesgue's integral is needed and fortunately it is found in a work of M. Morse and W. Transue. It is refined by D. K. Chang and the author (1986) to obtain a suitable form to use here. This is employed and with it we consider the isotropic random fields analysis which is detailed in Chapter 4. They include the corresponding (integral) representations of random fields for both harmonizable, Karhunen and Cramér classes that extend the former. Chapter 5 is devoted to some extensions of the preceding analysis if the indexing is an LCA group as well as an extension when it is a hypergroup. The work indicates many possibilities of the study when the indexing is considered on these structures that have both practical and theoretical consequences. There is also some analysis given in this chapter if the optimality criterion is not the usual squared loss but a general non-negative convex function that may be specialized for many applications.

The last chapter contains an analysis of bistochastic operators, their characterizations as well as their convergent properties. Also some accounts of isotropic analysis of random fields are included, and the area is active. It is also useful for readers to learn more about random measures with applications here. They are detailed in my earlier work in the volume entitled *Random and Vector Measures* (2012), especially in its last three chapters. This gives a good motivation as well as some new ideas on extending the present work in other directions, motivated by the work in (the long) Chapter 7 of the above book. An extended presentation of Cramér and Karhunen classes as a far-reaching generalizations of (weak) harmonizability (and weak stationarity in many respects) is also detailed in last two chapters that exhibit the extent of growth of the subjects with possible further applications and exploration for second order processes.

In the preparation of this monograph, I have been awarded a UCR Senate's Edward A. Dickson Professorship which has really assisted me in completing this volume. The material is composed using the LaTeX version with the great help of Ms. Ambika Vanchinathan for a period of over a year, from Chennai, India, and I am very grateful to her for this assistance. I am also much obliged to my friends Drs. Y. Kakihara and J. H. Park, for their great help in proofreading and effective corrections of the TeX version. Also from our Department office, I had been assisted by James Marberry, as well as Crissy Reising and Joyce Sphabmisai in some preparations of this work. I hope the material in this volume, along with its predecessors, will encourage new as well as seasoned researchers in the subject, to significantly advance this work into the future.

Riverside, CA
August 18, 2020   M. M. Rao

Contents

1

Harmonizability and Stochastic Analysis

In this chapter the well-known and powerful harmonizability technique will be introduced as a motivation for studying several intrinsic aspects of stochastic or random processes, and random fields. It is not a mere descriptive expression to say that Fourier analytic tools are among the most important techniques in such a detailed analysis of random functions and fields. First, it is natural to consider second order processes which indicate generalizations to many classes influenced by applications and develop an extended theory as well as the consequent analysis.

1.1 Second Order Processes and Stationarity

If an experiment is conducted and its outcome is observed denoting it by $\{X_t, t \in T\}$, during the time period T one may consider its moment structure, based on an underlying probability model, denoted by (Ω, Σ, P), whose first two moments, defined as $E(X_t)$ and $E(X_t^2)$, are assumed to exist and analyzed as a basic step in its analysis. Thus if $m_t = E(X_t)$ and $r(s,t) = E\big((X_s - m_s)(\overline{X_t - m_t})\big)$, where the X_t may be complex-valued, are called respectively the *means* and *covariances* of the process defined for a subset T of the reals $\mathbb{R}$, one may study the structural (or moment) properties of the process based on the mean $\{m_t, t \in T\}$ and the covariance function $\{r(s,t); s, t \in T\}$. To simplify writing, let $Y_t = X_t - m_t$, the centered process, $[E(Y_t) = 0]$ with the same covariances $r(s,t)(= E(Y_s \bar{Y}_t))$ which may be analyzed in detail without the interference of the means. But the covariance function r given by $r(s,t) = E(Y_s Y_t)[= E(Y_s \bar{Y}_t)]$ in the complex case], has the key property of *positive definiteness* in that $\sum_{i,j=1}^{n} a_i \bar{a}_j r(t_i, t_j) \geq 0$

and prompts its analysis. The important property of positive definiteness led by A. Ya. Khintchine (1934) to introduce and study the fundamental concept of *covariance stationarity* of the centered process $\{\tilde{X}_t = X_t - m_t, t \in T\}$ with two moments, implying that the covariance is invariant under translations, so that

$$r(s,t) = E\left[\left((X_s - E(X_s))\left(\overline{X}_t - E(\bar{X}_t)\right)\right)\right] = \tilde{r}(s - t). \qquad (1)$$

The continuity condition of r (and hence $\tilde{r}$) is natural for many applications. This immediately leads to a recognition of the following representation of $\tilde{r}$ (hence r) due to the fundamental Fourier integral characterization by S. Bochner (1933).

$$\tilde{r}(s - t) = \int_{\mathbb{R}} e^{i(s-t)\lambda} \, dF(\lambda), \quad s, t \in \mathbb{R}, \qquad (2)$$

where $F \colon \mathbb{R} \to \mathbb{R}^+$ is a *nondecreasing bounded function*. This result is also valid if $\tilde{r}(\cdot)$ is only a measurable function, noted by F. Riesz, after studying Bochner's theorem. In applications, the nondecreasing bounded function $F \geq 0$ is usually termed the *spectral distribution* of the process X_t. Prior to the representation (2) by Bochner, there existed a special result if $\mathbb{R}$ is replaced by the integers $\mathbb{Z}$, in which case (2) has a simpler form, discovered by G. Herglotz, so that the process $\{X_n, n \in \mathbb{Z}\}$ has the corresponding representation for $r(\cdot, \cdot)$ as,

$$r(m,n) = \tilde{r}(m - n) = \int_{-\pi}^{\pi} e^{i(m-n)\lambda} \, dF(\lambda), \quad m, n \in \mathbb{Z}, \qquad (3)$$

where $\mathbb{Z}$ denotes the integers and the function $F(\cdot)$ is, as before, nondecreasing and bounded. The continuity of the covariance is equivalent to the mean square continuity of the process, i.e., $E(|X(t) - X(s)|^2) \to 0$ as $s \to t$ in both cases.

Now the intervention of the Fourier analysis as integral in (2) and (3) above suggests a powerful motivation to weaken the hypothesis of mean continuity (i.e., the continuity of $r(\cdot, \cdot)$) as well as weakening of the hypothesis of stationarity of the process, so that $r(s, t)$ may not be $\tilde{r}(s - t)$ but merely retains the positive definiteness property. These ideas will be explored and extended in what follows.

The preceding discussion shows that the second moment, or more precisely its covariance function is the key parameter for an analysis of second order processes. But then one may ask if the first moment function $t \mapsto f(t) = E(X_t), t \in \mathbb{R}$, can be an arbitrary constant or may be adjusted by taking it to be zero after a subtraction for covariance? This point is not as simple as one may wish. Its nontriviality was shown by A. V. Balakrishnan (1959) and is instructive to consider as it will be of interest and a motivation for studying extensions of stationarity and this will be discussed in order to include some useful general classes of the basic stationary family that we started with. Such extensions are also useful for many applications. We then analyze the corresponding problem for families of nonstationarity processes below which are termed *Karhunen* and *Cramér classes* introduced by the researchers with these names.

This discussion implies and indicates that one may consider the process $t \mapsto X_t$ in $L^2(P)$ which satisfies the condition $\tilde{r}(s,t) = E(X_s \bar{X}_t)$, $(E(X_t) = 0))$ to be stationary if $\tilde{r}(s,t) = \tilde{r}(s - t)$ and apply the Khintchine–Bochner method so that $\tilde{r}$ satisfies (2) as a Fourier transform of a bounded measurable function $\tilde{F}$ and then study $\tilde{X}(t) = X(t) - m(t)$, the mean-corrected process so that $X(t)$ is the sum of a random and nonrandom (measurable) elements. Then the method used for $X(t)$ and $\tilde{X}(t)$ involves the (mean) function that also admits a Fourier transform. Its explicit form is determined first by Balakrishnan which is seen to be a nontrivial property of the Fourier representation of the first moment indicating that some extensions for higher moments are possible. First, we treat the representations of the mean and covariances, and then generalize stationarity, called "harmonizability", of the process and later discuss some other extensions and of higher moment studies.

1.2 Admissible Means for Stationary Processes and Extensions

Let $\{X_t, t \in \mathbb{R}\}$ be a covariance stationary process with mean $m \colon t \mapsto m(t) = E(X_t)$, and covariance $r(s,t)$ given by

$$r(s,t) = E(Y_s \bar{Y}_t) = \int_{\mathbb{R}} e^{i(s-t)\lambda} \, dF(\lambda), \ [= \tilde{r}(s - t)] \qquad (4)$$

where $Y_t = X_t - m(t)$, $m(t) = E(X_t)$, and $F(\cdot)$ as its "spectral distribution" to mean that it is a nonnegative and nondecreasing bounded function (left continuous by a convenient normalization), thus giving $r(\cdot)$, a Fourier integral representation. The precise form of the mean function

$m(\cdot)$ is later determined by Balakrishnan (1959), without demanding that it be a constant which was done in earlier times, as follows:

Theorem 1.2.1 *Let* $X = \{X_t, t \in \mathbb{R}\}$ *be a process with mean* $m(\cdot)$, *and covariance* $\tilde{r}(\cdot)$ *which is* $L^2(P)$-*stationary and continuous so that* $\tilde{r}(s-t) = E([X_s - m(s)]\overline{[X_t - m(t)]})$ *where* $m(t) = E(X_t)$. *Then for* $r(\cdot)$ *represented by* (4), $m(\cdot)$ *must also have a Fourier representation*

$$m(t) = \int_{\mathbb{R}} e^{it\lambda}\, d\mu(\lambda) = \int_{\mathbb{R}} e^{it\lambda} \left(\frac{d\mu}{dF}(\lambda) \right) dF(\lambda), \tag{5}$$

in order that $r(s,t) = \tilde{r}(s-t) - m(s)\bar{m}(t)$ *is a covariance in the sense that* $\mu \ll F$ *and* $\frac{d\mu}{dF}$ *lies in the unit ball of* $L^2(F)$, *where* F *is the measure representing* $r(\cdot, \cdot)$ *by* (4).

This result serves as a motivation for us to see that the mean function of the process is also important to consider along with the main parameter, namely the covariance function. [Note that both $r(\cdot, \cdot)$ and $\tilde{r}(\cdot)$ cannot be covariances of the same process.] The problem will be analyzed for a general class of processes in which harmonic (or Fourier) analysis plays a key role and subsumes the above result. The class we study here is the harmonizable family containing the stationary class that was raised in the earlier works which motivated these questions. To present a comprehensive analysis that includes random fields such as $\{X_t, t \in \mathbb{R}^n\} \subset L^2(P)$, $1 \le n < \infty$, let us replace $\mathbb{R}^n$ by G, a *locally compact abelian group*, denoted as $(G, +)$ and $\{X_t, t \in G\}$ be the new family, usually termed a *random field* (and *process* if $G = \mathbb{R}$).

The next extension from stationarity considered by Khintchine, is to study classes of nonstationary processes whose covariance functions $r(s,t) = E([X_s - m_s]\overline{[X_t - m_t]})$ admit integral representations relative to 'bimeasures' $\beta \colon \mathcal{B}(G) \times \mathcal{B}(G) \to \mathbb{C}$ so that $r(\cdot, \cdot)$ can be represented by a double integral relative to a "positive definite bimeasure" which, when concentrated on the diagonal of $G \times G$, gives back the (stationary) representation (4). Such an extension was considered (perhaps) independently by M. Loéve and Yu. A. Rozanov in early 1950's who then obtained a Fourier integral representation of the covariance, termed a harmonizable class when $G = \mathbb{R}^n$. We now describe this class in detail since in $\mathbb{R}^n$, $n > 1$, the integrals (4) are different accordingly as the new $F(\cdot, \cdot)$ has a finite *Vitali* or *Fréchet* variation based on the Lebesgue or other integrals.

These two concepts play key roles in our analysis, but there are several other "variations" which were also analyzed and the details are found in Clarkson and Adams (1933). We only employ here the result that the relation: $0 \leq \text{Fréchet}(F) \leq \text{Vitali}(F) \leq \infty$, with strict inequality between the first two variations when there is equality between the last two. An illustration will be given below. As is known from the standard measure theory, a complex measure on a σ-algebra is bounded, the (bi-) measure $\beta\colon \mathcal{B}(G) \times \mathcal{B}(G) \to \mathbb{C}$ is bounded as a complex measure and is nonnegative definite if

$$\sum_{i,j=1}^{n} \beta(A_i, A_j) a_i \bar{a}_j \geq 0, \quad n \geq 1, \ A_i \in \mathcal{B}(G).$$

Also if $r\colon G \times G \to \mathbb{C}$ is given by

$$r(s,t) = \int_{\hat{G}} \int_{\hat{G}} \langle s, x \rangle \overline{\langle t, y \rangle} \beta(dx, dy) \tag{6}$$

where $\hat{G}$ is the dual of G and $\langle s, \cdot \rangle$ is a function on $\hat{G}$, then $r(\cdot, \cdot)\colon G \times G \to \mathbb{C}$ is a bounded continuous positive definite (hence a covariance) function. This β of (6) plays a key role in extending the stationary case of (4) and hence a characterization of it is useful, but not simple, prompting Loéve to ask for its characterization. It will be sketched in the complements section later. We also discuss the problem of their means here. The simple example that $r(s,t) = f(s)\bar{f}(t)$ where $f(\cdot)$ is a continuous mapping that is not the Fourier transform of an integrable function (such functions are known to exist) shows that an $r(\cdot, \cdot)$ being positive definite (hence a covariance) does not admit a representation as (6) above. So it is nontrivial.

Any characterization of $r(\cdot, \cdot)$ of (6) depends on the integral used there since the bimeasure $\beta(\cdot, \cdot)$ need not define an absolute integral to invoke Lebesgue's integration properties. Along with this, we need to characterize the mean function as an extension of theorems above. For this one finds that bimeasure $\beta(\cdot, \cdot)$ of (6) has several variations, and the Vitali and Fréchet are the most relevant ones to use here. Following Clarkson and Adams two of these variations are given as follows and are stated for a set function:

Definition 1.2.2 *Let G be an LCA (= locally compact abelian) group and $\mathcal{B}(G)$ denote its Borel field. Let $\beta\colon \mathcal{B}(G) \times \mathcal{B}(G) \to \mathbb{C}$ be a bimeasure on the indicated product σ-algebra with values in the complex* *field* $\mathbb{C}$ *so that* $\beta(\cdot, A), \beta(A, \cdot)$ *are measures for each* $A \in \mathcal{B}(G)$. *Then* $\beta(\cdot, \cdot)$ *is said to have finite Fréchet variation on the product* σ-*algebra* $\mathcal{B}(G) \otimes \mathcal{B}(G)$ *if* $\|\beta\|(G, G)$ *is finite where, with* $a_i \in \mathbb{C}$,

$$\|\beta\|(G, G) = \sup \left\{ \left| \sum_{i,j=1}^{n} a_i \bar{a}_j \beta(A_i, A_j) \right| : A_i \in \mathcal{B}(G), |a_i| \leq 1 \right\} \quad (7)$$

for all disjoint A_i, $i = 1, \ldots, n$ $n \geq 1$. *The* $\beta(\cdot, \cdot)$ *here is of finite Vitali variation, if the* $a_i = 1$ *for all i in* (7) *and* $|\beta(A_i, A_j)|$ *replaces* $\beta(A_i, A_j)$, *thus moving the absolute value signs inside.*

It is clear that if $\beta(\cdot, \cdot)$ has finite Vitali variation, then it automatically has finite Fréchet variation but not true conversely. Further, if $\beta(\cdot, \cdot)$ has finite Vitali variation, denoted $|\beta|(G, G) < \infty$, then (3) is a familiar Lebesgue-Stieltjes integral. But in the more general case that $\|\beta\|(G, G) < \infty$ (and $|\beta|(G, G) = \infty$), the integral in (3) *cannot* be defined in the Lebesgue sense. The necessary weakening of Lebesgue's work is given by M. Morse and W. Transue (1956) (to be called the MT-integral here) and that will be recalled now. It reduces to the Lebesgue case if $|\beta|(G, G) < \infty$ but is weaker and hence more general.

Let $C_C(\Omega)$ be the standard complex continuous function space on a locally compact Hausdorff space Ω with $f \in C_C(\Omega)$ having compact support. Then each positive linear functional $I: C_C(\Omega) \to \mathbb{R}$ is uniquely representable as $I(f) = \int_\Omega f d\mu$ for a Baire measure μ on Ω, which is also denoted as $I(f) = \int_\Omega f dI$. If $I(\cdot)$ is complex valued it can be expressed as $I = I_1 - I_2 + i(I_3 - I_4)$, $I_j \geq 0$ and I is then termed a "$\mathbb{C}$-measure" by Morse and Transue who also considered bilinear forms.

If Ω_i, $i = 1, 2$ are locally compact spaces as above and $\wedge: C_C(\Omega_1) \times C_C(\Omega_1) \to \mathbb{C}$ is a bilinear form and if $\wedge(f, \cdot), \wedge(\cdot, g)$ are $\mathbb{C}$-measures then (f, g) is *Morse-Transue (or MT-)integrable* iff f is $\wedge(\cdot, g)$ and g is $\wedge(f, \cdot)$ integrable for each f and g, and both the numbers $\int_{\Omega_1} f(\omega_1) \wedge (d\omega_1, g), \int_{\Omega_2} g(\omega_2) \wedge (f, d\omega_2)$ exist and are equal. The common value is denoted by the double integral:

$$\int_{\Omega_1} \int_{\Omega_2} (f(\omega_1), g(\omega_2)) \wedge (d\omega_1, d\omega_2) = \int_{\Omega_1} f(\omega_1) \wedge (d\omega_1, g)$$

$$= \int_{\Omega_2} g(\omega_2) \wedge (f, d\omega_2). \quad (8)$$

The MT-integral is defined using the (equivalent) Daniell procedure than that of Lebesgue's and applies to [and admits] larger classes of functions. Further discussion is omitted here (but given in Chang and Rao (1986)). With this background, it is possible to present some useful second order processes more general than stationarity but utilize fully the benefits of Fourier analysis and its sharp forms. These are naturally termed *harmonizable classes*, and are defined as follows:

Definition 1.2.3 *(a)Let G be an LCA (locally compact abelian) group and $\mathcal{B}(G)$ be its Borel σ-algebra. Then $\beta\colon \mathcal{B}(G) \times \mathcal{B}(G) \to \mathbb{C}$ is called a bimeasure if $\beta(\cdot, B)$ and $\beta(A, \cdot)$ are complex measures on $\mathcal{B}(G)$, for each $A, B \in \mathcal{B}(G)$. [This does not imply that $\beta(\cdot, \cdot)$ is a $\mathbb{C}$-measure on the product σ-algebra $\mathcal{B}(G) \otimes \mathcal{B}(G)$.]*
(b)The bimeasure $\beta\colon \mathcal{B}(G) \times \mathcal{B}(G) \to \mathbb{C}$ is said to have finite Fréchet (or Vitali) variation as in Definition 1.2.2 above.
(c)A random family $X\colon G \to L^2(X)$ with mean functional $m(g) = E(X_g)$, and covariance $r(g_1, g_2) = E[(X_{g_1} - E(X_{g_1}))(X_{g_2} - E(X_{g_2}))^-]$ is weakly harmonizable if it admits a representation, for its covariance $r(\cdot, \cdot)$ with MT-integral, as:

$$r(s, t) = \int_{\hat{G}} \int_{\hat{G}} \langle s, \lambda \rangle \overline{\langle t, \lambda' \rangle} \beta(d\lambda, d\lambda') \qquad (9)$$

where the bimeasure $\beta(\cdot, \cdot)$ has finite Fréchet variation, and if $\beta(\cdot, \cdot)$ has finite Vitali variation, then it is called strongly harmonizable.
(d)A random family $X\colon G \to L^2(P)$ is termed weakly or strongly harmonizable accordingly as its covariance $r(\cdot, \cdot)$ of (9), is weakly or strongly harmonizable respectively.

It is desirable to have a usable or simpler criterion to study the structure of the harmonizable field $\{X_g, g \in G\} \subset L^2(P)$, and also a characterization of its mean functional as given in the above theorem. This will be obtained for the fields now. After characterizing the Khinchine stationarity, with S. Bochner's theorem on the Fourier integral representation of a positive definite function, a corresponding result by the same author is again available. It uses a new concept called variation (or V-) boundedness in two forms to be used for the weak and strong harmonizabilities.

Definition 1.2.4 *(a)On a locally compact abelian group G, a continuous positive definite function $r\colon G \times G \to \mathbb{C}$ is weakly V-bounded if $\|r\| < \infty$ where*

$$\|r\| = \sup \left\{ \left| \int_{G \times G} r(s,t) f(s) \bar{g}(t) d\mu(s) d\mu(t) \right| : \|\hat{f}\|_\infty \le 1, \right.$$

$$\left. \|\hat{g}\|_\infty \le 1, \ f, g \in L^1(G, \mu) \right\}. \quad (10)$$

Here μ is a Haar measure on G and $\hat{f}$ and $\hat{g}$ is the Fourier transform of f and g.

(b) *With the same notation as in (a) above let $\tilde{G} = G \times G$ be the direct product (with the product topology) which is an LCA group with its own Haar measure $\tilde{\mu}$ (see, for instance, Hewitt and Ross (1963), Chapter 6), then r is strongly V-bounded if*

$$\|r\|^\sim = \sup \left\{ \left| \int_{G \times G = \tilde{G}} r(\tilde{s}) \tilde{f}(\tilde{s}) \, d\tilde{\mu}(s) \right| : \|\tilde{f}\|_\infty \le 1 \right\} < \infty, \quad (11)$$

where $(\tilde{G}, \tilde{\mu})$ is the product (Haar) measure space, given in the above reference. Here $\tilde{f} \in L^1(\tilde{G}, \tilde{\mu})$ and is not merely the product type.

With these concepts we now can obtain characterizations of weak and strong harmonizable covariances and then of the means of the respective classes, extending Theorem 1.2.1 above given in the particular case of $G = \mathbb{R}$ and for $L^2(P)$-valued stationary processes. To help the reading of our work, it will be useful to briefly restate the Morse-Transue (or MT-) integral as applied to the present needs.

If the Ω_i, $i = 1, 2$, are locally compact and $C_C(\Omega_i)$ are complex continuous compactly based function spaces, a bilinear form $\wedge \colon C_C(\Omega_1) \times C_C(\Omega_2) \to \mathbb{C}$ defined through the Daniell–Bourbaki procedure, the desired MT-integral is obtained if $f \in C_C(\Omega_1)$ is $\wedge(\cdot, v)$ and $g \in C_C(\Omega_2)$ is $\wedge(u, \cdot)$-integrable for each $u \in C_C(\Omega_1)$, $v \in C_C(\Omega_2)$ and the (Lebesgue) integrals $\int_{\Omega_1} f(\omega_1) \wedge (d\omega_1, g(\omega_2))$ and $\int_{\Omega_2} g(\omega_2) \wedge (f(\omega_1), d\omega_2)$ exist as well as agree or equal; the common value is denoted: $\int_{\Omega_1} \int_{\Omega_2} (f, g)(\omega_1, \omega_2) \wedge (d\omega_1, d\omega_2)$. If $\beta \colon \mathcal{B}(\Omega_1) \times \mathcal{B}(\Omega_2) \to \mathbb{C}$ is a bimeasure, then a pair (f, g) is β-integrable (or MT-integrable for β), where $\mu_1 \colon A \mapsto \int_{\Omega_2} g(y) \beta(A, dy)$, $\mu_2 \colon B \mapsto \int_{\Omega_1} f(x) \beta(dx, B)$ exist as measures for $A \in \mathcal{B}(\Omega_1), B \in \mathcal{B}(\Omega_2)$ and are *equal*, so that the common value is denoted as:

$$\int_{\Omega_1} \int_{\Omega_2} (f(x), g(y)) \beta(dx, dy) = \int_{\Omega_1} f(x) \mu_1(dx) = \int_{\Omega_2} g(y) \mu_2(dy). \quad (12)$$

Since the integral is *not* absolute, a further condition is required for this work, somewhat resembling the needed restrictions in the Riemann (nonabsolute) integrals. We now include the desired versions, and work with them. These will be referred to respectively as the *MT*- and the *strict MT*-integrals. Thus a bimeasure $\mu : \mathcal{B} \times \mathcal{B} \to \mathbb{C}$, and Borel functions $f, g : B \to \mathbb{C}$ define an MT-integrable pair as above and is *strictly MT-integrable*, if for each pair $A, B \in \mathcal{B}$, one has

$$\int_A \int_B^* (f, g)(x, y) \mu(dx, dy) = \int_A f(x) \nu_1^B(dx) = \int_B g(y) \nu_2^A(dy) \quad (13)$$

where $\nu_2^A = \nu_2^f | \mathcal{B}(A); \nu_1^B = \nu_1^g | \mathcal{B}(B)$, for all $A, B \in \mathcal{B}$, $\mathcal{B}(A), \mathcal{B}(B)$ being the restricted (or trace) σ-algebra of $\mathcal{B}$ to A (and B). The strengthening ensures the validity of a dominated convergence theorem for such a bimeasure $\mu \colon \mathcal{B} \times \mathcal{B} \to \mathbb{C}$. This distinction is needed since the Fréchet variation of μ being finite, its Vitali variation can be infinite! (Cf. Exercise 1.) It is also discussed in the companion volume (2014), on Inference.

With this preparation let us present a characterization of weakly as well as strongly harmonizable covariances on an LCA group G and then describe the corresponding mean functions.

Theorem 1.2.5 *(a)A covariance function $r \colon G \times G \to \mathbb{C}$ is weakly harmonizable if and only if it is continuous and weakly V-bounded. When these conditions hold r is representable as a strict MT-integral relative to a unique bimeasure β as:*

$$r(s, t) = \int_{\hat{G}} \int_{\hat{G}}^* \langle s, x \rangle \overline{\langle t, y \rangle} \beta(dx, dy) \quad (14)$$

where $\hat{G}$ is the dual group of G.
(b)On the other hand, $r \colon G \times G \to \mathbb{C}$ is a strongly harmonizable covariance function if and only if it is strongly V-bounded and has the representation as a standard MT-integral

$$r(s, t) = \int_{\hat{G}} \int_{\hat{G}} \langle s, x \rangle \overline{\langle t, y \rangle} F(dx, dy) \quad (15)$$

for a positive definite F of finite Vitali variation which thus admits an extension to be a bounded scalar measure on the Borel σ-algebra $\mathcal{B}(\hat{G}) \otimes \mathcal{B}(\hat{G})$.

Remark 1. If $r(\cdot, \cdot)$ is a stationary covariance, then the function F of (15) concentrates on the diagonal $x = y$ of the product space so that the representation reduces to Bochner's classical result of a continuous positive definite function as desired. After establishing this result we characterize the means, extending the result of Theorem 1.2.1, and present some complements.

Proof. The sufficiency is easy. Indeed if $r(\cdot, \cdot)$ is weakly harmonizable then for any $f, g \in L^1(G, \mu)$, μ a Haar measure on G, we have

$$\left| \int_G \int_G r(s,t) f(s) \overline{g}(t) \, d\mu(s) d\mu(t) \right|$$

$$= \left| \int_G \int_G f(s) \overline{g}(t) \left[\int_{\hat{G}} \int_{\hat{G}}^{*} \langle s, x \rangle \overline{\langle t, y \rangle} \, d\beta(x, y) \right] d\mu(s) d\mu(t) \right|$$

$$= \left| \int_{\hat{G}} \int_{\hat{G}}^{*} \hat{f}(x) \bar{\hat{g}}(y) \, d\beta(x, y) \right|, \text{ by Fubini's Theorem,}$$

$$\leq \|\hat{f}\|_\infty \|\hat{g}\|_\infty |\beta|(\hat{G}, \hat{G}), \text{ using a property of the MT-integration.}$$

Since by hypothesis $|\beta|(\hat{G}, \hat{G}) < \infty$, this implies that r is weakly V-bounded, and the assertion follows in this direction.

For converse part, let r be weakly V-bounded. Consider the functional defined by the (Haar) integral:

$$l(f, g) = \int_G \int_G r(s,t) f(s) \overline{g}(t) \, d\mu(s) \, d\mu(t) \tag{16}$$

and set $T(f, g) = l(F^{-1}(\hat{f}), F^{-1}(\hat{g})) \colon C_0(G) \times C_0(G) \to \mathbb{C}$. Then T is well-defined since $F \colon f \mapsto \hat{f}, f \in L^1(G)$, being the Fourier transform, is one-to-one on $L^1(G) \to C_0(\hat{G})$. Now $T = l \circ (F^{-1}, F^{-1})$ is bilinear and the V-boundedness condition implies the boundedness of T:

$$\sup\{|T(\hat{f}, \hat{g})| : \|\hat{f}\|_\infty \leq 1, \|\hat{g}\|_\infty \leq 1\} = \|r\| < \infty. \tag{17}$$

Thus T is a bounded bilinear functional on $L^1(\hat{G}) \times L^1(\hat{G})$ and has a bound preserving extension to $C_0(\hat{G}) \times C_0(\hat{G})$. Then by the multidimensional extension of Riesz's representation theorem on $C_0(\hat{G})$, obtained by Dobrakov (1989), there exists a unique bounded bimeasure ν which is positive definite and is of finite Fréchet variation so that the following strict MT-integration holds:

$$T(\hat{f}, \hat{g}) = \int_{\hat{G}} \int_{\hat{G}}^{*} \hat{f}(x)\overline{\hat{g}(y)}\nu(dx, dy). \tag{18}$$

[For more details of bimeasure integrals, see also Chang and Rao (1986), p. 21 on these properties.] From (16) and (18) one has $l(f, g) = T(\hat{f}, \hat{g})$ which implies

$$\int_{G} \int_{G} r(s,t)f(s)\overline{g}(t)\, d\mu(s)\, d\mu(t)$$

$$= l(f, g)$$

$$= \int_{\hat{G}} \int_{\hat{G}}^{*} \left[\int_{G} \int_{G} \langle x, s\rangle f(s)\overline{\langle y, t\rangle g(t)}\mu(ds)\mu(dt) \right] \nu(dx\, dy). \tag{19}$$

It follows from (19) and Fubini's theorem that

$$\int_{G} \int_{G} \left[r(s,t) - \int_{\hat{G}} \int_{\hat{G}}^{*} \langle x, s\rangle \overline{\langle y, t\rangle}\nu(dx, dy) \right] f(s)\overline{g}(t)d\mu(s)d\mu(t) = 0.$$

Since $f, g \in L^1(G)$ are arbitrary, we get $[\ ] = 0$ a.e. $[\mu \otimes \mu]$ and by continuity of r, everywhere. This implies (14).

For the strongly harmonizable case, the direct product $\tilde{G} = G \times G$ will be an LCA group with its Haar measure $\tilde{\mu} = \mu \otimes \mu$ which is the usual product measure (cf. Hewitt and Ross (1963), e.g. (23.33)). Then if $x = (s, t) \in G \times G = \tilde{G}$, and $r : \tilde{G} \to \mathbb{C}$ is seen to be a strongly V-bounded mapping, so $r(\cdot)$ strongly V-bounded, gives:

$$\left| \int_{\tilde{G}} r(x)f(x)\, d\tilde{\mu}(x) \right| \leq K\|f\|_{\infty}$$

for all μ-integrable f with $\tilde{f}(x) = f(x)g(t), x = (s, t)$. Then by the classical Bochner representation theorem again one has

$$r(x) = \int_{\tilde{G}} \langle x, u\rangle dF(u)$$

for an F of bounded variation. With $\tilde{G} = G \times G$ and $\tilde{\mu} = \mu \otimes \mu$ it can be expressed (by the uniqueness in that representation) as:

$$r(x) = r(s, t) = \left[\int_{\tilde{G}} \langle x, y\rangle dF(y) = \right] \int_{\hat{G}} \int_{\hat{G}} \langle s, \lambda\rangle \overline{\langle t, \lambda'\rangle} dF(\lambda, \lambda'),$$

where now F is of bounded (Vitali) variation on $\hat{G} \times \hat{G}$. But since $r(\cdot, \cdot)$ is positive definite, it follows from this representation that F must be positive definite. This gives (15), and hence the proof of the theorem is complete. The essential distinction between weak and strong harmonizabilities is clarified. □

As a consequence of the above result, it is possible to present an *integral representation* of a (weakly or strongly) harmonizable process $\{X_t, t \in G\}$ on an LCA group G, generalizing the stationary case of the Bochner-Khintchine (classical) theorem as follows:

Theorem 1.2.6 *Let* $\{X_t, t \in G\} \subset L_0^2(\mathbb{P})$ *be a (mean) continuous strongly or weakly V-bounded process where G is an LCA group. Then it admits a unique (stochastic) integral representation as:*

$$X_t = \int_{\hat{G}} \langle t, u \rangle \, dZ(u), \quad t \in G, \tag{20}$$

relative to a vector measure $Z \colon \mathcal{B}(\hat{G}) \to L_0^2(P)$ whose spectral bimeasure $\beta \colon (A, B) \to E(Z(A)\overline{Z(B)})$ is weakly or strongly V-bounded.

Proof. Let $r(s, t) = E(X_s \overline{X}_t)$ be the covariance of the centered process X which is V-bounded with continuous paths on G. Then for $f, g \colon G \to \mathbb{C}$ bounded Borel functions with compact supports, consider

$$\int_G \int_G r(s,t)f(s)\overline{g}(t) \, d\mu(s) \, d\mu(t)$$

$$= E\left[\int_G X_s f(s) \, d\mu(s) \overline{\int_G X_t g(t) \, d\mu(t)} \right]$$

where μ is the Haar measure on G. Now using the V-boundedness hypothesis on X and taking $f = g$, the right side becomes:

$$E\left[\left| \int_G X_s f(s) \, d\mu(s) \right|^2 \right] = \int_G \int_G r(s,t)f(s)\overline{f}(t) \, d\mu(s) \, d\mu(t)$$

$$\leq K\|\hat{f}\|_\infty^2 \text{ using (10)},$$

for some $0 < K < \infty$. Hence

$$\left\| \int_G X_s f(s) \, d\mu(s) \right\|_{2,P} \leq \sqrt{K}\|\hat{f}\|_\infty.$$

Since a bounded part in $L^2(P)$ is relatively weakly compact, one has the set $\{\int_G X_s f(s) \, d\mu(s), f \in L^1(G), \|\hat{f}\|_\infty \leq 1\} \subset L^2(P)$ to be bounded.

Consider the mapping $T\colon C_0(\hat{G}) \to L^2(G)$ given by

$$T(\hat{f}) = l(f) = \int_G X_s f(s)\, d\mu(s) \tag{21}$$

which is bounded and linear, is weakly compact. So by the Riesz representation theorem, there exists a $Z\colon \mathcal{B}(\hat{G}) \to L_0^2(P)$, such that $E(Z(A)) = 0$ and represents T as

$$T(\hat{f}) = \int_{\hat{G}} \hat{f}(u)\, dZ(u) = \int_{\hat{G}} \int_G f(t)\langle t, u\rangle d\mu(t)\, dZ(u). \tag{22}$$

This may be written in detail as:

$$\int_G X_t f(t)\, d\mu(t) = T(\hat{f}) = \int_G \int_{\hat{G}} f(t)\langle t, u\rangle d\mu(t)\, dZ(u)$$

where the left is the Bochner and the right the Dunford-Schwartz (or Vector) integral. Hence by Fubini's theorem (in vector form)

$$\int_G \left[X_t - \int_{\hat{G}} \langle t, u\rangle\, dZ(u) \right] f(t)\, d\mu(t) = 0. \tag{23}$$

It follows that, from the arbitrariness of $f(\cdot)$, that $[\,] = 0$ holds a.e. and by the continuity of X_t, the μ-null set must be empty.

The same argument applies (with slight simplifications) to the strongly harmonizable case. The last statement on the spectral bimeasure of $Z(\cdot)$ is immediate. $\square$

So far the mean function is assumed to be zero, i.e. the harmonizable processes are centered. We now consider the possible mean function of such processes and establish a generalized form (i.e., subsuming the result of Theorem 1.2.1) of the means for such X_t.

Theorem 1.2.7 *Let $X = \{X_t, t \in G\}$ be a weakly harmonizable process with a continuous covariance function r having F as its spectral bimeasure. Then a mapping $m\colon G \to \mathbb{C}$ is the mean value of the random field X on the space (Ω, Σ, P) with r as its covariance if and only if there is a unique $g \in L^2(F)$ such that $(i)\,(g, g)_F \leq 1$ and $(ii)\, m(t) = \int_{\hat{G}} \langle t, u\rangle\, d\mu(u)$ where μ is a scalar measure given by*

$$\mu\colon A \mapsto \int_G^* \overline{g}(v) F(A, dv), \quad A \in \mathcal{B}(G). \tag{24}$$

Here $r(\cdot, \cdot)$ is the covariance of the X, and $F(\cdot, \cdot)$ is its (positive definite) spectral bimeasure and the integral with a star $(*)$ is the strict MT-integral.

Proof. First, observe that $L^2(F)$ is a semi-inner product space since F is a positive definite bimeasure. It is also true that the space is complete, and separable when G is separable. This fact is not entirely simple and can be tried as a problem.

Now using a few properties of the (MT)-integral, one can verify that the space $L^2(F)$ is complete for the inner product defined earlier using the (strict) MT-integral. The point here is that the MT-integral (and its strict version) is essential in this (generalized) analysis. Thus if $r(s, t) = E(X_s \overline{X}_t)$, which is now harmonizable, it admits an integral (vector or Dunford-Schwartz type) representation:

$$X_t = \int_{\hat{G}} \langle t, u \rangle \, dZ(u), \quad t \in G, \tag{25}$$

by (20) and $Z(A) \neq 0$ since $E(X_t) \neq 0$ by assumption, $A \in \mathcal{B}(G)$ and $m_t = E(X_t)$. Hence by a classical theorem of Hille (as in Dunford-Schwartz (1958), IV.10.8) one has from (25):

$$m_t = E(X_t) = \int_{\hat{G}} \langle t, u \rangle E(dZ(u)) = \int_{\hat{G}} \langle t, u \rangle \mu(du), \tag{26}$$

where $\mu(du) = E(Z(du))$, so μ is a scalar measure on $\mathcal{B}(\hat{G})$ and

$$m_t = l(f_t) = \int_G f_t(u) \mu(du), \quad f_t \in L^2(F), \tag{26'}$$

is a well-defined linear functional for some g of $l(f_t) = (f_t, g)$, and $(g, g)_F = \|l\| \leq 1$.

We now turn to the converse. Suppose $m = \int_{\hat{G}} f_t \, d\mu$ where $\mu(A) = \int_{\hat{G}} \bar{g}(v) F(A, dv)$, the $F(\cdot, \cdot)$ being a bimeasure determined by $Z(\cdot)$. It is to be verified that m_t is the mean of a harmonizable field X_t so that if $Y_t = X_t - m_t$, then Y_t is centered and $r(s, t) = E[(X_s - m_s)\overline{(X_t - m_t)}] = E(X_t \overline{X}_t) - m_s \overline{m}_t$. For this, it is enough to verify that $\tilde{r}$ is positive definite and continuous. Continuity being clear, it suffices to verify the positive definiteness.

Let $f(u) = \sum_{i=1}^n a_i \langle s_i, u \rangle$, $a_i \in \mathbb{C}$ and consider the strict MT-integrals:

$$\sum_{i,j=1}^{n} a_i \bar{a}_j r(s_i, s_j) = \int_{\hat{G}} \int_{\hat{G}}^{*} \sum_{i,j=1}^{n} a_i \bar{a}_j \langle s_i, u \rangle \overline{\langle s_j, v \rangle} F(du, dv)$$

$$- \left| \int_{\hat{G}} \sum_{i=1}^{n} a_i \langle s_i, u \rangle \, d\mu(u) \right|^2$$

$$= \int_{\hat{G}} \int_{\hat{G}}^{*} f(u) \bar{f}(u') F(du, du') - \left| \int_{\hat{G}} f(u) \, d\mu(u) \right|^2$$

$$= (f, f)_F - \left| \int_{\hat{G}} f(u) \int_{\hat{G}}^{*} \bar{g}(u') F(du, du') \right|^2$$

$$= (f, f)_F - \left| \int_{\hat{G}} \int_{\hat{G}}^{*} f(u) \bar{g}(u') F(du, du') \right|^2 \geq 0,$$

since the last integral is dominated by $(f, f)_F (g, g)_F \leq (f, f)_F$ by the CBS inequality and the fact that $\|g\|_F \leq 1$. So m_t is the mean of X_t. $\square$

In the strongly harmonizable case, the MT-integral is already strict and so the 'star' on the integral can be omitted. This may be stated for a convenient reference as:

Corollary 1.2.8 *Let $\{X_t, t \in G\}$ be a strongly harmonizable process with covariance $r(\cdot, \cdot)$. Then $m: G \to \mathbb{C}$ is its mean function, so that $m_t = E(X_t)$, if and only if there is a function $g \in L^2(F)$ such that (i) $(g, g)_F \leq 1$ and (ii) $m(t) = \int_G (t, u) \, d\mu_g(u)$, where $\mu_g: A \to \int_G \bar{g}(v) F(A, dv)$, $A \in \mathcal{B}(G)$, the integral being the (ordinary not strict) MT-integral.*

Remark 2. 1. We now record some consequences. If the process is second order stationary so that $r(s, t) = r(s - t)$, then the 'spectral function' F concentrates on the diagonal so that $\mu_g(\cdot)$ of the above corollary satisfies $\mu_g(A) = \int_A \bar{g}(u) \, dH(u)$ where $H(\cdot)$ is the (bounded) spectral measure of r. So $\frac{d\mu_g}{dH} = g \in L^2(H)$ and $\int_{\hat{G}} |g|^2 dH \leq 1$.

In the stationary case, one has

$$r(s, t) = \tilde{r}(s - t) = E[X_s \overline{X_t}] - m_s \bar{m}_t$$

so that (with $s = t$), $|m_t|^2 = E[|X_t|^2] - \tilde{r}(0)$ is a constant. Thus one has $|m_t|^2 = E[|X_t|^2] - \tilde{r}(0)$ and $m_t = a(t, m_0)$ for some $m_0 \in G$, and $t \in \mathbb{R}$.

2. If r and $\tilde{r}$ are a pair of general covariances (not necessarily even harmonizable), the fact that $(s, t) \mapsto m_s \bar{m}_t$ is always positive definite implies that $\tilde{r} - r$ is positive definite, and hence by the well-known reproducing kernel Hilbert space theory due to Aronszajn, the resulting Hilbert spaces $\mathcal{H}_r, \mathcal{H}_{\tilde{r}}$ also often called Aronszajn spaces (see below) satisfy the inclusion $\mathcal{H}_{\tilde{r}} \subset \mathcal{H}_r$ corresponding to the positive definiteness property of $\tilde{r} - r$, and hence $\| \cdot \|_{\tilde{r}} \geq \| \cdot \|_r$, from which one obtains $\|m\|_r \leq \|m\|_{\tilde{r}} \leq 1$, if m_t is the mean of a harmonizable process with covariance r. This fact holds also for some non harmonizable process, e.g. for a pair of Brownian motion processes which are not necessarily harmonizable. That was observed by Ylvisaker (1961), where $G = [0, 1]$ with addition (mod 1) as group operation and he then noted that $m \colon G \to \mathbb{R}$ is a possible mean value of a process with r as covariance if and only if $\int_G \left| \frac{dm_t}{dt} \right|^2 \leq 1$. This does not yet give the result of Theorem 1.2.8, since the corresponding characterization of $\mathcal{H}_r$ for the processes is not at hand.

3. A function $g \colon G \to \mathbb{R}$ can be positive definite and nonmeasurable. For instance let $f \colon \mathbb{R} \to \mathbb{R}$ be a nonmeasurable function such that $f(x + y) = f(x) + f(y)$ and $g(x) = e^{-tf(t)}$ for $t > 0$, so that f is conditionally positive definite. Then 'g' is positive definite and even nonmeasurable. Existence of such an f is well-known. (See Exercise 3 for more detail.)

4. In place of Fourier transform in the above work, one may consider other one-one mappings such as Mellin and a certain "L-transform" discussed by Kawata (1965), and the work can be extended for such classes also. Some analogs for unimodular groups will be considered later in this book.

If $r \colon S \times S \to \mathbb{C}$ is a positive definite function (or a covariance) then the *Aronszajn space*, denoted $\mathcal{H}_r$, defined on S of complex functions, is a Hilbert space satisfying the conditions:

1. $r(\cdot, t) \in \mathcal{H}_r, t \in S$, and
2. $f \in \mathcal{H}_r \Rightarrow (f, r(\cdot, t)) = f(t)$.

An easy consequence of the above definition is the fact that $m \colon S \to \mathbb{C}$, $r(s, t) = m(s)\bar{m}(t)$, is a covariance and this $\mathcal{H}_r$ is the (vector) space of all multiples of m, with norm $\|m\|_{\mathcal{H}_r} = 1$. The basic property established by Aronszajn (1950), useful in the analysis of this space, is given by:

Theorem 1.2.9 *If $r_i \colon S \times S \to \mathbb{C}$, $i = 1, 2$ are covariances such that r_1 dominates r_2 (i.e., $r_1 - r_2$ is positive definite), we have then the inclusion $\mathcal{H}_{r_1} \supset \mathcal{H}_{r_2}$ in that $\|f\|_{\mathcal{H}_{r_2}} \leq \|f\|_{\mathcal{H}_{r_1}}$, $f \in \mathcal{H}_{r_2}$ arbitrary.*

This result based on Aronszajn's theory (cf. Aronszajn (1950)) implies the following extension of corollary 1.2.8, due to Ylvisaker (1959), and is of interest here.

Theorem 1.2.10 *Let $\{X_t, t \in T \subset L^2(P)\}$ be a process with the mixed product moment $K : T \times T \to \mathbb{C}$, defined by $K(s,t) = E(X_s \bar{X}_t)$, and $m \colon T \to \mathbb{C}$ be a function. Then $(s,t) \to K(s,t) - m(s)\bar{m}(t)$ is the covariance function of the X_t-process, if and only if $m \in \mathcal{H}_K$ of unit norm, i.e., belongs to the unit ball of $\mathcal{H}_K$.*

These two results indicate adjuncts to the theory considered above, and will be of interest in extension of the earlier analysis. We omit the details here, since this aspect is not related to harmonizable analysis of the process.

1.3 Positive Definiteness as a Basis of Stochastic Analysis

In a substantial part of stochastic analysis, the concept and properties of positive definiteness play fundamental roles. This has been basic both for stationary as well as harmonizable analyses. Here we shall sketch some of the key roles of it for the concept of Brownian motion.

An English botanist named Robert Brown, observed in 1826 that a particle suspended in fluid makes irregular spontaneous movements caused by the molecular impacts on it by those of the medium. There appeared to be negligible bonds between the particles and the surrounding medium. The impacts on the particle ω are irregular and the displacement $X^\omega(t+s) - X^\omega(t)$ is the sum of a large number of (centered) impacted displacements. [A similar phenomenon was noticed later around 1900 by a French researcher named L. Bachelier termed 'speculation theory', about stock market movements.] Thus the $X(t)$-process has independent (centered) increments, and the already available central limit theorem suggests that the random element $X(t+s) - X(t), s > 0$, is (approximately) *Gaussian* (or *Laplacian* as termed in France). This conclusion is also due, in 1905, to A. Einstein in Germany (and to M. von Smoluchawski in Poland). The result was proved mathematically rigorously by N. Wiener in 1923 in the U.S., and so it is also termed a Wiener process. Thus we have a process $\{X_t, t \geq 0\}$ which is termed Brownian motion, or the Wiener process, such that $(X_{t_k} - X_{t_{k-1}})$ is Gaussian with mean zero and variance $c(t_k - t_{k-1}) > 0$ for any $0 < t_1 < \cdots < t_n$ and has stationary independent increments. This process formed a powerful motivation for an extension to second order processes whose covariances lead to weakly stationary classes as a first generalization and then to (weakly) harmonizable classes as its natural extension.

The crucial fact in this extension is the *positive definiteness property* of the covariance $r(s, t) = E(X_s \bar{X}_t)$, where $s, t \in T(= G)$ which has a semi-group structure under addition. This process motivates a study of the more general harmonizable classes where the index G is allowed to be a locally compact group.

Recall that $r : T \times T \to \mathbb{C}$ is *positive definite* if for $t_i \in T, a_i \in \mathbb{C}$

$$\sum_{i,j=1}^{n} r(t_i, t_j) a_i \bar{a}_j \geq 0, \quad n \geq 1.$$

However, there is no direct or simple method of verifying this property. An integral representation of $r(\cdot, \cdot)$ is obtained by various mathematicians including S. Bochner, H. Cramér, P. Lévy, F. Riesz and others. This is really a characterization of functions that are Fourier transforms, or those that admit just vector integral representations using V-boundedness again. This is restated, more generally, as follows and used *to characterize vector Fourier transforms* of interest here.

Definition 1.3.1 *If G is an LCA group and $\mathfrak{X}$ a Banach space, a mapping $f : G \to \mathfrak{X}$ is V-bounded if (i) $f(G) \subset \mathfrak{X}$ is bounded, (ii) f is strongly measurable relative to the Borel σ-algebras of G and $\mathfrak{X}$, (iii) the following set $\bar{W}$ in $\mathfrak{X}$ is weakly compact, where*

$$W = \left\{ \int_G f(t) g(t)\, dt : \|\hat{g}\|_\infty \leq 1, \ g \in L^1(G, \mu) \right\} \subset \mathfrak{X}. \tag{27}$$

[Here 'dt' denotes a Haar measure on G.]

Theorem 1.3.2 *Let G be an LCA group and $\mathfrak{X}$ a reflexive Banach space. A mapping $X : G \to \mathfrak{X}$ is a Fourier transform of a vector measure $\nu : \mathcal{B}(\hat{G}) \to \mathfrak{X}$ if and only if X is V-bounded and weakly continuous. Thus*

$$X_g = \int_{\hat{G}} \langle g, g' \rangle\, dZ(g'), \qquad g \in G \tag{28}$$

where $Z(\cdot)$ is the unique vector measure on $(\hat{G}, \mathcal{B}(\hat{G}))$ so $\{X_g, g \in G\}$ is a weakly harmonizable random field, valued in $\mathfrak{X}(= L^2(P)$ is possible).

Proof. Let $X: G \to \mathfrak{X}$ be V-bounded and weakly continuous. If $L^1(G)$ denotes the (Lebesgue) space of integrable functions relative to the Haar measure, denoted dt, and $f \in L^1(G)$, let $\hat{f}(\cdot)$ be its Fourier transform, then $T_X: \hat{f} \to \int_G X_t f(t)\, dt$ is well-defined and linear on $L^1(G)$ to $\mathfrak{X}$, satisfying, for some constant $K > 0$:

$$\|T_X(\hat{f})\|_{\mathfrak{X}} \le K\|\hat{f}\|_{\infty}, \qquad \text{by (27).} \tag{29}$$

Also as a consequence of the Riemann–Lebesgue lemma, $T_X(\hat{f})$ vanishes at infinity, i.e., $T_X(\hat{f}) \in C_0(G, \mathfrak{X})$ and the V-boundedness hypothesis allows us to invoke the (vector) Riesz representation theorem so that for a unique vector measure $\nu: \mathcal{B}(\hat{G}) \to \mathfrak{X}$ we have

$$T_f(\hat{g}) = \int_G \hat{g}(s)\, d\nu(s), \quad g \in L^1(G). \tag{30}$$

This may be rewritten on using the fact that $T_f(\hat{g})$ is also an integral of f, g, i.e., $T_f(\hat{g}) = \int_G f(t)g(t)\, dt$, and one can invoke the classical Fubini theorem for the scalar (signed) measures $x^* \circ \nu, x^* \in \mathfrak{X}^*$, to have:

$$\int_G x^*(f(t))g(t)\, dt = x^*(T_f(\hat{g})) = \int_{\hat{G}} \hat{g}(s)\, d(x^* \circ \nu)(s). \tag{31}$$

The left integral can be simplified as:

$$x^*\left(\int_G f(t)g(t)\, dt\right) = x^*\left[\int_G g(t)\left(\int_{\hat{G}} d\hat{\nu}(s)\right)dt\right]. \tag{32}$$

This implies that $\hat{f} = \hat{\nu}$, a.e., so that the V-bounded f is the Fourier transform of $\nu: \mathcal{B}(\hat{G}) \to \mathfrak{X}$.

Conversely, if f is the Fourier transform of ν, then the weak compactness of the set (27) is to be established, since the function is clearly bounded, $(\|f(t)\| \le \|\nu\|(G) < \infty)$. The weak compactness of W of (27) is also clear since it is bounded by $\|\hat{g}\|_{\infty} \le 1$ and ν has finite variation, and that in a reflexive space bounded sets have weakly compact closures. This completes the argument. $\square$

Remark 3. The boundedness hypothesis in the above result is used in concluding the compactness of the closure of W in (27). However, the result holds for nonreflexive X also, with the rest of the hypothesis on using some abstract analysis which invokes the Eberlein–Šmulian Theorem. The detail is sketched in the author's book (Rao (2004), Measure Theory and Integration, 2nd Ed. Exercise 1, on p. 558), and some related extensions were also discussed there for interested readers.

We shall return later to consider some extensions and applications of the general analysis touched on above. *It is essential to discuss the positive definiteness property of the second order random fields, starting with the Brownian motion when the indexing is* $\mathbb{R}^n, n > 1,$ *as it depends on some new ideas.*

From the earlier discussion of this section, a Brownian motion, BM, $\{X_t, t \geq 0\}$ is a *centered* Gaussian process with stationary independent increments so that from the elementary identity $(X_s - X_t)^2 = X_s^2 + X_t^2 - 2X_s X_t$, the cov $(X_s, X_t) = E(X_s X_t)$ implies

$$C_X(s,t) = E(X_s \bar{X}_t) = \frac{1}{2}[E(|X_s|^2) + E(|X_t|^2) - E(|X_s - X_t|^2)]$$

and hence for the BM the covariance must satisfy:

$$\left(\min(s,t) = \right) C_X(s,t) = \frac{1}{2}[\|s\| + \|t\| - \|s - t\|], \qquad s, t \in \mathbb{R}^+. \quad (33)$$

Here $\| \cdot \|$ stands for the norm of $\mathbb{R}$.

Since we consider the processes $\{X_t, t \in T\}$ with $T \subset \mathbb{R}^n, n \geq 1$, and even $T = G$ an LCA group in our study, it becomes essential to know if $C_X(\cdot, \cdot)$ of (33) defines a covariance, i.e., a *positive definite function* on $T \times T$. This is not easy and P. Lévy was able to construct a Gaussian process $\{X_t, t \in \mathbb{R}^2\}$ and show that its covariance is given by (33) so that *it is positive definite*, but an independent proof of the latter property could not be found by him! We present a general solution.

Definition 1.3.3 *Let* $(T, \| \cdot \|)$ *be a normed vector space and let* $X = \{X_t, t \in T\}$ *be a real Gaussian process on* (Ω, Σ, P), *a probability space. The process is called a Lévy–Brownian motion with* T *as its parameter set if (i)* $X_0 = 0$, *a.e., (ii)* $E(X_t) = 0$, $t \in T$, *and for* $t, t' \in T$, $E\big((X_t - X_{t'})^2\big) = \|t - t'\|$.

Clearly if $T = \mathbb{R}$ this is the BM considered above, and the case that T is of higher dimension is not an easy extension and new ideas are needed. We present two (different) methods, if $T = \mathbb{R}^n$, and later if T is more general.

Thus let $T = \mathbb{R}^n, n > 1$. Consider a family of complex functions on $\mathbb{R}^n$, as $\phi_s \colon \mathbb{R}^n \to \mathbb{C}$, $n > 1$ defined by

$$\phi_s(x) = C_n |x|^{-(n+1)/2}(e^{i(s,x)} - 1), \ s \in \mathbb{R}^n, \ (s,x) = \sum_{i=1}^{n} s_i x_i. \quad (34)$$

Observe that the complex function $\phi_s \in L^2(\mathbb{R}^n, dx)$, the classical Lebesgue space. The class $\{\phi_s, s \in \mathbb{R}^n\}$ forms a dense set in the Lebesgue space $L^2(\mathbb{R}^n, dx)$, and $\phi_0(x) = 0$. Note that for suitable C_n

$$\int_{\mathbb{R}^n} |\phi_s - \phi_t|^2(x) dx = C_n^2 \int_{\mathbb{R}^n} |e^{i(s,x)} - e^{i(t,x)}|^2 \frac{dx}{|x|^{(n+1)}} = |s-t|. \quad (35)$$

A computation by evaluating this integral shows that setting $X_t = \phi_t$, $t \in \mathbb{R}^n$, is a process with $X_0 = 0$, and for $s, t \in \mathbb{R}^n$, one has

$$E[(X_s - X_t)^2] = E[|\phi_s - \phi_t|^2]$$
$$= \int_{\mathbb{R}^d} |\phi_s(x) - \phi_t(x)|^2 dx = |s - t|.$$

But the left side is the real process which on expansion gives $|s| + |t| - 2E(X_s X_t)$ so that the covariance is:

$$\mathrm{cov}\,(X_s, X_t) = \frac{1}{2}[|s| + |t| - |s - t|].$$

This gives (33) which thus depends on several tricks and it is adopted from Neveu (1968). We state this result as:

Proposition 1.3.4 *For each integer $n \geq 1$ there exists a real Brownian motion $\{X_t, t \in \mathbb{R}^n\}$ such that $X_0 = 0$ a.e., and*

$$E[(X_s - X_t)^2] = |s - t| \left(= \sqrt{\sum_{i=1}^{n}(s_i - t_i)^2}\right), \ and$$

$$E(X_s X_t) = \frac{1}{2}[|s| + |t| - |s - t|]. \quad (36)$$

A different argument is needed if $\mathbb{R}^n$ is replaced by an infinite-dimensional Hilbert space. This will now be given as it is also important and useful. For some concrete work below, it is useful to know that an abstract Hilbert space can isometrically be realized as a subspace of the familiar $L^2(P)$ on a suitable probability space (Ω, Σ, P). The desired statement is as follows:

Theorem 1.3.5 *If $\mathcal{H}$ is a nontrivial Hilbert space, then it is always realizable isometrically and isomorphically as a closed subspace of $L^2(P)$ on a probability space (Ω, Σ, P), even with P as Gaussian.*

Proof. We sketch the (well-known) argument here for completeness. Let $\{e_i, i \in I\} \subset \mathcal{H}$ be a complete orthonormal set which exists. Note that such a set is obtained and is maximal (i.e., cannot be enlarged) by using Zorn's lemma. Thus if $\Omega_i = \mathbb{R}, \Sigma_i = \mathcal{B}$ (Borel σ-algebra of $\mathbb{R}$) and $P_i: A \to \frac{1}{\sqrt{2\pi}} \int_A e^{\frac{-t^2}{2}} dt, A \in \Sigma_i$, let $(\Omega, \Sigma, P) = \otimes_{i \in I}(\Omega_i, \Sigma_i, P_i)$. If $\Pi_i: \Omega \to \Omega_i$ is the coordinate projection, then $P \circ \Pi_i^{-1} = P_i$ and $(\Omega, \Sigma, P) = \otimes_{i \in I}(\Omega_i, \Sigma_i, P_i)$, gives the desired probability space, and $P \circ \Pi_i^{-1} = P_i$ and if $\{\phi_i, i \in I\}$ is a complete orthonormal set in $\mathcal{H}$, let $\{X_i, i \in I\} \subset L^2(\Omega, \Sigma, P) = L^2(P)$ be an orthonormal set so that $\tau(\phi_i) = X_i, i \in I$. Then $(\tau\phi_i, \tau\phi_j) = (X_i, X_j)_{L^2(P)} = (\phi_i, \phi_j)_I = \delta_{ij}$ and an isomorphism between the spaces $\mathcal{H}$ and $L^2(P)$ is obtained. Because of the maximality (and Zorn's lemma), the space $\mathcal{H}$ and $L^2(P)$ are isomorphic (or $\mathcal{H}$ is identifiable with a closed subspace of $L^2(P)$), so that the abstract space $\mathcal{H}$ is realizable as an $L^2(P)$. □

A comparison of this result with the preceding representation in which the indexing of the process is restricted and thereby demanding the corresponding class of the (Gaussian) process—here a Brownian motion class—is found to be more intricate than expected and we need to consider the processes (or random fields) with special (multiple) indexing again as emphasized by P. Lévy. It is also crucial that the covariance given by (33) is *real-valued* in addition to it is being a multidimensional domain. The detail depends on the basic works of L. Schwartz and I. J. Schoenberg on real positive definite forms which will be presented here, following the simplification by Cartier (1971).

Recall that for a standard Brownian Motion $\{X_t, t \geq 0\}$, it is seen that $E(X_t) = 0, E[(X_t - X_{t'})^2] = E(X_t^2) + E(X_{t'}^2) - 2E(X_t X_{t'})$, which thus can be written as:

$$E(X_t X_{t'}) = \frac{1}{2}[E(X_t^2) + E(X_{t'}^2) - E(X_t - X_{t'})^2]$$
$$= \frac{1}{2}\big[|t| + |t'| - |t - t'|\big] \ (= \text{cov}_X(t, t')), \qquad (37)$$

for the classical Brownian motion with the time $T = \mathbb{R}^+$, and the product moment gives a positive definite function. If $T = \mathbb{R}_+^n$, the positive orthant, the fact that (37) is positive definite is not so easy. P. Lévy has shown this to be true for $n = 2$ and the problem for $n > 2$ was non-trivial. It was solved independently by L. Schwartz and I. J. Schoenberg and later more abstractly simplified by P. Cartier (1971) whose proof we now present. It is valid for T as a Hilbert space and is definitely sharper than the preceding result.

Theorem 1.3.6 *For each real Hilbert space T, there exists a Lévy-Brownian motion $X = \{X_t, t \in T\}$, indexed by T, on a probability space $(\Omega, \Sigma, P), E(X_t) = 0$ and the mapping $(t, t') \mapsto E(X_t X_{t'})$ satisfies (37), as its covariance expression.*

Proof. By the preceding theorem an abstract Hilbert space can be realized isomorphically as $L^2(\Omega, \Sigma, P)$ on a probability space where P is a Gaussian (probability) measure. Thus for the present proof, it is enough to show the existence of a function $f \colon T \to \mathcal{H}$, and a Hilbert space $\mathcal{H}$, such that $f(0) = 0, \|f(t) - f(t')\|_{\mathcal{H}}^2 = \|t - t'\|_T$ since T and $\mathcal{H}$ are different spaces. But such an f exists iff $(t, t') \overset{g}{\mapsto} \|t - t'\|_T$ is negative definite in that for any reals $(a_1, \ldots, a_n), \sum_{i=1}^{n} a_i = 0$, one has

$$g(t_i, t_i) = 0, \quad \sum_{i,j=1}^{n} g(t_i, t_j) a_i a_j \leq 0, \quad g(-t_i, -t_j) = g(t_i, t_j).$$

This ensures that $C(t, t') = E(X_t X_{t'})$ of (37) is positive definite. Indeed,

$$\sum_{i,j=2}^{n} C(t_i, t_j) a_i a_j = -\frac{1}{2} \sum_{i,j=1}^{n} g(t_i, t_j) a_i a_j \geq 0, \qquad (38)$$

where $t_1 = 0$ and $\sum_{i=1}^{n} a_i = 0$. Thus an inner product on T can be defined with $C(\cdot, \cdot)$, and let $\mathcal{H}_2$ be the completion of T using $C(\cdot, \cdot)$. Define a mapping $h \colon T \to \mathcal{H}_2$ with $h_1(0) = 0, (h(t), h(t')) = C(t, t')$, so that $g(t, t') = \|h(t) - h(t')\|_{\mathcal{H}_2}^2 = \|t - t'\|_T$. Then $g(\cdot, \cdot)$ is (termed) conditionally negative definite because:

$$\sum_{i,j=1}^{n} g(t_i, t_j) a_i a_j = \sum_{i,j=1}^{n} (h(t_i) - h(t_j), h(t_i) - h(t_j)) a_i a_j$$

$$= 2 \sum_{i,j=1}^{n} h(t_i)^2 a_i a_j - 2 \sum_{i,j=1}^{n} h(t_i) h(t_j) a_i a_j \leq 0$$

since $\sum_{i=1}^{n} a_i = 0$ making the first term on the right vanish, and h is positive definite. We now exhibit such a g on any (finite) subspace.

Let $T_0 = \text{sp}\{t_i, 1 \le i \le d\} \subset T$, a d-dimensional subspace which may be identified as $\mathbb{R}^d$ with a Haar measure, corresponding to the Lebesgue measure on $\mathbb{R}^d$, denoted by μ and let $\mathcal{H}_1 = L_0^2(T_0, \mu)$ with inner product $\langle g, h \rangle = \int_{T_0} \text{Re}\,(g(t)\overline{h(t)})\, d\mu(t)$, so that $\mathcal{H}_1$ is a Hilbert space. For $\alpha, t \in T_0 - \{0\}$, let $\langle \alpha, t \rangle = \sum_{i=1}^{d} \alpha_i t_i$ and

$$h_\alpha(t) = \|t\|^{-(d+1)/2}(1 - e^{i\langle\alpha,t\rangle}), t \in T_0 - \{0\}, \quad \alpha \in T_0. \tag{39}$$

To see that $h_\alpha \in \mathcal{H}_1$, note that

$$|h_\alpha(t)|^2 \le \|\alpha\|^2\|t\|^{1-d}\left(\frac{\sin(\langle\alpha,t\rangle/2)}{\langle\alpha,t\rangle/2}\right)^2. \tag{40}$$

Here we used the result that $|1 - e^{i\langle\alpha,t\rangle}|^2 = 2(1 - \cos(\langle\alpha,t\rangle))$. Thus from (40) we note that $|h_\alpha(t)|^2$ is integrable so that $h_\alpha \in \mathcal{H}_1$ as desired. If $V(\alpha) = \int_{T_0} |h_\alpha(t)|^2 d\mu(t)$, then $V(0) = 0$, and $V(a\alpha) = |a|V(\alpha)$. Also for any automorphism $U\colon T_0 \to T_0$, $V(U\alpha) = V(\alpha)$. Then it is known from analysis that such a mapping V must be of the form $V(\alpha) = C_\alpha\|\alpha\|$ for some $C_\alpha > 0$. Hence integrating:

$$\|h_\alpha - h_{\alpha'}\|_{\mathcal{H}_1}^2 = \|h_{\alpha-\alpha'}\|_{\mathcal{H}_1}^2 = V(\alpha - \alpha') = C_{\alpha-\alpha'}\|\alpha - \alpha'\|_T. \tag{41}$$

Letting $f_\alpha = C_\alpha^{-\frac{1}{2}} h_\alpha \colon T_\alpha \to \mathcal{H}_1$, it is seen that $f\colon \alpha \to f_\alpha$ verifies all the requirements, and hence gives the desired assertion. $\square$

If T is a subset of $\mathbb{R}^d, d > 1$, then a specialization of the above result in such a case is of interest for some applications and will be a companion of Proposition 1.3.4 above, established somewhat differently, by deducing it from the above, with the Dunford-Schwartz method. Thus we let $Z\colon \mathcal{B}_0(T) \to L^2(T)$ be a vector measure, $T \subset \mathbb{R}^d$ and use the D-S integration as indicated above, for $f \in L^2(T, \mu)$. So $f \mapsto \int_T f(t)dZ(t)$ is defined. We thus will state the following:

Proposition 1.3.7 *If $T = \mathbb{R}^k, 1 \le k < \infty$, then the Lévy-Brownian motion $\{X_t, t \in T\}$ admits the following vector integral representation:*

$$X_t = \int_T f_t(u)\, dZ(u), \quad t \in T(= \mathbb{R}^k),$$

where $\{Z(A), A \in \mathcal{B}_0\}$ is the noise process noted, and $\mathcal{B}_0$ is the δ-ring of bounded Borel sets of T, $f_t(\cdot) = C_t^{-\frac{1}{2}} h_t(\cdot)$ where C_k and h_t are given respectively in (39) and (41) above.

Another way of stating Lévy's Brownian Motion (BM), useful for some applications, is as follows: A gaussian random field $\{X_t : t \in \mathbb{R}^N\}$ is termed Lévy's BM, if the following conditions hold:

(i) $E(X_t) = 0$, (ii) $X_0 = 0$, a.e., and (iii) $E[(X_s - X_t)^2] = r(s, t)$, and its covariance $\Gamma(s, t) = E(X_s X_t) = \frac{1}{2}[r(0, s) + r(0, t) - r(s, t)]$. Note that $t \mapsto X_t$ is continuous with probability one. The following is a class considered for some analysis by P. Lévy himself.

Let $S_N(t)$ be a sphere in $\mathbb{R}^N$ centered at the origin and radius $t > 0$. If $\sigma_t(\cdot)$ is the uniform probability distribution on $S_N(t)$, define a new (Gaussian) process $M_N = \{M_N(t), t \geq 0\}$ by the path integral:

$$M_N(t) = \int_{S_N(t)} X(u) \sigma_t(du), \quad t > 0, \tag{42}$$

called *Lévy's $M_N(t)$-process.* Note that $E(M_N(t)) = 0$, and

$$\Gamma_N(s, t) = E(M_N(s) M_N(t)) = \frac{1}{2}(s + t - \rho_N(s, t)),$$

where the function $\rho_N : (s, t) \mapsto \int_{S_N(s)} \int_{S_N(t)} r(u, v) \sigma_s(du) \sigma_t(dv)$. An explicit evaluation of $\rho_N(\cdot, \cdot)$ is difficult, but for some special cases, these functions have been evaluated and their properties discussed in the monograph by Hida and Hitsuda (1993) where some other interesting properties of Gaussian random fields have been considered to which we refer the interested readers.

We consider certain *other* aspects of Lévy's B.M. of several parameters as discussed by R. Gangolli (1967) indicating new directions. The preceding result motivates the ensuing analysis. Also, the classical Schur observation that the pointwise product of a pair of Hermitian nonnegative matrices is a nonnegative Hermitian matrix is of particular interest in our analysis on the Lévy-Schoenberg kernels which uses as essential elements of the Lévy-B.M. theory. A generalized form is thus:

Proposition 1.3.8 *Let S be a topological space and $r : S \times S \to \mathbb{R}$ be a kernel such that $r(a, b) = r(b, a), r(0, 0) = 0$ for a distinguished point $0 \in S$, and let $f(a, b) = \frac{1}{2}[r(a, 0) + r(b, 0) - r(a, b)], a, b \in S$. Then f is positive definite if and only if the kernel $\theta : S \times S \to \mathbb{R}$ defined by $\theta_t(a, b) = \exp\{-t \cdot r(a, b)\}$ is positive definite, where $t > 0$.*

Proof. Now $f(a, b) = \frac{1}{2}[r(a, 0) + r(b, 0) - r(a, b)]$ being real valued, its positive definiteness follows if for any signed measure $\mu(\cdot)$, compactly supported, one has

$$\int_S \int_S f(x, y)\, d\mu(x)\, d\mu(y) \geq 0. \tag{43}$$

Since $f(a, 0) = f(b, 0) = 0$, it can be assumed that $\mu(S) = 0$ to prove (43). Hence this becomes

$$\int_S \int_S f(a, b)\, d\mu(a)\, d\mu(b)$$
$$= \frac{1}{2} \int_S \int_S (r(a, 0) + r(b, 0) - r(a, b))\, d\mu(a)\, d\mu(b)$$
$$= -\frac{1}{2} \int_S \int_S r(a, b)\, d\mu(a)\, d\mu(b). \tag{44}$$

Then (43) is equivalent to verifying

$$\int_S \int_S r(a, b)\, d\mu(a)\, d\mu(b) \leq 0. \tag{45}$$

Now in the forward direction $\theta_t(\cdot, \cdot)$ is positive definite, so

$$0 \leq \int_S \int_S \exp\{-t \cdot r(a, b)\}\, d\mu(b)\, d\mu(a)$$
$$= \int_S \int_S (1 - t \cdot r(a, b) + o(t^2))\, d\mu(a)\mu(b)$$
$$= -t \int_S \int_S r(a, b)\, d\mu(a)\, d\mu(b) + o(t^2), \quad t \downarrow 0, \tag{46}$$

where $o(t^2)$ is uniformly small on compact subset supports of μ, as $t > 0$. This gives (45) and the direct part of the result follows.

For the converse; suppose the given $f(\cdot, \cdot)$ is positive definite. Then $e^{t \cdot f}$ is positive definite for $t \geq 0$. Hence for each $n \geq 1, a_n \in \mathbb{C}$

$$\sum_{i,j=1}^n a_i \bar{a}_j \exp\{-t \cdot r(a_i, a_j)\}$$
$$= \sum_{i,j=1}^n a_i \bar{a}_j \exp\{-t[r(a_i, 0) + r(a_j, 0) - 2f(a_i, a_j)]\}$$

by hypothesis. Set $\beta_i = \alpha_i e^{-t \cdot r(a_i, 0)}$ in the above so that

$$\sum_{i,j=1}^{n} \beta_i \bar{\beta}_j e^{2t \cdot f(a_i, a_j)} \geq 0, \tag{47}$$

since $e^{2t \cdot f}, t \geq 0$, is positive definite so that $\theta_t(a, b)$ is also, as asserted.

$\square$

Remark 4. This result has a complex as well as an operator valued version with essentially the same argument, as observed by P. Masani (1973). It will not be detailed here.

The weak stationarity of a square integrable process has been formulated in an extended and very useful form by S. Bochner (1955), initially termed $L^{2,2}$-bounded processes, whose structural analysis is found to be very useful both for applications and an advancement of the theory including the stationary and harmonizable classes. The $L^{2,2}$-class is first discussed for a motivation as well as initial applications.

Definition 1.3.9 *Let* $\{X_t, t \in I \subset \mathbb{R}\} \subset L^2(P)$ *be a process. It is called* $L^{2,2}$*-bounded if for each simple function* $f : I \to \mathbb{R}, f = \sum_{i=1}^{n} a_i \chi_{(t_i, t_{i+1}]}$, *the integral* $\int_I f(t) \, dX_t$, *defined as the usual sum* $\tau f = \sum_{i=0}^{n} a_i(X_{t_{i+1}} - X_{t_i})$, *satisfies for some* $C > 0$,

$$E\left(|\int_I f(t) \, dX_t|^2\right) \leq C \int_I |f(t)|^2 \, dt. \tag{48}$$

A standard example of an $L^{2,2}$-bounded process is the Brownian Motion with variance parameter $\sigma^2 (= C,$ here) and independent centered increments so that (48) becomes

$$E\left(|\tau f|^2\right) = \sigma^2 \int_I |f(t)|^2 \, dt. \tag{49}$$

Since a stochastic measure on a ring into a Banach space is a vector measure, it has several types of variations, as detailed by Clarkson and Adams (1933), we recall here two of the most useful ones called the Fréchet and Vitali concepts that are of immediate use. Thus if $\mathcal{B}(I)$ is the Borel σ algebra of an interval $I \subset \mathbb{R}$, $\mathcal{X}$ is a Banach space with norm $\|\cdot\|$, and the mapping $Z : \mathcal{B}(I) \to \mathcal{X}$ is σ-additive (or a vector measure), then consider the variations:

$$\|Z\|(A) = \sup\left\{\|\sum_{i=1}^{n} a_i Z(A_i)\| : |a_i| \leq 1, A_i \in \mathcal{B}(I), A_i \subset A, \text{disjoint}\right\}$$

called the semi-variation (or Fréchet variation) and another:

$$|Z|(A) = \sup\left\{ \sum_{i=1}^{n} \|Z(A_i)\| : A_i \in \mathcal{B}, \text{disjoint}, A_i \subset A \right\}$$

called the *Vitali* (or the usual) variation of Z. It can be seen from the above pair of authors work that the $|Z|(\cdot)$ is always σ-additive but often nonfinite while $\|Z\|(\cdot)$ is always finite but only σ-subadditive. If $f_n \colon I \to \mathbb{R}$ is a simple $\mathcal{B}(I)$-measurable function, $f_n \to f$ pointwise, and $\{\int_I f_n \, dZ, n \geq 1\} \subset \mathcal{X}$ is Cauchy, then $\lim_{n\to\infty} \int_I f_n dZ = \int_I f \, dZ$ is well-defined, $f \mapsto \int_I f \, dZ$ is linear, and $\| \int_I f \, dZ \|_{\mathcal{X}} \leq \|f\|_\infty \|Z\|(I)$ also holds. The vector integrals here are taken in the Dunford-Schwartz sense. However, the properties of Lebesgue's integration are not always valid. We present the following result to clarify matters, and the last observation, noting the distinction, is due to P. A. Meyer (1985).

Proposition 1.3.10 *Let $X = \{X_t, t \in I\}$ be an $L^{2,2}$-bounded process. Then X induces a vector measure $Z \colon \mathcal{B}(I) \to L^2(P)$ such that the Fréchet and Vitali variations of Z agree, and $\tau \colon f \mapsto \int_I f \, dX$ can be expressed as $\tau(f) = \int_I f(t) X(t) \, dt$, (a Bochner integral) so that the induced measure $Z(\cdot)$ can be treated as a Lebesgue-Stieltjes set function only if $X(\cdot)$ has finite variation on I in the classical sense.*

Proof. The mapping $\tau \colon L^2(I, dt) \to L^2(P)$ defined as $\tau(f) = \int_I f \, dX$, is a continuous linear operator (cf. (49)). Since $L^2(P)$ is reflexive, τ is weakly compact and by the Riesz-Dunford theorem there is a unique vector measure $Z \colon \mathcal{B}(I) \to L^2(P)$, with $\tau(f) = \int_I f dZ, f \in L^2(I, du)$ and Z is the induced measure of X. Now treat $I = (a, b)$ as an LCA group under addition mod $(b - a)$, and let $\hat{I}$ be its dual. So by the Plancherel theorem, we have

$$E\left(|\tau f|^2\right) \leq C \int_I |f|^2 dt = C \int_{\hat{I}} |\hat{f}| \, d\mu$$

where μ is the dual measure on $\hat{I}$ and $\hat{f} = \mathcal{F}(f)$ is the Fourier transform of f, the μ is also (a constant multiple of) the Lebesgue measure. Hence if $T = \tau \circ \mathcal{F}^{-1}$ (so $T(\hat{f}) = \tau(f)$), then T is a continuous linear mapping on $L^2(\hat{I}, dt)$ and there is a vector measure $\tilde{Z} \colon \mathcal{B}(I) \to L^2(P)$, such that

$$T(\hat{f}) = \int_I \hat{f}(x) \tilde{Z}(dx), \quad f \in L^2(\hat{I}, d\mu).$$

Now using Fubini's theorem, we get

$$
\tau(f) = T(\hat{f}) = \int_{\hat{I}} \left(\int_I e^{ixt} f(t) \, dt \right) \tilde{Z}(dx)
$$

$$
= \int_I f(t) \left[\int_{\hat{I}} e^{ixt} \tilde{Z}(dx) \right] dt
$$

$$
= \int_I f(t) Y(t) dt \quad \left(= \int_I f(t) \, dX_t \right). \tag{50}
$$

Here $Y(t)$ is the Fourier transform of the vector measure $\tilde{Z}(\cdot)$, and the integral is in Bochner's sense, so $Y(\cdot)$ is weakly harmonizable.

Note that in (50) we cannot write $dX_t = Y(t) \, dt$ if X is possibly a Brownian motion which has no (Lebesgue) derivative anywhere, so that the stochastic integral is not of Lebesgue-Stieltjes type.

For the last part, let $Z(\cdot)$ be the induced measure of X, and the integral can be taken in the Stieltjes sense. We now show that $Z(\cdot)$ must then have finite variation. Indeed consider $\pi_n : a = t_0 < t_1 < \ldots < t_n = b$ on $I = [a, b]$ and for $f : I \to \mathbb{R}$, a bounded Borel mapping S_n is defined by

$$
S_n(f) = \int_I f_n \, dZ = \sum_{i=0}^{n-1} f(t_i)[Z(t_{i+1}) - Z(t_i)] \in L^2(P) \subset L^1(P),
$$

where $f_n = \sum_{i=0}^{n-1} f(t_i) \chi_{[t_i, t_{i+1}]}(t)$. Then $S_n(f) \to \tau(f)$ as the partitions π_n are refined and $\sup_n \|S_n(f)\|_X < \infty$ for each $f \in B(I)$, space of bounded function on I. By the uniform boundedness principle, it follows that $\sup_n \|S_n\| = \alpha_0 < \infty$, and the set $\{S_n, n \geq 1\}$ is uniformly bounded. If we put

$$
h_n^\omega = \sum_{i=0}^{n-1} \operatorname{sgn}\left(Z(t_{i+1}) - Z(t_i)\right)(\omega) \chi_{[t_i, t_{i+1}]}, \quad \omega \in \Omega,
$$

then $h_n^\omega \in B(I, \mathcal{B}_I)$, (bounded Borel), $\|h_n^\omega\|_\infty \leq 1$, and

$$
S_n(h_n^\omega) = \sum_{i=0}^{n} |Z(t_{i+1}) - Z(t_i)|(\omega) \leq \|S_n\| \cdot \|h_n^\omega\|_\infty \leq \alpha_0 < \infty.
$$

Integrating this on Ω, one sees that $Z : \mathcal{B}(I) \to L^1(P)$ must have finite variation so that $Z(\cdot)$ can be treated as a Lebesgue-Stieltjes measure. $\square$

Remark 5. It should be noted that $L^{2,2}$-boundedness and weak (or K-) stationarity are *not* the same, although there are some important processes that belong to both. It can be seen that Brownian motion process belongs to both classes.

For the following work, it will be convenient to extend $L^{2,2}$-*bounded* to $L^{\rho,p}$-*bounded*, where $\rho > 0, p > 0$ for an additive $X \colon \mathcal{S} \to L^p(P)$ where $\mathcal{S}$ is a semi-ring of half-open intervals of $\mathbb{R}^n$. If there is a constant $C(= C_{\rho,p} > 0)$ such that for real $\mathcal{S}$-simple ϕ on $\mathbb{R}^n$ one has, relative to the Lebesgue measure μ:

$$\left\| \int_{\mathbb{R}^n} \phi \, dX \right\|_p^p \leq C_{\rho,p} \int_{\mathbb{R}^n} |\phi|^p \, d\mu. \tag{51}$$

If $p = \rho = 2$ we have the $L^{2,2}$-boundedness already considered. Now $\tau \colon \phi \mapsto \int_{\mathbb{R}^n} \phi \, dX$ is a bounded linear mapping on simple functions ϕ and has an extension to $L^p(P)$; denoted as $\tau(\phi)$ or $F(\phi) = \int_{\mathbb{R}^n} \phi \, dX$. It will be useful for our analysis to see if the above stochastic integral can be given a vector integral form of the Dunford-Schwartz type so that we can use much of their (by now classical) vector integration arguments.

Proposition 1.3.11 *Let $\mathcal{S}$ be a semi-ring of $\mathbb{R}^n$ of half-open intervals and $X \colon \mathcal{S} \to L^2(P)$ be an $L^{2,2}$-bounded additive function. Then there is a vector (or stochastic) measure $Z \colon \mathcal{B} \to L^2(P)$ such that the $F(\phi) = \int_{\mathbb{R}^n} \phi(t) \, dX(t)$ can be represented by another vector integral (as given in Dunford-Schwartz)*

$$F(\phi) = \int_{\mathbb{R}^n} \hat{\phi}(\lambda) Z(d\lambda), \tag{52}$$

where $\hat{\phi}$ is the Fourier transform of $\phi \in L^2(\mu)$, and $\mathcal{B}$ above is the Borel σ-algebra of $\mathbb{R}^n$ and the integral in (52) is the classical vector integral of Dunford-Schwartz type.

Proof. By the $L^{2,2}$-boundedness hypothesis and the classical fact that the Fourier transform on $L^2(\mathbb{R}^n, \mu)$ is an onto isometry, we have

$$\left\| \int_{\mathbb{R}^n} \phi \, dX \right\|_2^2 \leq C_1 \int_{\mathbb{R}^n} |\phi(t)|^2 \, d\mu(t) = C_1 \int_{\mathbb{R}^n} |\hat{\phi}(\lambda)|^2 \, d\mu(\lambda),$$

where $\hat{\phi} = \mathcal{F}(\phi)$, the Fourier transform of $\phi \in L^2(\mu)$ and $\tau(\phi) = \int_{\mathbb{R}^n} \phi \, dX$. Hence $T = \tau \circ \mathcal{F}^{-1}$ with $T(\hat{\phi}) = \tau \circ \mathcal{F}^{-1}(\hat{\phi}) = \tau(\phi)$, is a continuous linear mapping on $L^2(\mu) \to L^2(P)$, and it is well-known that $\tau(\cdot)$ is representable as

$$\tau(\phi) = T(\hat{\phi}) = \int_{\mathbb{R}^n} \hat{\phi}(\lambda) Z(d\lambda), \quad \phi \in L^2(\mu)$$

for a unique vector measure $Z : \mathcal{B} \to L^2(P)$, (cf. Dinculearu's 'Vector Measures' (1967), p. 259). This gives (52) as desired. $\square$

More detailed and general concepts coming from $L^{2,2}$-bounded concept and the basic Khintchine stationarity, to be detailed in the following chapters, we present another useful but fundamental extension, noted by K. Karhunen (1947), and develop its fundamental properties and applications later. Thus if $r(s,t)$ is a (continuous) stationary covariance so that $r(s,t) = \tilde{r}(s - t)$ has the Fourier representation with a spectral density $f(\lambda) = F'(\lambda) \geq 0$.

$$r(s,t) = \tilde{r}(s-t) = \int_{\mathbb{R}} e^{i(s-t)\lambda} \, dF(\lambda) = \int_{\mathbb{R}} (e^{is\lambda}\sqrt{f(\lambda)})\overline{(e^{it\lambda}\sqrt{f(\lambda)})} \, d\lambda$$

which may be expressed by Plancherel's theorem if we express $g(a, \cdot) = e^{ia(\cdot)}\sqrt{f(\cdot)} \in L^2(\mathbb{R}, d\lambda)$, and hence as:

$$r(s,t) = \tilde{r}(s - t) = \int_{\mathbb{R}} g(s, \lambda)\bar{g}(t, \lambda)d\lambda, \quad s, t \in \mathbb{R}. \tag{53}$$

Taking this as a motivation we can consider a general class of processes valued in $L_0^2(P)$, called of *Karhunen type* if $\{X_t, t \in T\}$, $X_t \in L_0^2(P)$, and if there is a family $\{g(t, \cdot), t \in T\} \subset L^2(S, \mathcal{S}, \nu)$, with

$$r(s,t) = E(X_s \bar{X}_t) = \int_S g(s, u)\bar{g}(t, u) \, d\nu(u), \quad s, t \in T. \tag{54}$$

The class $\{g(s, \cdot), s \in S\} \subset L^2(S, \mathcal{S}, \nu)$ and the class $\{X_t, t \in T\} \subset L_0^2(P)$ can be quite general and subsumes the Klintchine (stationary) process. Since by Theorem 1.3.5 above every (abstract) Hilbert space can be realized isometrically as a subspace of $L^2(P)$ on a probability space (Ω, Σ, P), the general analysis of Karhunen processes will be useful in our study and its detailed treatment will be included later.

1.4 Important Remarks on Abstract and Concrete Versions of Hilbert Spaces

It is natural that many practical problems are first formulated concretely on a space of square summable sequences and their properties investigated for a basic understanding, and then abstraction. This presented the numerical sequence l^2-spaces, generalized to $L^2(\mu)$-classes of Lebesgue's, and on noting the key concept and its properties of positive definiteness led J. von Neumann to formulate an abstract space axiomatically, calling it an abstract Hilbert space, which helped an enormous growth of the subject, both in theory and applications. On the other hand, Theorem 1.3.5 above shows that an abstract Hilbert space can always be realizable concretely as a (sub-) space of $L^2(P)$ on some probability triple (Ω, Σ, P). Thus this dichotomy is what makes our subject and the treatment so essential and useful for a study. An abstract version presents an overview of the subject suggesting newer and far-reaching aspects of the problems to be studied, while the concrete version leads us to detail the intricate aspects of the underlying structures, and thus both points are essential for us. The concrete and abstract idea mix was originated by Kolmogorov (1941) himself to indicate how useful and natural it is to develop the subject. Thus if $X_t \in L_0^2(P)$, with means zero and stationary covariance $r(s,t) = E(X_s \bar{X}_t) = r(s + h, t + h), s, h \in \mathbb{R}$, then $U_t : X_s \mapsto X_{s+t}$, defines a linear operator on the closed subspace $\mathcal{L} = \overline{\text{sp}}\{X_t, t \in \mathbb{R}\} \subset L_0^2(P)$, the Hilbert space of centered square integrable random variables X_t, so that the shift mapping $U_t : X_s \mapsto X_{s+t}$ is well-defined on the linear space of $\{X_t, t \in \mathbb{R}\} \subset L_0^2(P)$, satisfying

$$\|U_t X_s\|_2^2 = \|X_{s+t}\|_2^2 = r(s+t, s+t) = r(s,s) = \|X_s\|_2^2,$$

so that $\|U_t X_s\|_2 = \|X_s\|_2$. A similar computation shows:

$$(U_t X_s, X_h) = (X_{s+t}, X_h) = r(s+t, h) = r(s, h-t)$$
$$= (X_s, X_{h-t}) = (X_t, U_{-t} X_h), \quad h, s \in \mathbb{R}, \quad (55)$$

so that the linear operator U_t satisfies $U_t^* = U_{-t}$, where U_t^* is the adjoint of U_t. Then (55) implies that $\{U_t, t \in \mathbb{R}\}$, defined on $L_0^2(P)$, is a weakly continuous unitary group of operators and $X_t = U_t X_0$, so that using the spectral representation of the U_t-group one has

$$X_t = U_t X_0 = \left(\int_{\mathbb{R}} e^{it\lambda} \, dE_\lambda \right) X_0 = \int_{\mathbb{R}} e^{it\lambda} \, dZ(\lambda), \quad (56)$$

where $Z\colon A \mapsto E_A X_0, A \in \mathcal{B}(\mathbb{R})$, the Borel σ-algebra, and the integral is well-defined. Here $Z(\cdot)$ is orthogonally valued when the X_t-process is stationary in which case $(Z(A), Z(B)) = \alpha(A \cap B)$ for a finite positive measure $\alpha\colon \mathcal{B}(\mathbb{R}) \to \mathbb{R}^+$, called the spectral distribution of the process. In general $\beta(A, B) = (Z(A), Z(B))$ defines just a bimeasure and β on $\mathcal{B}(\mathbb{R}) \otimes \mathcal{B}(\mathbb{R})$ need not have finite variation and one has to turn to bimeasure integration theory of M. Morse and W. Transue (briefly seen above but) for a detailed analysis of the subject; that study will start in the next chapters under the more general class called *weakly harmonizable* family. The bimeasure $\beta\colon (A, B) \mapsto (Z(A), Z(B))$ which is thus σ-additive in A and B separately for disjoint families A_n and B_n, and $\beta(\cdot, \cdot)$ can have finite *Vitali* or only finite *Fréchet* variation, as given in Definition 1.2.2 above. The importance of these concepts is that we can enlarge the study of second order classes from the Khintchine stationary family to much larger sets of classes of interest in filtering, sampling and other applications. The following is a useful and motivational illustration.

Theorem 1.4.1 *Let $\{X_t, t \in T\} \subset L_0^2(P)$ be a random field, centered, and whose covariance $r(s, t)$ is representable as:*

$$r(s, t) = E(X_s \bar{X}_t) = \int_S g(s, \lambda)\overline{g(t, \lambda)}\, d\sigma(\lambda), \quad s, t \in T, \qquad (57)$$

for some $\{g(s, \cdot), s \in T\} \subset L^2(S, \mathcal{S}, \sigma(\cdot))$. Then there exists $\{B_t, t \in T\}$ a class of bounded commuting family of operators on $L_0^2(P)$ such that one has

$$X_t = U_t X_{t_0} = \int_S g(t, \lambda)\, dZ_{X_{t_0}}(\lambda), \quad t \in T, \qquad (58)$$

for a unique orthogonally valued random measure $Z_{X_{t_0}}\colon \mathcal{S} \to L_0^2(P)$, such that $E(Z_{X_{t_0}}(A)\bar{Z}_{X_{t_0}}(B)) = \alpha(A \cap B), A, B \in \mathcal{S}$, the vector integral in (58) being in the usual Dunford-Schwartz sense.

In this result, one can have if $T = \mathbb{R}^+, S = \mathbb{R}^n$ and $X_{t_0} = X_0$ then the family $\{U_t, t \geq 0\}$ can be a closed densely defined set of linear operators, as noted by Getoor (1956). The function g of (58) is called the *kernel* representing the process. If it belongs to some larger classes, the problem leads to some other generalized classes of interest which may be indicated here.

An extension of stationarity is a shift $V_s\colon X_t \mapsto X_{t+s}$ and it to be linear, one must have $\sum_{j=1}^n a_j X_{t_j} = 0 \Rightarrow \sum_{j=1}^n a_j X_{t+s} = 0$ for $n \geq$

1 and $\|V_s \left(\sum_{j=1}^{n} a_j X_{t_j} \right) \|_2 \leq C \| \sum_{j=1}^{n} a_j X_{t_j} \|_2$ for $n \geq 2$ for some $C > 0$. As a shift one should also have $V_{s_1} V_{s_2} = V_{s_1+s_2}, s_1, s_2 \geq 0$, with $V_0 = $ identity. If $V_s = \tau^s$ for a bounded linear $\tau : \mathcal{L} \to \mathcal{L}$ and $Y_i \in \mathcal{L} \subset L^2(P)$, linear subspaces, one must have for $Y_0, Y_1, \ldots, Y_t$ in $\mathcal{L}$, either

$$\sum_{m,n=0}^{p} (\tau^n Y_m, \tau^m Y_n) = \sum_{m,n=0}^{p} (Y_{n+m}, Y_{m+n}) \geq 0, \tag{59}$$

or better

$$\sum_{m,n=0}^{p} (\tau^{n+1} Y_n, \tau^{m+1} Y_n) \leq C_0 \sum_{m,n=0}^{p} (\tau^n Y_m, \tau^m Y_m). \tag{60}$$

Such τ is called a *subnormal* operator and $C_0 = \|\tau^2\|$. It was shown by J. Bram (1955) that this operator τ can be extended to an inclusive Hilbert space $\tilde{\mathcal{L}} \supset \mathcal{L}$ on which τ and its adjoint τ^* commute (i.e. $\tau \tau^* = \tau^* \tau$) and one can take the probability space (Ω, Σ, P) rich enough so $\tilde{\mathcal{L}} \subset L_0^2(P)$. Then under (60), $\tau^s = V_s, s > 0$, the constants C_s replace C_0 in (60) and $\{V_s, s \geq 0\}$ will be a semi-group of normal operators on $\tilde{\mathcal{L}} \subset L_0^2(P)$, giving a nonstationary process generalization of the unitary group of the stationary class considered earlier, to include the Karhunen class when the semi-group is weakly continuous and without further restrictions. Then this process $\{X_t = V_t X_0, t \geq 0\} \subset L_0^2(P)$, though nonstationary, can be detailed for an integral representation, that includes the stationary processes, since $\{V_t, t \geq 0\}$ forms a weakly continuous normal semi-group. We can now present the general representation of the process as:

Theorem 1.4.2 *Let $\{X_t, t \geq 0\} \subset L_0^2(P)$ be a (covariance) continuous process admitting a right translation operator $\tau_s : X_t \mapsto X_{t+s}$ which is a bounded linear subnormal mapping, on the span of the process and hence has a normal extension to $L_0^2(P)$, using an enlargement of the probability space if necessary after which (60) holds. If the class $\{V_s, s \geq 0\}$ forms a strongly continuous semi-group then $\{X_t = V_t X_0, t \geq 0\}$ is a Karhunen process which admits a unitary spectral representation whose covariance has the integral formulation, on a Borel set $\Delta \subset \mathbb{R}^2$, as*

$$r(s,t) = \iint_\Delta e^{s\lambda + t\bar{\lambda}} \, d\beta(\lambda) \tag{61}$$

where $\beta(A) = E(|Z_{X_0}(A)|^2), \beta \colon \mathcal{B}(\Delta) \to \mathbb{R}^+$. For a normal semi-group one has the decomposition $\tau_t = R_t U_t$ where $\{R_t, t \geq 0\}$ is a positive self-adjoint semi-group commuting with the unitary group $\{U_t, t \in \mathbb{R}\}$ given by $X_t = V_t X_0 = R_t U_t X_0, \mathcal{B}(\Delta) = \mathcal{B}|\Delta$, restriction.

Proof. Since by (60), the process $\{X_t = V_t X_0, t \geq 0\}$ in $L^2(P)$ and the family $\{V_t, t \geq 0\}$ forms a weakly continuous semi-group of bounded normal operators on $L_0^2(P)$ one can invoke the spectral theorem for such operators (cf. Hille and Phillips (1957), Theorem 22.4.2), to obtain the representation:

$$X_t = V_t X_0 = \int \int_\Delta e^{t\lambda} E(d\lambda) X_0 = \int \int_\Delta e^{t\lambda} \, dZ_{X_0}(\lambda), \qquad (62)$$

where $Z_{X_0} \colon \mathcal{B}(\Delta) \to L_0^2(P)$ is an orthogonally valued measure. This represents a Karhunen field with covariance given by

$$r(s,t) = \int \int_\Delta e^{s\lambda - t\lambda'} \, d\beta(\lambda) \qquad (63)$$

where $\beta(A) = E(|Z_{X_0}(A)|^2)$, so that (63) is similar to the weakly (or K-) stationary case. This can be refined using the fact that $V_t = R_t U_t = U_t R_t$ with $\{U_t, t \in \mathbb{R}\}$ being a unitary group where $U_{-t} = U_t^*, t > 0$, and $\{R_t, t > 0\}$ a *positive* self-adjoint semi-group of operators and R_t, U_t commute. The semi-group property of the V_t yields the following:

$$V_{t+s} = R_{t+s} U_{t+s} = V_t V_s = R_t R_s U_t U_s = R_{t+s} U_{t+s}. \qquad (64)$$

It follows that $\{R_t, t \geq 0\}$ and $\{U_t, t \geq 0\}$ are both strongly continuous semi-groups of operators on $L_0^2(P)$, and the R_t-family is positive in addition. This implies, along with U_t, R_t commuting, the following:

$$\begin{aligned}
r(s,t) &= (X_s, X_t) = (V_s X_0, V_t X_0) = (R_s U_s X_0, R_t U_t X_0) \\
&= (U_t X_0, R_s R_t U_t X_0) = (U_t^* U_s X_0, R_{s+t} X_0) \\
&= (U_{s-t} X_0, R_{s+t} X_0) = \overline{(R_{s+t} X_0, U_{s+t} X_0)}, s + t > 0, s - t \in \mathbb{R} \\
&= E(\bar{X}_{s-t}, \bar{\bar{X}}_{s+t}) = \tilde{r}(s-t, s+t), \quad \text{(say)}.
\end{aligned} \qquad (65)$$

Thus $\tilde{r}$ is positive definite and defined on the cone $\{(s,t) : |t| \leq s\}$ and

$$\tilde{r}(s,t) = r\left(\frac{s+t}{2}, \frac{s-t}{2}\right), \quad 0 < t < s. \qquad (66)$$

This shows that $\tilde{r}$ is positive definite on convex sets and its integral representation is not a direct consequence of the Bochner theorem. Such functions were discussed by Devinatz (1954) which in our case leads to the following, with (63), for $(s,t) \in C = \{(s,t) \in \mathbb{R}^+ \times \mathbb{R} : |t| \leq \frac{s}{2}\}$:

$$\sum_{i,j=1}^{n} a_i \bar{a}_j \tilde{r}(s_i + s_j, t_i - t_j) = \sum_{i,j=1}^{n} a_i \bar{a}_j \left(R_{s_i+s_j} X_0, U_{t_i-t_j} X_0\right)$$

$$= \sum_{ij=1}^{n} a_i \bar{a}_j \left(R_{s_i} U_{t_i} X_0, R_{s_j} U_{t_j} X_0\right),$$

$$\text{since } V_t = R_t U_t \text{ etc.}$$

$$= \sum_{i=1}^{n} (a_i R_{s_i} U_{s_i} X_0, a_i R_{s_i} U_{s_i} X_0) \geq 0.$$

(67)

The key converse implication is a consequence of Devinatz's theorem. With the commutativity of R_s and U_s and their spectral representations yield for (62) the following:

$$X_t = V_t X_0 = R_t U_t X_0 = \int_{\mathbb{R}^+} \lambda^t dE_{1\lambda} \int_{\mathbb{R}} e^{i\lambda'} dE_{\lambda'} X_0 \quad (R_t = R_1^t)$$

$$= \int_{[\text{Re } \lambda > 0]} e^{t(\log \lambda + \lambda')} d\tilde{E}_{\lambda\lambda'} X_0 \quad (\tilde{E}_{\lambda\lambda;} = E_{1\lambda} E_{\lambda'}), \quad (68)$$

where the commutativity of $E_{1\lambda}$ and $E_{\lambda'}$ is used. This alternative method of representation (62) for normal operators is valid for the closed densely defined case as well. □

Remark 6. This result adapted from the author (cf. Rao (2008)) is extendable also for semi-groups $\{\tau_t, t \geq 0\}$ defining $X_t = \tau_t X_0, t \geq 0$, which are closed and densely defined (see Getoor (1957)).

1.5 Complements and Exercises

1. This problem shows that a positive definite bimeasure has finite Fréchet but not Vitali variation. It is a modification of one given in Clarkson and Adams (1933) showing the difference from Lebesgue's theory which cannot be used here in the harmonizable treatments in

general. Consider an $n \times 2^{n-1}$ matrix $A_n = (a_{ij}^n, 1 \leq i \leq n, 1 \leq j \leq 2^{n-1}, a_{ij} = \pm 1)$ for $n = 2m + 1, m \geq 0$ with all combinations of ± 1, except that $a_{nj}^n = 1, j = 1, \ldots, 2^{n-1}$. Define $\|A_n\| = \sup\{|\sum_{i=1}^n \sum_{i=1}^{2^{n-1}} \sum_{j=1}^{2^{n-1}} a_{ij}^n b_i c_j| : b_i \in \mathbb{C}, |b_i| \leq 1, |c_j| \leq 1\}$. Set $|A_n| = \sum_{i=1}^n \sum_{j=1}^{2^{n-1}} |a_{ij}^n| = n2^{n-1}$. Verify next (after detailed computation) that $\|A_n\|/|A_n| = r_n \to 0$ as $n \to \infty$ with $x = 2m + 1$. Now consider the bimeasure β, defined by $\beta(A, B) = \{\sum \beta_t(P) : P \in A \times B\}$, where $\beta_k(P) = (-1)^k a_i^k / t_k 2^{t_k - 1}$ if $P = P_{ij}^{t_k}$, and 0 otherwise. Then $\beta_k(\cdot, \cdot)$ is well-defined on $\mathcal{B}(\mathbb{R})$. $|\beta_k|(R, R) = 1$, and $\beta(\cdot, \cdot)$ is a bimeasure and has finite Fréchet variation. However $|\beta|(R, R) \geq \sum_{j=1}^k |\beta_j|(R, R) = k \to \infty$, so that its Vitali variations infinite. [The details are not simple, and are in Chang and Rao (1986). This is a somewhat modified version of the original Clarkson-Adams work.]

2. Let G be an LCA group, ξ be a random variable on (Ω, Σ, P) with mean zero and $E(\xi^2) = 1$, and $X_t = \xi f(t), t \in G$. Verify that $\{X_t, t \in G\}$ need not be harmonizable even if $f : G \to \mathbb{C}$ is continuous and $f(\infty) = 0$. If $g(x) = e^{-t \cdot f(x)}, t > 0, f(x+y) = f(x) + f(y)$, verify that $g : G \to \mathbb{R}$ is positive definite and nonmeasurable.

3. This problem shows the role of V-boundedness in our analysis. Let $X_t = f(t)\xi$ where $f : G \to \mathbb{C}$ is a mapping and ξ is a random variable with mean zero and unit variance. Then the process $\{X_t, t \in G\}$ is not necessarily harmonizable even if f is continuous and vanishes at infinity. It is V-bounded only if f is the Fourier transform of an integrable function (i.e., $f = \hat{g}$).

4. This problem presents an estimation of a parameter, extending a result of Balakrishnan's (1959) from weekly stationary errors to the (weakly) harmonizable case. So let $Y_t = \alpha m_t + X_t, t \in \mathbb{R}$, and $\alpha \in \mathbb{R}$ as signal and X_t the noise having mean zero and a harmonizable covariance r with a spectral bimeasure F. The previous study was when the noise was taken stationary. The problem is to estimate α, as a weighted linear unbiased estimator on observing the Y_t-process on an interval, $[-B, B]$, using a weight function $p(\cdot)$ of bounded variation, thus by $\hat{\alpha}_B = \int_{-B}^{B} Y_t dp(t)$, so that $E(\hat{\alpha}_B) = \alpha$, the unbiasedness condition. Typically one should get the "best" *lower-bound* for the variance of this 'unbiased estimator' $\hat{\alpha}_B$ since the actual distribution function of $\hat{\alpha}_B$ is quite difficult, to

study its behavior. The *lower bound analysis* is the next best thing and is thus of interest. [See Rao (1961) on the problems and its philosophy of use in many such studies.]

5. This result presents how classes of Lévy-Schoenberg kernels can be generated, supplementing Proposition 1.3.8. As before let our G be a separable topological group and $K \subset G$ a closed subgroup. If $f : G/K \to \mathbb{R}$ is a Lévy-Schoenberg kernel as defined in Proposition 1.3.8, then the kernel defined by:

$$r(a, b) = f(a, a) + f(b, b) - 2f(a, b), a, b \in G/K,$$

satisfies $r(a, b) = r(b, a) : r(a, a) = 0$, and $r(xa, xb) = r(a, b)$, for $a, b \in G/K$ and $x \in G$, and the kernel $\theta_t : (a, b) \to \exp\{-t \cdot r(a, b)\}$ is positive definite for $t \geq 0$. On the other hand if $r(a, b) \in \mathbb{R}$, where $a, b \in G/K$ and satisfying the preceding conditions then for any $0 \in G/K$ the function $f : (a, b) \mapsto \frac{1}{2}[r(a, 0) + r(b, 0) - r(a, b)]$, defines a Lévy-Schoenberg kernel. Thus there are plenty of Lévy-Schoenberg kernels on infinite dimensional (very) general topological spaces, and hence the Lévy-BM classes are large.

6. This problem presents a condition for a process, or random field $\{X_t, t \in G\} \subset L^p(P), 1 \leq p \leq 2$, to have a Fourier integral representation, to be termed a *strictly harmonizable random field*. Thus $\{X_t, t \in G\}$ is strictly harmonizable if it satisfies the *V-boundedness condition* (G is a compact abelian group)

$$\left\| \sum_{i=1}^{n} a_i X_i \right\|_T \leq C \sup \left\{ \left| \sum_{i=1}^{n} a_i \chi_i \right| : \chi_i \in \tilde{G}, a_i \in \mathbb{C} \right\}$$

for some $0 < C < \infty$, and χ_i are characters of G i.e., $\chi_i \in \tilde{G}$ (dual of G). Then the representation that $X_n = \int_G e^{int} dZ(t)$, holds if $G = \mathbb{Z}$, the integer group (so $\hat{G} = (\pi, \pi]$). (See Hosoya (1982).)

7. The concept of 'weak harmonizability' can be extended for the $L^p(P), p \geq 1$, spaces (and even Orlicz spaces). Thus a sequence $\{X_n, n \in \mathbb{Z}\}$ is L^p-*harmonizable* if it admits a representation:

$$X_n = \int_I e^{inu} d\mu(u), \quad n \in \mathbb{Z}, I \subset \mathbb{R},$$

where $\mu : \mathcal{B}(I) \to L^p(P)$ is a *vector measure* and $\mathcal{B}(I)$ is a Borel σ-algebra.

Here μ is often termed a spectral stochastic measure. The series $\{X_n, n = 0, \pm 1, \pm 2, \ldots\}$ is called an L^p-*harmonizable process*. I. Kluvánek (1967) has shown that the following precise statement holds: [Let $\mathcal{X}$ be a Banach space and G an LCA group, $X : G \to \mathcal{X}$ is a mapping. Then X is a generalized Fourier transform of a regular vector measure $v : \mathcal{B}(\hat{G}) \to \mathcal{X}$, so $X(g) = \int_{\hat{G}} \langle g, s \rangle v(ds), g \in G$, where $\hat{G}$ is the dual group of the LCA group G, iff $X(\cdot)$ is weakly continuous and V-bounded, in that the set $\{\int_G f(t)X(t)dt : \|\hat{f}\|_\infty \le 1, f \in L^1(R)\}$ as a subset of $\mathcal{X}$ is relatively weakly compact in $\mathcal{X}$.] (For a detailed background of the work, see the another's somewhat long paper (Rao (1982)), and for a good background and application, see Dehay (1991) discussing products of pairs of harmonizable processes.)

1.6 Bibliographical Notes

The work of this volume starts with analyses of the second order processes as the ideas really give a familiar feeling for many others, motivate a study and explore new areas in this research. The basis for a large part of our analysis is the positive definiteness property appearing in different forms. Its characterization under minimal conditions, seen as a Fourier transform of a suitable measure obtained by S. Bochner, has aided enormous growth of stochastic processes, starting with A. I. Khintchine's analysis and his characterization of (weakly) stationary classes, detailing the covariance structure. Many developments were based on the centered second order processes. The mean functions, if not taken to be zero (for 'convenience') is a problem that depends on some serious (Fourier) analysis was shown by A. V. Balakrishnan (1959), and it prompted further analysis that is included in this chapter. The original positive definiteness property and its characterization leading to harmonizable classes, first indicated briefly by Loéve and Rozanov, has been extended much further again with the key concept of V-boundedness introduced by S. Bochner himself which gave the weak and strong harmonizabilities of the covariances of second order processes, and the consequent work on means of the processes that were characterized. All this is included here.

Some of the deeper analysis of harmonizable classes, in contrast with the stationary processes demands some advanced but interesting techniques, to be used and extended in the following chapters. Here a use of M. Morse and W. Transue analysis of (multiple) integrals which are weaker than Lebesgue's are needed and their use will continue in these chapters.

The 'positive definite analysis' leads to Lévy's Brownian motions opening up new aspects of stochastic analysis, which takes to characterizations of abstract Hilbert spaces isometrically as certain Lebesgue spaces $L^2(P)$ on a Gaussian measure space, connecting and extending aspects of abstract Hilbert space analysis with concrete $L^2(P)$-type theory. This combination is benefitting in both areas.

In the following chapters some specializations of the analysis as well as extensions, bringing in several ideas and applications of abstract harmonic analysis will be made. Since the Fourier analysis plays such a key role here, it may be of interest to record the following somewhat unusual experience for others to read and reflect. This is about knowing and retelling the actual evolution of our subject. For this I have noticed an AMS publication on the 'History of Mathematics' by a Greek distinguished mathematician, Nicolas K. Artemiadis, who obtained his Doctor d'Etat (like our Ph.D) in Paris and taught mathematics as a Professor at the University of Wisconsin and Southern Illinois University, and later returned to his native Greece and wrote this scholarly book in Greek language which later was translated and published in English by the AMS in 2004. I bought the volume and found among others an interesting account on Fourier and his problems on publishing it in the French Academy for many years (nearly 20 years) delayed before that classic work was finally published. [It seems that the merit of the ideas, was judged and appreciated by the Stalwarts, Lagrange, Laplace, and Legendre, but the work was not accepted for not being mathematically rigorous! His persistence and final publication in early 1820s saved us the Fourier analysis.] This historical account, published in Greek language originally quoted a passage from M. Klein's book on Math History, published by a British publisher in England, was paid for and given in the Greek version, but the original passage was given in the AMS version. Somebody in the U.S. has thought it was "plagiarization" and made the AMS to withdraw the book, although the author produced letters of permission and purchase of a passage for the Greek edition! When I saw the note on withdrawing the English version by the AMS, to avoid court litigation, I have written a letter in 2005, to the AMS notices in which the book's withdrawal was noted asking them to have the book made available, and mentioning that any historical account reported by various writers naturally will be the same and could not be "plagiarized", and after all the Klein account could not be the original one about an ancient instance! (Cf. AMS Notices, 52 (2005), p. 1174.) After seeing my letter in the notices, Prof. Artemiadis wrote me a letter, dated October 20, 2005, stating "your letter to the NOTICES was for me an unexpected ray of life (literally). I cannot find strong enough words to express my deepest and most profound gratitude." It was nice to see a scholar, who was much disturbed for withdrawing his cherished work's circulation, to feel somewhat relieved. In a later issue of the Notices (a couple of years later), I have read the sad news that Prof. N. K. Artemiadis has passed away. It was some relief to think that the improper understanding of his work is slightly abated.

The important role that the Fourier analysis, in its modern forms, playing in the following chapters will be self-evident. The random processes $\{X_t, t \in T\}$ where $T \subset \mathbb{R}^n, n \geq 1$, subgroup, and T an LCA group will be seen as central to much of what follows.

2

Harmonic Approaches for Integrable Processes

As seen in Chapter 1, stochastic analysis is deepened as well as enriched, both in content and form for theory as well as applications, by a use of the methods of harmonic analysis and related integration methods. We start with an analysis of second order processes that consider together with the first and second moments, as shown in Chapter 1, implying that the usual assumption of centering at their means is somewhat restricting the general structural analysis of the (second order) processes. *Also included here are the integral determination of all local functionals, and also **a probability solution of the classical Riemann hypothesis**. Thus the chapter contains some **key aspects** of the subject.

2.1 Morse-Transue Integration Method and Stochastic Analysis

We start with necessary integrability properties of second order processes, their basic structures and a general analysis for classes motivated by (and centering around) harmonizable classes, as seen in Chapter 1. It depends on explaining, detailing and employing seriously, the Morse-Transue (or MT-) integration method which is *weaker* than the classical Lebesgue's version, but is important for our work.

This was briefly considered in Chapter 1. A few related results and some applications will be discussed here for quick reuse. Thus let S_i be locally compact, $\mathcal{K}(S_i)$, the space of continuous compactly based complex functions, $i = 1, 2$ and $B \colon \mathcal{K}(S_1) \times \mathcal{K}(S_2) \to \mathbb{C}$ be a bilinear mapping, called a *complex* (or $\mathbb{C}$) *bimeasure* on $S_1 \times S_2$ if $B(\cdot, v) \colon \mathcal{K}(S_1) \to \mathbb{C}$ and $B(u, \cdot) \colon \mathcal{K}(S_2) \to \mathbb{C}$ are relatively bounded linear functionals for each $u \in \mathcal{K}(S_1)$ and $v \in \mathcal{K}(S_2)$. Here $B(\cdot, v)$ is said to be *relatively bounded* if for each compact $K \subset S_1$ and $\mathcal{K}(K) \subset \mathcal{K}(S_1)$ is considered, then $B(\cdot, v)\colon \mathcal{K}(K) \to \mathbb{C}$ is bounded. Similarly $\Lambda(u, \cdot), u \in \mathcal{K}(S_1)$, has an analogous meaning for $B(u, \cdot)$. This is designed to use the MT-work. Similarly, we associate a set-theoretical function $\beta\colon \mathcal{S}_1 \times \mathcal{S}_2 \to \mathbb{C}$, where $\mathcal{S}_i$ is σ-Borel algebra of S_i, and call it a *bimeasure* if $\beta(\cdot, F), \beta(E, \cdot)$ are σ-additive for each $E \in \mathcal{S}_1$ and $F \in \mathcal{S}_2$. Their *Fréchet* and *Vitali* variations are defined and denoted by $\|\beta\|(\cdot, \cdot)$ and $|\beta|(\cdot, \cdot)$, where (Vitali)

$$|\beta|(S_1, S_2) = \sup \left\{ \sum_{i,j=1}^{n} |\beta(A_i, B_j)| : A_i \in \mathcal{S}_i, B_j \in \mathcal{S}_j \right\}, \quad (1)$$

and (for the Fréchet),

$$\|\beta\|(S_1, S_2) =$$
$$\sup \left\{ \left| \sum_{i,j=1}^{n} a_i \bar{a}_j \beta(A_i, B_j) \right| : A_i \in \mathcal{S}_i, B_j \in \mathcal{S}_j, |a_i| \le 1, |a_j| \le 1 \right\}. \quad (2)$$

Then β has *finite Vitali variation* if $|\beta|(S_1, S_2) < \infty$, and has finite *Fréchet variation* if $\|\beta\|(S_1, S_2) < \infty$, so that $\|\beta\|(S_1, S_2) \le |\beta|(S_1, S_2)$. There will actually be strict inequality if the Vitali variation is infinite.

For the following analysis, it will be useful to include a few more details of the *Morse-Transue integration* as it extends the Lebesgue integral by weakening its absolute integration condition. The concept of MT-integration is set down as follows.

Definition 2.1.1 *Let $f_i\colon S_i \to \mathbb{C}, i = 1, 2$ be Baire functions, and Λ be a $\mathbb{C}$ bimeasure on $S_1 \times S_2$. Then (f_1, f_2) is Morse-Transue (or MT)-integrable if (i) f_i is $\Lambda(\cdot, g_2)$ integrable and f_2 is $\Lambda(g_1, \cdot)$ integrable for $g_i \in \mathcal{K}(S_i)$, $i = 1, 2$, the linear functionals $f_1 \to \Lambda(f_1, \cdot)(g_2)$ and $f_2 \to \Lambda(\cdot, f_2)(g_1)$ are Radon measures, and that the integrals satisfy $\Lambda(f_1, \cdot)(g_2) = \Lambda(\cdot, f_2)(g_1)$ so that the common value is denoted by $\Lambda(f_1, g_2) = \Lambda(g_1, f_2)$:*

$$\Lambda(f, g) = (MT) \int_{S_1} \int_{S_2} (f, g) d\Lambda = \Lambda(f, \cdot)(g) = \Lambda(\cdot, g)(f). \quad (3)$$

This formulation is a *slightly extended version* of the original MT-definition, applicable to measurable functions (not only the continuous ones), as shown (extended) by Erik Thomas (1970), and we use this extended form in the following applications as needed.

The desired result for us is given as follows:

Theorem 2.1.2 *Let S_i be a locally compact space with $\mathcal{S}_i$ as its Borel σ-algebra and $\Lambda\colon \mathcal{S}_1 \times \mathcal{S}_2 \to \mathbb{C}$ be a bimeasure. Let $f_i \in \mathcal{K}(S_i)$ be such that (f_1, f_2) is Λ-integrable as in (3) so that the pair (f_1, f_2) is Λ-integrable and $(f_1, f_2) \mapsto \int_{S_1} \int_{S_2} (f_1, f_2) d\Lambda$ defines a $\mathbb{C}$-bimeasure on $\mathcal{K}(S_1) \times \mathcal{K}(S_2)$. Also any Λ-integrable pair (f_1, f_2) is MT-integrable, and both integrals agree.*

Proof. We sketch the argument for comprehension of the MT-method of the (weaker) integration in which one utilizes some results (detailing the MT-method by Ylinen (1978)) for understanding certain points, which help shortening the usual detailed analysis.

Thus let $\Lambda\colon \mathcal{S}_1 \times \mathcal{S}_2 \to \mathbb{C}$ be a measure and let (f, g) be Λ-integrable. Note that the concept of Λ-integrability implies that each pair of bounded Baire functions (u, v) is so integrable and $|\Lambda(f, g)| \le \|\Lambda\|\|(S_1, S_2)\|\|f\|_\infty\|g\|_\infty$, the latter symbols on $\mathcal{K}(S_i)$ are the 'sup' norms, so that $\Lambda(\cdot, \cdot)$ is a bounded $\mathbb{C}$-bimeasure and that $\Lambda(\cdot, g), \Lambda(f, \cdot)$ define bounded Radon measures on $\mathcal{B}$ by the usual measure theory analysis. Thus one has:

$$\Lambda(f, \cdot)(g) = \Lambda(\cdot, g)(f) = \int_S f(x)\Lambda_g(dx, S) = \int_S g(y)\Lambda_f(S, dy), \quad (4)$$

using the known properties of the MT-integrals (Morse-Transue, 1956). From this and the standard arguments, (4) gives since (f, g) is Λ-integrable:

$$\int_S \int_S (f, g) d\Lambda = \int_S g(y)\Lambda(f, dy)$$
$$= \Lambda(f, \cdot)(g). \quad (5)$$

Analogous argument shows that $\Lambda(\cdot, g)(f)$ and $\Lambda(f, \cdot)(g)$ exist and are equal, i.e., $\Lambda(f, \cdot)(g) = \Lambda(\cdot, g)(f)$ so that (f, g) is MT-integrable.

Regarding the last statement, let Λ be a bounded $\mathbb{C}$-bimeasure on $\mathcal{K}(S) \times \mathcal{K}(S)$, then by the MT-theory each pair of bounded Baire functions (f, g) is MT-integrable, since Λ is a *bounded* $\mathbb{C}$-bimeasure. By the theory of these authors, $\beta\colon \mathcal{B}(S) \times \mathcal{B}(S) \to \mathbb{C}$ defined by $\beta(E, F) = \Lambda(\chi_E, \chi_F)$ will be seen to be a complex bimeasure so that it is bounded. If $\Lambda'\colon (h, k) \mapsto \int_S \int_S (h, k) d\beta$, is defined for a pair of bounded Baire (h, k), then by the earlier part above Λ' is a $\mathbb{C}$-bimeasure agreeing with Λ on (χ_E, χ_F) and so $\Lambda'(u, v) = \Lambda(u, v)$ for $u, v \in \mathcal{K}(S)$. It then follows by the extension method, that $\Lambda = \Lambda'$ as asserted. $\square$

Remark 7. As Ylinen (1978) shows by example that the hypothesis of boundedness of Λ after extension to $\mathcal{K}(S_1) \times \mathcal{K}(S_2)$ is essential to employ the result for more (not always continuous) functions. Thus the MT-integral is not absolute, contrasting it with that of Lebesgues. We include an example to this effect in the Complements below (cf. Ex. 1).

For our further analysis, a restriction of the MT-integral is desirable. That is called a *strict bimeasure integral* (or *strict MT-integral*). It is defined as follows.

Definition 2.1.3 *Let* $(\Omega_i, \Sigma_i), i = 1, 2$ *be a measurable pair and* $f_i \colon \Omega_i \to \mathbb{C}$ *be measurable and* $\beta \colon \Sigma_1 \times \Sigma_2 \to \mathbb{C}$, *a bimeasure. Then the pair* (f_1, f_2) *is called* strictly β-integrable *if we have:*

(a) f_1 *is* $\beta(\cdot, F)$ *and* f_2 *is* $\beta(E, \cdot)$ *integrable (Lebesgue), so that* $_{f_1}\beta(E, B) = \int_E f_1(\omega_1)\beta(d\omega_1, B)$, *and* $\beta_{f_2}(A, F) = \int_F f_2(\omega_2)$, $\beta(A, d\omega_2)$, *for* $A \in \Sigma_1, B \in \Sigma_2$ *are complex integrals for each* $E \in \Sigma_1, F \in \Sigma_2$, *and*

(b) f_1 *is* $\beta_{f_2}(\cdot, F)$ *and* f_2 *is* $_{f_1}\beta(E, \cdot)$-*integrable for each* $E \in \Sigma_1, F \in \Sigma_2$ *and* $\int_E f_1(\omega_1)\beta_{f_2}(d\omega_1, F) = \int_F f_2(\omega_2)_{f_1}\beta(E, d\omega_2)$ *holds. When these conditions obtain, the double integral is given by:*

$$\int_E \int_F^* (f_1, f_2)d\beta = \int_{S_1} \int_{S_2} (\chi_E f_1, \chi_F f_2)\, d\beta = \int_E f_1(\omega_1)\beta_{f_2}(d\omega_1, F).$$
(6)

It can be verified that each strictly integrable pair for β, is naturally β-integrable as defined before, both have the same values and that the β-integrability results in Ylinen (1978) are valid for our strict β-integrals. The following *change of variables formula* can be established with some slightly modified (usual) arguments.

Proposition 2.1.4 *Let* (f, g) *be strictly* β-integrable on $(\Omega_i, \Sigma_i), i = 1, 2$ *for a bimeasure* $\beta \colon \Sigma_1 \times \Sigma_2 \to \mathbb{C}$, *and let* $\mu \colon (A, B) \to \int_{\Omega_1} \int_{\Omega_2} (\chi_A f, \chi_B f)d\beta$ *for* $A \in \Sigma_1, B \in \Sigma_2$. *If* $h \colon \Omega_1 \to \mathbb{C}$ *is bounded measurable and* $k \colon \Omega_2 \to \mathbb{C}$ *is similarly obtained, then writing* $\mu = \mu_{f,g}$, *we get*

$$\int_A \int_B^* (fh, gk)\,d\beta = \int_A \int_B^* (h, k)\,d\mu, \quad A \in \Sigma_1, B \in \Sigma_2. \quad (7)$$

Proof. Since μ is evidently a bimeasure on $\Sigma_1 \times \Sigma_2$, define the linear functional on $\Sigma_1 \times \Sigma_2 \to ca(\Omega_2, \Sigma_2)$ as:

$$k'_F(\lambda) = \int_F k(\omega_2)\lambda(d\omega_2), \quad F \in \Sigma_2, \lambda \in ca(\Omega_2, \Sigma_2),$$

$k(\cdot)$ being that given in the statement. But then there exist step functions $k_n \to k$ pointwise, $|k_n| \le |k|$, and we denote a typical k_n as $k_n = \sum_{j=1}^n b_j^n \chi_{B_j^n}$. Then for each $E \in \Sigma_1$, one has

$$\begin{aligned}
\beta_{gk}(E, F) &= \int_F (gk)(\omega_2)\beta(E, d\omega_2), \text{ by definition,}\\
&= \lim_{n\to\infty} \int_F (gk_n)(\omega_2)\beta(E, d\omega_2), \text{ by dominated convergence,}\\
&= \lim_{n\to\infty} \sum_{j=1}^n b_j^n \beta_g(E, F \cap B_j^n)\\
&= \lim_{n\to\infty} \int_F k_n(\omega_2)\beta_g(E, d\omega_2) = \int_F k(\omega_2)\beta_g(E, d\omega_2)\\
&= k'_F(\beta_g(E, \cdot)).
\end{aligned}$$

Now the standard Dunford-Schwartz theory gives:

$$\begin{aligned}
\int_F (gk)(\omega_2)_f \beta(E, d\omega_2) &= k'_F \left(\int_E g(\omega_2)_f \beta(\cdot, d\omega_2) \right)\\
&= k'_F \left(\int_E f(\omega_1)\beta_g(d\omega_1, \cdot) \right)\\
&= \int_E f(\omega_1)\beta_{gk}(d\omega_1, F).
\end{aligned}$$

Thus (f, gk) is strictly β-integrable and moreover,

$$\int_E \int_F^* (g, k)\,d\beta = k'_F(\mu(E, \cdot)) = \int_F k(\omega_2)\mu(E, d\omega_2) = \tilde{\mu}_E(E, F).$$

In a similar way, one shows that (fh, gk) is strictly β-integrable and deduces

$$\int_E \int_F^* (fh, gk)d\beta = \int_E h(\omega_1)\tilde{\mu}_E(d\omega_1, F) = \int_E \int_F^* (h, k)d\mu,$$

which completes the proof. $\square$

The corresponding statement for $\mathbb{C}$-bimeasure and MT-integrals is:

Proposition 2.1.5 *Let* $(S_i, \mathcal{B}_i), i = 1, 2$ *be* σ-compact *Borel spaces, and* f, g *be* $\mathcal{B}_1, \mathcal{B}_2$-*measurable scalar functions. If* Λ *is a bounded* $\mathbb{C}$-*bimeasure on* $\mathcal{B}_1 \times \mathcal{B}_2$, *then* (fh, gk) *is MT-integrable for* Λ *and for all bounded scalar functions* h, k *on* S_1, S_2 *iff* $(\chi_A f, \chi_B g)$ *is MT-integrable for all* $A \in \mathcal{B}_1$ *and* $B \in \mathcal{B}_2$.

The proof in one direction is immediate, and the converse is established as in the preceding proposition, with an interplay of (Radon) measures and linear functionals given in Bourbaki, and the details are left to the reader.

   The following consequence is often useful for applications.

Corollary 2.1.6 *Let* $(S_i, \mathcal{B}_i), i = 1, 2$ *be measurable spaces and* $\beta \colon \mathcal{B}_1 \times \mathcal{B}_2 \to \mathbb{C}$ *be a bimeasure. Then we obtain:*

 (i) *A pair* (f, g) *of complex functions is* β-*integrable strictly if and only if the pair* $(|f|, |g|)$ *is so integrable.*
 (ii) *If* $f_i \colon S_i \to \mathbb{C}, i = 1, 2$ *are measurable and* $|f_1| \le |f|, |f_2| \le |g|$ *and* (f, g) *is strictly* β-*integrable, then so is* (f_1, f_2).
 (iii) *If* (f_m, g_n) *is a sequence of measurable functions,* $f_n \to \tilde{f}, g_n \to \tilde{g}$ *pointwise and* $|f_n| \le |f|, |g_n| \le |g|, (f, g)$ *as in (i) then* $(\tilde{f}, \tilde{g})$ *is also strictly* β-*integrable and one has*

$$\int_E \int_F^* (\tilde{f}, \tilde{g})d\beta = \lim_{\substack{m \to \infty \\ n \to \infty}} \int_E \int_F^* (f_n, g_m)d\beta, \quad E \in \mathcal{B}_1, F \in \mathcal{B}_2.$$

The result can be established with the standard argument familiar in Real Analysis, and need not be repeated. If $(\chi_A f, \chi_B g)$ is MT-integrable relative to a bimeasure, the result holds for them also. Hereafter when the result holds for strict as well as the general case of a bimeasure, the distinction will be omitted in the statements.

   The class of second order processes that is covered by our work includes the harmonizable families, but the methods are adapted to some general classes introduced first by Karhunen (1947) and independently by Loéve (1948). Soon after, H. Cramér (1951) has given a formulation that includes both, amplifying the structure. We analyze these classes to clarify and enhance the application-potential of the classes, all of which are of second order. The existence of these classes is an immediate consequence of the fundamental Kolmogorov-Bochner projective limit theorem discussed and detailed in the initial part of our trilogy (cf. Rao (1995) Chapter I). Also it is noted that each positive definite continuous function $r\colon G \times G \to \mathbb{C}$ on an LCA group G is the covariance function of a (even a Gaussian) centered process (= random field) with this r as its covariance function. Then for such r we may also suppose the existence of a positive definite function $F\colon G \times G \to \mathbb{C}$ of locally finite Fréchet variation $\|F\|(K) < \infty$, in that on each compact product set $K \times K \subset G \times G$, for each set of (scalar) Baire functions $\{g(s, \cdot), s \in G\}$ one has (G being an LCA group as always)

$$\int_G \int_G g(s, \lambda)\bar{g}(s, \lambda')F(d\lambda, d\lambda') < \infty, \quad s \in G. \tag{8}$$

For such a family we can introduce the general Cramér and Karhunen classes of processes that subsume the stationary and then the harmonizable classes enhancing the study of our stochastic analysis in many ways.

The preceding account on bimeasure integration (of MT-type) is utilized crucially in obtaining integral representations of harmonizable as well as Karhunen and Cramér classes of processes which form a substantial part of the second order (continuous parameter and other) processes of great interest in applications.

We have already introduced the Karhunen class in second order processes in Chapter 1 (Section 3), as the second order class with means zero, and covariance $r(\cdot, \cdot)\colon (s, t) \mapsto E(X_s \bar{X}_t), X_t \in L_0^2(P)$, given by

$$r(s, t) = \int_S g(s, u)\bar{g}(t, u)d\nu(u), \quad s, t \in T, \tag{9}$$

where $\{g(s, \cdot), s \in S\} \subset L^2(S, \mathcal{S}, \nu)$ for some measure ν on $(S, \mathcal{S})$. Thus the class of second order processes $\{X_s, s \in S\} \subset L_0^2(P)$ whose covariance is representable as (9) is termed the Karhunen class which properly includes the (centered) Khintchine or weak stationary class. But we have also seen that the (weakly) stationary class is included properly in the (weakly) harmonizable class. Similarly, Cramér (1951) has extended the Karhunen class in the following (natural) way.

Definition 2.1.7 *Let* $\{X_t, t \in \mathbb{R}\} \subset L_0^2(P)$ *be a process with covariance* $r\colon (s,t) \mapsto E(X_s \bar{X}_t)$ *that is representable as,*

$$r(s,t) = \int_{\mathbb{R}} \int_{\mathbb{R}} g(s,u)\bar{g}(t,v)\beta(du, dv), \quad s,t \in \mathbb{R}, \qquad (10)$$

where β *is a positive definite set function (bimeasure) of bounded Vitali variation on* $\mathbb{R}^2$, *where the integral in* (10) *is in the standard Lebesgue-Stieltjes sense in the complex plane.*

Note that if $g(s,u) = e^{isu}$ in (10), this reduces to the Loève definition of a (in the present terminology) *strongly hormonizable* process. Clearly this goes over to the stationary case if $g(s,u) = e^{isu}$ and $\beta(\cdot, \cdot)$ concentrates on the line $u = v$ in $\mathbb{R}^2$. It also extends the Karhunen class, if $g(\cdot, \cdot)$ is more general but $\beta(\cdot, \cdot)$ concentrates on $u = v$. For a clear understanding of the extension involved, *we first characterize the (all important) Karhunen fields.*

Theorem 2.1.8 *Let* $\{X_t = N_t X_0 \colon t \in T\}, X_0 \in L_0^2(P)$, *be a process generated by a family of bounded commuting set of normal operators on* $L_0^2(P)$, *of a centered random variable* $X_0 \colon \Omega \to \mathbb{R}$, $N_0 = id.$, *and* T *to be a separable topological space. Then there exist a continuous function* $b\colon T \times [0,1] \to \mathbb{R}$, *and an orthogonally valued* σ-additive $Z\colon \mathcal{B}(I) \to L_0^2(P), I = [0,1]$, *such that we have the (vector) integral representation:*

$$X_t = \int_I b(t, \lambda) dZ(\lambda), \quad t \in T, \qquad (11)$$

and its covariance function $r\colon (s,t) \mapsto E(X_s \bar{X}_t)$ *is given by*

$$r(s,t) = \int_I b(s, \lambda)\bar{b}(t, \lambda) d\alpha(\lambda), \quad s,t \in T, \qquad (12)$$

where the measure $\alpha\colon \mathcal{B}(I) \to \mathbb{R}^+$ *satisfies* $\alpha(A) = \|Z(A)\|_2^2$ *for all Borel sets* $A \subset I; \mathcal{B}(I)$ *being the Borel* σ-algebra of I.

Proof. The original argument depended on a classical theorem due to J. von Neuman, but we give an elegant short modern proof due to B. R. Gelbaum (1964) of interest.

Let $\mathfrak{a}$ be the set of continuous linear operators, made into a closed (sub) algebra of (continuous) linear mappings from $B(L_2^2(P))$ where $\mathfrak{a}$ is determined by the set $\{I, N_s^*, N_t, s, t \in T\}$ under the uniform (or operator) norm. Since T is a separable set, the family $\mathfrak{a}$ is a separable Banach*-algebra with identity and is isomorphic to the space $C(M)$, the space of continuous complex functions on the compact metric space M which is the space of "maximal ideals" of $\mathfrak{a}$. This is a consequence of I. M. Gel'fand's representation theorem for Banach*-algebras (cf. Loomis (1953), Sec. 26). It is the new or modern application of the *abstract Harmonic Analysis*, that gives a clearer structure of the problem although we need to use the abstract ideas. [If $\mathfrak{a}$ is not separable, then M will not be metric, and still compact, but our argument applies to both cases.]

The compact M is thus a complete separable metric (or Polish) space but M is uncountable. Now one can invoke a well-known theorem, due to Kuratowski of 1930, stating that there exists an $f: I \to M$, which is one-to-one and onto so f, f^{-1} are Borel functions, called a Borel equivalence (cf. e.g. Royden (1988), p. 406). Since a continuous complex function is uniformly approximable by (complex) simple functions (by the Stone-Weierstrass theorem), if $a < t_k < 1$, and $I_k = [0, t_k] \uparrow I$, their correspondents denoted E_{t_k} in $\mathfrak{a}$, form a resolution of the identity, and the isomorphism between $\mathfrak{a}$ and $C(M)$ gives for each $B \in \mathfrak{a}$, a unique $b_i \in C(M)$ (or $B_c \leftrightarrow b_i$), so that $\sum_{i=1}^{n} \lambda_i (\chi_{I_i} - \chi_{I_{i-1}}) = b_i^n$ satisfies, for the uniform norm,

$$\|b_t - b_t^n\|_\infty < \epsilon, \tag{13}$$

and this gives (using operator norm on β_i) by Gel'fand's isomorphism

$$\|B_t - B_t^n\| < \epsilon \tag{14}$$

for each $\epsilon > 0$. Using the Riemann-Stieltjes approximation one gets

$$B_t = \int_I b(t, \lambda) dE_\lambda, \quad t \in T. \tag{15}$$

It now follows, on taking $B_t = M_t$, for an element $X_{t_0} \in L_0^2(P)$,

$$X_t = N_t X_{t_0} = \int_I b(t, \lambda) dz(\lambda), \quad t \in T.$$

Thus $Z(\cdot)$ is a vector measure satisfying the conditions of the theorem and $\|Z(A)\|_2^2 = \alpha(A)$ gives ($\alpha(\cdot)$ is a finite measure):

$$r(s, t) = \int_I b(s, \lambda) \bar{b}(t, \lambda) d\alpha(\lambda). \tag{16}$$

This is (12) and the result follows. $\square$

Remark 8. 1. The key function $b(s, \cdot)$ above can be given a concrete construction, also following Gelbaum (1964) who refined Kuratowski's method. It is noted that there is a Cantor set $S \subset I = [0, 1]$ which is (isometrically) equivalent to M, and let $f: S \to M$ be an onto isomorphism. Now set $M_s = f([0, s) \cap S) \subset M, s \in I$ and let χ_{M_s} correspond, under isometric equivalence, to $E_s \in \mathfrak{a}$. Then $\{E_s, s \in I\}$ is a resolution of the identity and the collection $B_s \leftrightarrow b(s, \cdot)$ defined as: $b(t, u) = B_t(f(a))\chi_S(u) + \alpha b(t, u_1) + \beta b(t, u_2)$, if $u = \alpha u_1 + \beta u_2$, where (u_1, u_2) is a deleted interval for S in $I, \alpha \in I$, $\beta = 1 - \alpha$. It can be seen that this $b(t, \cdot)$ satisfies (16). A more detailed computation is given, with discussion in the above noted interesting Gelbaum paper.

2. If T is not separable, then the Kuratowski theorem used above is not applicable. Then by a type of *counting method* Gelbaum shows that the above work needs a drastic change and (16) does not hold if card $(T) > 2^{\mathrm{card}(M)}$. The corresponding result will be sketched here which is still of Karhunen's class.

A generalized (less sharp) version can be given as follows:

Theorem 2.1.9 *On a Lebesgue space $L_0^2(P)$ of centered random variables on (Ω, Σ, P), and an indexed set $\{B_t, t \in T\}$ of a commuting normal collection of bounded linear operators on $L_0^2(P)$, if $X_0 \in L_0^2(P)$ is a centered random variable, consider $X_t = B_t X_0 \in L_0^2(P), t \in T$. Then $\{X_t, t \in T\}$ forms a **Karhunen field**, in that there exists a compact set M, a stochastic measure $Z: \mathcal{B}(M) \to L_0^2(P)$ of orthogonal values, and a continuous family of functions $f(t, \cdot): M \to \mathbb{C}$, relative to which the following stochastic integral representation holds:*

$$X_t = B_t X_{t_0} = \int_M f(t, \lambda)dZ(\lambda), \quad t \in T, \qquad (17)$$

$Z(\cdot) = Z_{X_0}(\cdot)$, and the covariance function r of the X_t-field is representable as:

$$r(s, t) = (X_s, X_t) = \int_M f(s, \lambda)\bar{f}(t, \lambda)d\alpha(\lambda), \quad s, t \in T, \qquad (18)$$

where $\alpha(A) = \|Z(A)\|_2^2, A \in \mathcal{B}(M), Z(\cdot)$ being orthogonally valued.

Proof. The argument uses an interesting theorem due to N. Dunford (cf. Dunford–Schwartz, Part III (1963), Theorem X.2.1) which is an extension of a result due to J. von Neumann and uses some abstract analysis.

Thus let $\mathfrak{a}$ be a B^*-algebra generated by $\{I, B_s^*, B_t, s, t \in T\}$. Then by the Gel'fand–Neumark theorem $\mathfrak{a}$ is isomorphic and *-isometric to $C(M)$, the continuous complex function space on the set M of maximal ideals of $\mathfrak{a}$, which is a compact Hausdorff-space in the 'hull-kernel' topology. Thus for each $t \in T$, there exists an $f_t = f(t, \cdot) \in C(M)$, so that $f \leftrightarrow B_t(f_t = \hat{B}_t)$. With this, we can invoke Dunford's *generalized spectral theorem* [detailed in Dunford-Schwartz (1963), Part II of this book, Theorem X.2.1] so that there is a unique spectral measure $E(\cdot): \mathcal{B}(M) \to \mathfrak{a}$ such that (the construction of $E(\cdot)$ being the new element):

$$B_t = \int_M f(t, \lambda) dE(\lambda), \quad t \in T,$$

and then

$$X_t = B_t X_{t_0} = \int_M f(t, \lambda) dE(\lambda) X_{t_0} = \int_M f(t, \lambda) dZ_{X_0}(\lambda),$$

which gives (7), the projection $E(\lambda)$ commuting with all elements of $\mathfrak{a}$, $E(\cdot)$ being strongly σ-additive and orthogonally valued. Then (17) and (18) are immediate consequences. $\square$

The point of the Karhunen field represented by the above theorem is to place the second order process or field in an abstract Hilbert space context, taking most second order processes into its fold. To make this idea clear, by showing how several of the random fields are brought under its fold, we present these points in a general setting, and show how H. Cramér gave an essential "ultimate" formulation, making again the harmonizable class a key part here.

Theorem 2.1.10 *Let $\{X_t, t \in T\} \subset L_0^2(P)$ be a process or field with mean zero and covariance $r(s, t)$ which has the integral representation:*

$$r(s, t) = E(X_s \bar{X}_t) = \int_S g(s, u) \bar{g}(t, u) d\alpha(u) \tag{19}$$

for some measure space $(S, \mathcal{S}, \alpha)$ and a square integrable set $\{g(s, \cdot), s \in T\} \subset L^2(\alpha)$ so that the process (or random field) is of Karhunen class. Then it has an operator representation, relative to a set of commuting operators $\{B_r, B_s^*, r, s \in T\} \subset B(L_0^2(P))$ so that

$$X_t = B_t X_{t_0} = \int_S g(t, \lambda) dZ_{X_{t_0}}(\lambda), t \in T, \tag{20}$$

for a unique orthogonally valued measure $Z_{X_{t_0}} : S \to L_0^2(P)$ such that $\alpha(A) = \|Z_{X_{t_0}}(A)\|_2^2, A \in S$. On the other hand, every abelian family $\{B_t, t \in T\} \subset B(L_0^2(P))$ of normal operators defines a Karhunen field $\{X_t = B_t X_{t_0}, t, t_0 \in T\}$, by the preceding theorem.

Remark 9. This result can be given a somewhat generalized form with the associated concept introduced by H. Cramér, to be detailed below and this set of ideas connects nicely the above vector measures on $L^2(P)$.

It may be noted in passing that the general class of our Karhunen fields is representable by an abelian set of bounded operators on $L_0^2(P)$ whose generalized spectrum (in the Gel'fand sense) is locally compact. If the process $\{X_t, t \in \mathbb{R}\}$ also admits a bounded shift operator and is weakly continuous, then there is a strongly continuous semi-group $\{N_t, t \in \mathbb{R}\}$ of normal operators such that $\{X_t = N_t X_0, t > 0\}$ and a spectral set $\Lambda \subset \mathbb{R}$. If the process is stationary, then $X_t = U_t X_0$, where $\{U_t, t \in \mathbb{R}\}$ is a unitary group on $L_0^2(P)$. Thus there is always an operator representation and the process admits a stochastic integral formulation with an orthogonally valued measure on $\mathcal{B}(\Lambda)$ into $L_0^2(P)$. This result specialized to Karhunen processes was obtained by Getoor (1956) using a different method. If $T = \mathbb{R}$ (or an LCA group) and X_t is weakly continuous, then one has $X_t = A_t U_t Y_0$ where A_t commutes with U_t, and A_s is a closed densely defined mapping, $\{U_t, t \in \mathbb{R}\}$ being a unitary group. [See Mizel and Rao (2009) for some extensions.]

As in the harmonizable case, the orthogonally valued condition on $Z(\cdot)$ can also be relaxed for the Karhunen class as follows:

Proposition 2.1.11 *Let the mean continuous Karhunen process $\{X_t, t \in T\} \subset L_0^2(P)$ with representation $X_t = B_t X_0$ be as in (20) above. If the abelian class $\{B_t, t \in T\}$ forms a group, T being an LCA group, then the X_t admits a representation*

$$X_t = \int_T \langle t, \lambda \rangle dZ(\lambda), t \in T, \quad (\langle t, \cdot \rangle \in \hat{T}), \tag{21}$$

where the measure is $Z : \mathcal{B}(\mathbb{R}) \to L_0^2(P)$, so that the field is weakly harmonizable. ($\hat{T}$ is the dual of T.)

Proof. The result is a slight refinement of the preceding consideration, as it depends on a modification due to Wermer (1954, and see Dunford-

Schwartz, Part III, Lemma XV.6.1) on commuting normal operators, according to which there exists a bounded self-adjoint operator C with a bounded inverse so that $U_t = CB_tC^{-1}$ is unitary. Consequently, there is a spectral resolution of U_t giving the representation (by a classical result due to Hille, used before),

$$B_t = C^{-1}U_tC = \int_{\hat{T}} \langle t, \lambda \rangle dC^{-1}E_\lambda C, \quad t \in T, \tag{22}$$

and hence we have

$$X_t = B_t X_{t_0} = \int_{TR} e^{it\lambda} dZ(\lambda), t \in T, \tag{23}$$

where $Z \colon \mathcal{B}(\mathbb{R}) \to L_0^2(P)$ is σ-additive but not orthogonally valued. Hence $\{X_t, t \in T\}$ is just weakly harmonizable. $\square$

An extension of the preceding results based on Karhunen's work was given by Cramér (1951), and it will now be discussed as it presents a completion of this set of ideas leading to a generalized harmonizability. Thus Cramér considers a centered second order process $X_t, t \in \mathbb{R}$, whose covariance $r \colon (s, t) \to E(X_s \bar{X}_t)$ is representable as:

$$r(s, t) = \int_{\mathbb{R}} \int_{\mathbb{R}} g(s, u)\bar{g}(t, v)\beta(du, dv), \quad s, t \in \mathbb{R}, \tag{24}$$

where $\beta(\cdot, \cdot)$ is a positive definite set function of *bounded Vitali variation* so that (24) is defined as a Lebesgue-Stieltjes integral.

It is natural to follow the basic format of the (weakly) harmonizable process and its integral representation, extending the ideas and methods of the harmonizable classes and we can present the generalized representation for the (weak) Cramér class itself. It is useful to restate the matter to avoid ambiguity. Thus if $Z \colon \mathcal{B}(\mathbb{R}) \to L_0^2(P)$ is an orthogonally valued (vector) measure and $V \colon L_0^2(P) \to L_0^2(P)$ is a bounded linear operator then the function $\tilde{Z} = VZ \colon \mathcal{B}(\mathbb{R}) \to L_0^2(P)$ is σ-additive and β defined as $\beta(C, D) = (\tilde{Z}(C), \tilde{Z}(D))$ is a bimeasure, and the representation (21) becomes in this case:

$$Y_t = VX_t = \int_{\mathbb{R}} g(t, \lambda)d\tilde{Z}(\lambda), \quad t \in \mathbb{R}. \tag{25}$$

Then the Y_t-process is termed a *weak Cramér process* with $\beta(\cdot, \cdot)$ as its bimeasure of *finite-Fréchet* (not Vitali) variation. Its covariance is then given by (for $s, t \in \mathbb{R}$)

$$r(s,t) = (Y_s, Y_t) = \int_{\mathbb{R}} \int_{\mathbb{R}} g(s,u)\bar{g}(t,u')\beta(du,du'). \qquad (26)$$

Here the bimeasure $\beta(\cdot,\cdot)$ of (26) is not of (even locally) finite Vitali variation, and the integral now should be defined in *strict MT-sense*, to be useful in the next chapter.

Here we include a dilation result connecting the general weak Cramér and Karhunen processes somewhat extending the corresponding harmonizable cases to round out the present analysis. This will clarify the underlying structures based on Fourier or harmonic analysis that is at the base of this problem.

Theorem 2.1.12 *Let $\{X_t, t \in T\}$ be a centered process of weak Cramér class relative to a family $\{g(t,\cdot), t \in T\}$ of strictly β-integrable functions defining its covariance r that is representable as:*

$$r(s,t) = (X_s, X_t) = \int_S \int_S g(s,u)\bar{g}(t,u')\beta(du,du'), s,t \in T. \quad (27)$$

Then there exists a stochastic measure $Z \colon \mathcal{B}(S) \to L_0^2(P)$ such that

$$X_t = \int_S g(t,\lambda)dZ(\lambda), t \in T, \qquad (28)$$

holds as a Dunford-Schwartz vector integral, where $E(Z(C)\bar{Z}(D)) = \beta(C,D), C, D \in \mathcal{S}_0$, the δ-ring of bounded Borel sets from S contained in compact sets. Conversely, every process defined by (28) is of weak Cramér class with covariance (27) as a strict MT-integral.

The result is detailed with necessary background material on *(strict) Morse-Transue integrals* in the article entitled 'Bimeasures and Nonstationary Processes' by Chang and Rao (1986), going over 100 pages and hence it will be referred to, but not reproduced here. That every Cramér process may be 'dilated' to a Karhunen process on a super-Hilbert space will now be stated. This is similar to (but extends) the result that a (weakly) harmonizable process can be dilated to a stationary one on an inclusive Hilbert space which is analogous to, and using, the classical result of Naĭmark's with its extension by Sz.-Nagy which in our case implies that a (weak) Cramér family can be dilated to a Karhunen class on a larger Hilbert space. In the classical context, this says that a harmonizable class can be dilated to a stationary family on a super Hilbert space, just guides to further work.

The desired dilation of Cramér process to the Karhunen class on a larger $L^2(P)$-space can be given as follows. [There are however some differences between this and the harmonizable case.]

Theorem 2.1.13 *Let* $\{X_t, t \in T\} \subset L_0^2(P)$ *be a process of weak Cramér class so that it is representable as:*

$$X_t = \int_S g(t, \lambda) dZ(\lambda), \quad t \in T, \tag{29}$$

as in the preceding result. Then there exists a dilation to a large Hilbert space $\mathcal{K}$ *containing the vector space* $\mathcal{L}$ *generated by the* X_t*-process, denoted as* $\{Y_t, t \in T\}$*, which is a Karhunen process relative to a measure* μ *on* $(S, \mathcal{S})$*, such that* $X_t = QY_t$ *where* Q *is the orthogonal projection of* $\mathcal{K}$ *onto* $\mathcal{L}$ *determined by the* X*-process, the space spanned by the weak Cramér process,*

$$Y_t = \int_S g(t, u) d\tilde{Z}(u), \quad t \in T, \tag{30}$$

where the $\tilde{Z}(\cdot)$ *has orthogonal increments so that* $(\tilde{Z}(C), \tilde{Z}(D)) = \mu(C \cap D), C, D \in \mathcal{S}$*. But a* Y_t*-process given by (30) defined as* $X_t = VY_t, t \in T$ *is always of a weak Cramér process whose covariance can be given as an integral for* $s, t \in \mathbb{R}$*:*

$$r(s, t) = (Y_s, Y_t) = \int_{\mathbb{R}} \int_{\mathbb{R}} g(s, u) \bar{g}(t, v) \beta(du, dv), \tag{31}$$

only when $\{g(t, \cdot), t \in T\}$ *is strictly MT-integrable.*

We omit the details of proof as it uses some works that are needed for the preceding result, and are available in the same reference. The work shows that the Cramér and Karhunen classes are the abstractions of the harmonizable and the stationary families that opened the probabilistic analysis and use extensions of Fourier methods. It will be seen below that several other areas are opened up by this abstraction which enriches and deepens the analysis further. A few related results will be included in the Complements at the end of this chapter. Several of these ideas admit interesting Hilbert space extended operator analysis with probabilistic insights, leading to studies of new problems of use with stochastic applications.

The above discussion implies the following easy (abstract) characterization of harmonizable processes in the way stationarity was conceived by Khintchine in the earlier times:

Theorem 2.1.14 *Let $\{X_t, t \in T \subset \mathbb{R}\} \subset L_0^2(P)$ be a centered second order process. Then it is weakly harmonizable iff*

(i) $E(|X_t|^2) \leq M < \infty, t \in T$, whose covariance $r(\cdot, \cdot)$ is continuous on $T \times T$ and

(ii) $\sup\{E(|\int_T f(t)X_t dt|^2): \|\hat{f}\|_\infty \leq 1\} < \infty$, where $\hat{f} : t \mapsto \int_{\mathbb{R}} e^{it\lambda} f(\lambda) d\lambda$, is the Fourier transform of f. Here in (ii) the symbol is the (vector or) Bochner integral.

The following result illustrates the above theorem and gives a feeling for the subject. Thus let $\{S(t), t \geq 0\}$ be a contraction semi-group of linear operators on $L_0^2(P)$. This means $\|S(t)\| \leq 1$, (operator norm), $S(u + v) = S(u)S(v), S(0) = id, u, v \geq 0$, and $\|S(u)f - f\| \to 0$ as $u \to 0^+$, where $\|f\|^2 = E(|f|^2)$. Now let $Y(t) = S(t)X_0$, if $t \geq 0$, and $= S^*(-t)X_0$ if $t < 0$, where $S^*(t)$ is the *adjoint operator* of $S(t)$. Hence

$$E((S(u)f), \bar{g}) = E((f, S^*(u)\bar{g})), f, g \in L_0^2(P).$$

Setting $S(-u) = S^*(u)$, the process $\{Y(t), t \in \mathbb{R}\}$ becomes weakly harmonizable, and the family $\{S(u), u \in \mathbb{R}\}$ is positive definite:

$$\sum_{i=1}^{n} \sum_{j=1}^{n} E((S(u_i - u_j)f_i, \bar{f}_j)) \geq 0, f_i \in L_0^2(P),$$

for each set $\{u_1, \ldots, u_n\} \subset T$. This is easy if $T = \mathbb{Z}$ and the case $T = \mathbb{R}$ can then be reduced to the former using some standard analysis.

The relation between the covariance function and its measure for (weakly) harmonizable processes, an analog of the classical inversion theorem, can be given which is of interest in this analysis. The result was considered by Loève for the strongly harmonizable case and its analog for the weak harmonizability is not obvious. We include the result as it motivates further analysis and insight in this theory.

The following inversion formula, an analog of the classical Lévy case and of Loève's for the strongly harmonizable result, is obtainable with some standard computations and some modifications. Here is the formula:

Proposition 2.1.15 *Let $r(\cdot, \cdot)$ be a weakly harmonizable covariance function of $\{X_t, t \in \mathbb{R}\} \subset L_0^2(P)$ with $F(\cdot, \cdot)$ as its spectral bimeasure. If $A = (a_1, a_2), B = (b_1, b_2)$ are continuity intervals of F, then*

$$F(A, B) = \lim_{\substack{a \to \infty \\ b \to \infty}} \int_{-a}^{a} \int_{-b}^{b} \frac{e^{-ia_2 s} - e^{-ia_1 s}}{-is} \cdot \frac{e^{+ib_2 t} - e^{+ib_1 t}}{it} r(s, t) ds dt.$$

It will be of interest to find an estimator of F "optimally" using some extended statistical analogs of the stationary case. This has not been done, and may be a useful problem to consider.

The preceding discussion and analysis on the importance of a weak (and strong) harmonizable process and its dependence on Fourier analysis, shows a need to present characterizations of the key weak harmonizability, extending the classical (weak) stationarity in the next section and then use it in the following analysis.

2.2 V-Boundedness, Weak and Strong Harmonizabilities

Since V-boundedness plays a key role in our analysis, it may be useful to present some of its basic properties and results here and use them freely later on. Recall that a centered process $X_t, t \in \mathbb{R}$, is *(strongly) harmonizable* if its covariance r admits the integral representation as a Fourier transform relative to a bimeasure F, so that

$$r(s, t) = \int_{\mathbb{R}} \int_{\mathbb{R}} e^{is\lambda - it\lambda'} dF(\lambda, \lambda'), \quad s, t \in \mathbb{R}, \tag{32}$$

where F is of bounded Vitali variation. However, this is not really broad enough for applications. In fact consider $\{X_n, n \in Z\}$, $X_n \in L_0^2(P)$, and be an orthonormal family so that the set is stationary, and $r(m, n) = \tilde{r}(m - n) = \frac{1}{2\pi} \int_{-\pi}^{\pi} e^{i(m-n)\lambda} d\lambda, m, n \in \mathbb{Z}$, but the truncated series $\tilde{X}_m = X_n, n > 0, = 0$ for $n \leq 0$, has the property that their covariance $r : (m, n) \to E(\tilde{X}_m \bar{\tilde{X}}_n) = 1$ if $m = n$, and $= 0$ otherwise, so that $\tilde{r}$ cannot be expressed as (32), by a nontrivial extension of a theorem of F. and M. Riesz, due to S. Bochner, that $\tilde{r}$ must be absolutely continuous for all $m = n$, and $\tilde{r}(m - n) \to 0$ as $|m| + |n| \to \infty$ which does not hold here. Hence $\tilde{r}$ cannot be represented as (32), using the Lebesgue integral.

In order to advance the subject, we need to go for a *weaker integral*, due to M. Morse and W. Transue (1955/56) which is found suitable here. We shall see that the (Morse–Transue or) MT-*integral* recalled in the last section, and a slight modification of it, will play an important role in our analysis as well as in many other applications.

Let Ω_1, Ω_2 be locally compact Hausdorff spaces and $\Lambda\colon C_c(\Omega_1) \times C_c(\Omega_2) \to \mathbb{C}$ be a bilinear form on the product compactly based continuous scalar function spaces such that $\Lambda(f, \cdot)$ and $\Lambda(\cdot, g)$ are linear and are C-measures (or that each is a continuous complex linear functional). Thus by the usual Daniell formulation we have, $I = I_1 - I_2 + i(I_3 - I_4)$ and each I_j corresponds to a "C-measure".

As a motivation for the main characterization of weakly harmonizable processes, we recall the following earlier result due to Niemi (1975) which gives a feeling for the subject.

Theorem 2.2.1 *Let $X\colon \mathbb{R} \to L_0^2(P)$ be a process which is weakly continuous and valued in a ball, so that $\|X(t)\|_2 \leq M < \infty, t \in \mathbb{R}$. Then the process X is weakly harmonizable relative to a covariance bimeasure F of finite semi-variation, if and only if there exists a stochastic measure $Z\colon \mathcal{B}(\mathbb{R}) \to L_0^2(P)$ with $F(\cdot, \cdot)$ as its bimeasure $F\colon (A, B) \mapsto (Z(A), Z(B))$ and representation.*

$$X(t) = \int_{\mathbb{R}} e^{it\lambda} Z(d\lambda), \quad t \in \mathbb{R}, \tag{33}$$

as a Dunford-Schwartz integral, and $\|Z\|(\mathbb{R}) < \infty$. Particularly $X(\cdot)$ is strongly harmonizable iff the bimeasure F of Z in (33) is of bounded variation (in Vitali's sense) in $\mathbb{R}^2$. In any case the harmonizable process $X\colon \mathbb{R} \to L_0^2(P)$ is uniformly continuous, and is represented as a vector integral given by (33).

This result contains an important spectral representation due to Rozanov (1959), and Niemi (1975) has given an independent demonstration. It will be presented here with a different demonstration using an important new concept naturally due to Bochner (1956), called V-boundedness of great interest. It was also considered by Phillips (1950) in a slightly different context, and the ideas are used here which illustrate the key role of the Fourier analysis in our subject.

Definition 2.2.2 *Let $X = \{X(t), t \in \mathbb{R}\} \subset \mathfrak{X}$, where $\mathfrak{X}$ is a Banach space. It is V-bounded (V for variation) if the set $\{X(t)\colon t \in \mathbb{R}\} \subset \mathfrak{X}$ is in a ball, $X(\cdot)$ is strongly measurable (i.e., $X(\mathbb{R})$ is separable and $X^{-1}(B) \in \mathcal{B}$, for each Borel set $B \subset \mathbb{R}$), and the set C is relatively weakly compact in $\mathfrak{X}$ where*

$$C = \left\{ \int_{\mathbb{R}} f(t)X(t)dt\colon \|\hat{f}\|_u \leq 1, f \in L(\mathbb{R}) \right\} \subset \mathfrak{X} \tag{34}$$

and $\hat{f}\colon t \mapsto \int_{\mathbb{R}} e^{it\lambda} f(\lambda) d\lambda$, *the integral of the vector function* $f(\cdot)$ *is in Bochner's sense (or the standard Bochner integral).*

A fundamental characterization of weak harmonizability due to Bochner (1956), with some earlier analysis by Phillips (1950), is given by the following key result when $\mathcal{X} = L_0^2(P)$:

Theorem 2.2.3 *A stochastic process* $X \colon \mathbb{R} \to L_0^2(P)$ *is weakly harmonizable iff it is V-bounded, and weakly continuous.*

Proof. First, let X be weakly continuous and V-bounded, so

$$\left\| \int_{\mathbb{R}} f(t) X(t) dt \right\|_2 \leq c \|\hat{f}\|_u, \quad f \in L^1(\mathbb{R}), c > 0, \tag{35}$$

by definition of V-boundedness. Also $\mathcal{Y} = \{\hat{f} \colon f \in L^1(\mathbb{R})\} \subset C_0(\mathbb{R})$ the continuous complex function space each $\hat{f}$ vanishing at infinity, by the Riemann-Lebesgue lemma. Since $\mathcal{Y}$ is a real algebra in $C_0(\mathbb{R})$ and separates points of $\mathbb{R}$, it is uniformly dense by the classical Stone-Weierstrass theorem. If $\mathcal{F} \colon f \mapsto \int_{\mathbb{R}} f(\lambda) e^{-it\lambda} d\lambda, t \in \mathbb{R}$, then $\mathcal{F} \colon L^1(\mathbb{R}) \to C_0(\mathbb{R})$ is one-to-one and contractive. The mapping $T \colon \mathcal{Y} \to \mathcal{X} = L_0^2(P)$, as $T(\hat{f}) = \int_{\mathbb{R}} f(t) X(t) dt \in \mathcal{X}$ is well-defined. The following diagram is commutative where $T_1(\cdot)$ is given by:

$$T_1(f) = \int_{\mathbb{R}} f(t) X(t) dt \in \mathcal{X}.$$

By hypothesis T is bounded and since $\mathcal{Y} \subset C_0(\mathbb{R})$ is dense, it has a norm-preserving extensions $\tilde{T}$ to $C_0(\mathbb{R})$. Since $\mathcal{X}$ is reflexive, the mapping $\tilde{T}$ is weakly compact and by a classical theorem in representing such operators (cf., Dunford-Schwartz (1958), VI.73), it can be given an integral form as follows: However, details are unfortunately unavailable, in order to make the work (result) clearer. Details are sketched here. Thus let $\mathcal{L} = \mathbb{R}$ and $\tilde{\mathcal{L}}$ be the 'one point' compactification of $\mathcal{L}$

and consider the space $C(\tilde{\mathcal{L}})$. Now $C_0(\mathcal{L})$ can be identified as a subspace of $C(\tilde{\mathcal{L}})$ whose elements vanish at '∞'. Since $\tilde{T}: C_0(\mathcal{L}) \to \mathcal{X}$ is continuous and $C_0(\mathcal{L})$ is an "abstract M-space", there is a continuous extension of $\tilde{T}$, agreeing with the original of course on $C_0(\mathcal{L})$. [This uses the well-known theorem due to Kelley-Nachbin-Goodner, and then we have $T = \tilde{T} = Q$.] All this work implies that there is a $\tilde{Z}: \tilde{\mathcal{L}} \to \mathcal{X}$, a vector measure, such that

$$T(f) = \int_{\tilde{\mathcal{L}}} f(t)\tilde{Z}(dt), \quad f \in C(\tilde{\mathcal{L}}), \tag{36}$$

and $\|\tilde{T}\| = \|\tilde{Z}\|(\tilde{\mathcal{L}})$, using the D-S integral on the right. If we set $Z: A \to \tilde{Z}(\mathcal{L} \cap A), A \in \mathcal{B}(\mathcal{L})$, then $Z(\cdot)$, a vector measure, satisfies $\|Z\| \leq \|\tilde{Z}\|$, and if $f_0 = f|_{\mathcal{L}}$, we have:

$$\tilde{T}(f) = \int_{\mathcal{L}} f_0(t)dZ(t) + \int_{\{\infty\}} f(t)\tilde{Z}(dt), \quad f \in C(\tilde{\mathcal{L}})$$
$$= \bar{T}(f_0), \quad \text{since } f(\infty) = 0.$$

Hence $\tilde{T}(f) = \bar{T}(f), f \in C_0(\mathcal{L}), \|\tilde{T}\| \leq \|\bar{T}Q\| \leq \|\tilde{T}\|$, and

$$\tilde{T}(f) = \int_{\mathcal{L}} f(t)Z(dt), \quad f \in C_0(\mathcal{L}). \tag{37}$$

Thus writing $\mathbb{R}$ for $\mathcal{L}$, it follows now that

$$\|\tilde{T}\| = \sup\left\{ \left\| \int_{\mathbb{R}} f(t)Z(dt) \right\| : f \in C_0(\mathbb{R}), \|f\| \leq 1 \right\}$$
$$= \|Z\|(\mathbb{R}) = \|\tilde{Z}\|(\mathbb{R}), \tag{38}$$

and $T \Leftrightarrow Z$ corresponds uniquely. Also $\|T\| = \|Z\|(\mathbb{R})$.

Let $l \in \mathcal{X}^*$ and applying it to (37) with $\hat{f}$ in place of f, we get

$$\int_{\mathbb{R}} \hat{f}(t)(l \circ Z)(dt) = \int_{\mathbb{R}} f(t)l \circ X(t)dt, f \in L'(\mathbb{R}), \tag{39}$$

since $\tilde{T}|\mathcal{Y} = T$ so that $T(\hat{f}|\mathbb{R}) = \int_{\mathbb{R}} f(t)X(t)dt$ for all $f \in L^1(\mathbb{R})$. It follows from (39) with properties of Lebesgue integrals that

$$\int_{\mathbb{R}} f(t)dt \int_{\mathbb{R}} e_t(\lambda)l \circ Z(d\lambda) = \int_{\mathbb{R}} f(t)l \circ X(t)dt.$$

Hence on rearrangement, we have

$$\int_{\mathbb{R}} f(t) l \left(\int_{\mathbb{R}} e_t(\lambda) Z(d\lambda) - X(t) \right) dt = 0, f \in L(\mathbb{R}). \qquad (40)$$

Since f is arbitrary, and $l \in \mathfrak{X}^*$ is also unrestricted, (40) implies

$$X(t) = \int_{\mathbb{R}} e_t(\lambda) Z(d\lambda) = \int_{\mathbb{R}} e^{it\lambda} Z(d\lambda), t \in \mathbb{R}. \qquad (41)$$

This shows that X is weakly harmonizable by (32).

In the opposite direction, let $X: \mathbb{R} \to L_0^2(P)$ be weakly harmonizable so that it can be expressed as (41), with $\|Z\|(\mathbb{R}) < \infty$. Hence $\|X(t)\|_2 \leq M_0 < \infty, t \in \mathbb{R}$, and $(l \circ X)(\cdot)$ is the Fourier transform of $l \circ Z, l \in \mathfrak{X}^*$. So $X(\cdot)$ is weakly continuous. Then the following computation is valid, using the vector integration properties:

$$l \left(\int_{\mathbb{R}} f(t) X(t) dt \right) = \int_{\mathbb{R}} f(t) l \left(\int_{\mathbb{R}} e_t(\lambda) Z(d\lambda) \right) dt$$

$$= \int_{\mathbb{R}} \int_{\mathbb{R}} f(t) e_t(\lambda) l \circ Z(d\lambda) dt = \int_{\mathbb{R}} \hat{f}(\lambda) l \circ Z(d\lambda)$$

$$= l \left(\int_{\mathbb{R}} \hat{f}(\lambda) Z(d\lambda) \right). \qquad (42)$$

Since $l \in \mathfrak{X}^*$ is arbitrary, (42) implies

$$\int_{\mathbb{R}} f(t) X(t) dt = \int_{\mathbb{R}} \hat{f}(\lambda) Z(d\lambda) \in \mathfrak{X}. \qquad (43)$$

It follows from this and the preceding computations that

$$\left\| \int_{\mathbb{R}} f(t) X(t) dt \right\|_2 \leq \|\hat{f}\|_u \|Z\|(\mathbb{R}) = C\|\hat{f}\|_u, f \in L^1(\mathbb{R}), \qquad (44)$$

where $C = \|Z\|(\mathbb{R}) < \infty$. Hence the following set $\{ \int_{\mathbb{R}} f(t) X(t) dt : \|\hat{f}\|_u \leq 1 \}$ is bounded for $f \in L^1(\mathbb{R})$. The reflexivity of $\mathfrak{X}$ implies that X is then V-bounded, completing the converse. $\square$

Some consequences and extensions of the above result will be presented below both for applications and some of its extensions which also are of interest. It should be observed that the above result (and proof) extends if $\mathbb{R}$ is replaced by an LCA (= locally compact abelian) group G, so $G = \mathbb{R}^n$ included.

In particular, if $T\colon L_0^2(P) \to L_0^2(P)$ is a bounded linear operator and $Y(t) = TX(t), t \in \mathbb{R}$, where X is weakly harmonizable, then the abstract analysis so far used implies, for $T \in B(\mathcal{X})$,

$$Y(t) = T\left(\int_{\mathbb{R}} e^{it\lambda} Z(d\lambda)\right) = \int_{\mathbb{R}} e^{it\lambda} (T \circ Z)(d\lambda)$$

where $\tilde{Z} = T \circ Z$ is a stochastic measure, $\|\tilde{Z}\|(\mathbb{R}) \leq \|T\| \|Z\|(\mathbb{R}) < \infty$. Thus,

Corollary 2.2.4 *The linear span of weakly harmonizable processes forms a module over the class of all bounded linear transformations on* $\mathcal{X} = L_0^2(P)$.

As seen already in the earlier discussion that on an orthonormal sequence $\{X_n, -\infty < n < \infty\}$ which is trivially (weakly) stationary, has its image $\{X_n, n \geq 0\}$ by an orthogonal projection need not be stationary again, and not even strongly harmonizable, but is weakly harmonizable by the above corollary; this is of interest in both theory and applications. We have the following approximation, due to Niemi (1975) that is also of interest here. [For another *type* of standard representation, see Theorem 2.2.10 below.]

Theorem 2.2.5 *Let* $\{X_t, t \in \mathbb{R}\} \subset L_0^2(P)$ *be a weakly harmonizable process. Then there exists a sequence* $\{X_n(t), t \in \mathbb{R}\}_{n=1}^{\infty} \subset L_0^2(P)$ *of strongly harmonizable processes such that* $X_n(t) \to X(t)$, *in* $L_0^2(P)$ *as* $n \to \infty$, *uniformly in* t *on compact subsets of* $\mathbb{R}$. *The same conclusion holds if* $\mathbb{R}$ *is replaced by an LCA group* G *and then the sequence is to be replaced by a net of such processes (or fields).*

Proof. By the preceding (or earlier) analysis the process X_t can be given an integral representation as

$$X_t = \int_{\mathbb{R}} e_t(\lambda) dZ(\lambda), \quad t \in \mathbb{R} \tag{45}$$

for a stochastic (or vector) measure $Z\colon \mathcal{B}(\mathbb{R}) \to L_0^2(P)$. Let $\mathcal{X} = \bar{sp}\{X_t \colon t \in \mathbb{R}\} \subset L_0^2(P)$ which is a separable subspace. Hence there exists a complete orthonormal sequence $\{\phi_n, n \geq 1\} \subset \mathcal{X}$ so that

$$X_n(t) = \sum_{k=1}^{n} \phi_k (X_k(t), \phi_k), t \in \mathbb{R}. \tag{46}$$

We assert that $\{X_n(t), t \in \mathbb{R}, n \geq 1\}$ is a strongly harmonizable sequence that approximates (strongly), the weakly harmonizable process $X(t)$. It is clear that $X_n(t) \to X(t)$ in $L_0^2(P) = \mathfrak{X}$ for each $t \in \mathbb{R}$. To see that $X_n(t)$ is strongly harmonizable, let $l_k \colon X \mapsto (X, \phi_k)$, so $l_k \in \mathfrak{X}^*$, where $\mathfrak{X}^*$ is the adjoint space of $\mathfrak{X}$. Also since X_t is weakly harmonizable,

$$
\begin{aligned}
X_n(t) &= \sum_{i=1}^{n} \phi_k \cdot l_k(X(t)) = \sum_{k=1}^{n} \phi_k \cdot l_k \left(\int_{\mathbb{R}} e_t(\lambda) Z(d\lambda) \right) \\
&= \sum_{k=1}^{n} \phi_k \int_{\mathbb{R}} e_t(\lambda) \cdot (l_k \cdot Z)(d\lambda) = \int_{\mathbb{R}} e_t(\lambda) \zeta_k(d\lambda),
\end{aligned}
$$

where $\zeta_k(d\lambda)$ is the sum $\sum_{k=1}^{n} \phi_k l_k \circ Z(d\lambda)$ here. Also $G_n(A, B) = (\zeta_n(A), \zeta_n(B))$ has finite total variation since $(l_k \circ Z)(\cdot)$ is a signed measure. It then follows that $X_n(t)$ is strongly harmonizable.

Since $X(\cdot)$ is weakly harmonizable, it is seen to be strongly continuous. Also if $K \subset \mathbb{R}$ is compact then the image $X(K) \subset L_0^2(P)$ is also (norm) compact. Now $X_n(t)$ in $L_0^2(P)$ for each t, and since $X(K) \subset \mathfrak{X} = L_0^2(P)$ is compact implies (by the metric approximation property of Hilbert spaces) $X_n(t) \to X(t)$ in $L_0^2(P)$-uniformly in $t \in K \subset \mathbb{R}$. This implies all the assertions of the theorem. $\square$

It should be noted that the class of weakly harmonizable process forms a proper subset of bounded continuous processes in $L_0^2(P)$. We give an example in the complements section below to exemplify this. The following approximation (from above) is thus of interest in our study.

Theorem 2.2.6 *Let* $X \colon \mathbb{R} \to L_0^2(P)$ *be a weakly harmonizable process with* $Z \colon \mathcal{B}(\mathbb{R}) \to L_0^2(P)$ *as its representing measure given in Theorem 2.2.1 above. Then there exists a sequence of regular Borel measures* $\beta_n \colon \mathcal{B} \to \mathbb{R}^+$ *such that for each* $f \in C_0(\mathbb{R})$, *one has:*

$$
\left\| \int_{\mathbb{R}} f(t) Z(dt) \right\|_2 \leq \liminf_n \| f \|_{2, \beta_n}. \tag{47}
$$

Proof. By the preceding result there exists a strongly harmonizable sequence $X_n \to X$ uniformly on compact subsets of $\mathbb{R}$. If now ζ_n represents X_n, so both $\zeta_n, Z \colon \mathcal{B} \to L_0^2(P)$, we have

$$\int_{\mathbb{R}} f(\lambda)Z(d\lambda) = \lim_{n\to\infty} \int_{\mathbb{R}} f(\lambda)\zeta_n(d\lambda), \qquad (48)$$

for all trigonometric polynomials $f(\cdot)$ which are uniformly dense in $C_0(\mathbb{R})$ and separate point of $\mathbb{R}$. It then follows that (48) holds for all $f \in C_0(\mathbb{R})$, by the standard reasoning. Hence for $f \in C_0(\mathbb{R})$,

$$\left\|\int_{\mathbb{R}} f(\lambda)dZ(\lambda)\right\|_2^2 = \lim_{n\to\infty} \left\|\int_{\mathbb{R}} f(\lambda)\zeta_n(\lambda)\right\|_2^2$$

$$= \lim_{n\to\infty} \int_{\mathbb{R}}\int_{\mathbb{R}} f(\lambda)\overline{f(\lambda')}dF_n(\lambda, \lambda')$$

where $F_n(s, t) = (\zeta_n(-\infty, s), \zeta_n(-\infty, t))$ is a covariance of bounded variation for each n. Let $|F_n|(\cdot, \cdot)$ denote the Vitali variation measure of F_n. Then the hermitian property of F_n gives, $|F_n|(A, B) = |F_n|(B, A)$, and if $\beta_n(A) = |F_n|(A, \mathbb{R})$, it then defines a finite Borel measure, and

$$\int_{\mathbb{R}} f(\lambda)\beta(d\lambda) = \frac{1}{2}\left[\int_{\mathbb{R}}\int_{\mathbb{R}} f(s)\left[|F_n|(ds, dt) + f(t)|F_n|(ds, dt)\right]\right],$$
$$(49)$$

and since F_n is positive semi-definite, we have

$$0 \le \int\int_{\mathbb{R}^2} f(s)\overline{f(t)}F_n(ds, dt) \le \int\int_{\mathbb{R}^2} |f(s)\overline{f(t)}||F_n|(ds, dt)$$

$$\le \frac{1}{2}\int\int_{\mathbb{R}^2} (|f(s)|^2 + |f(t)|^2)|F_n|(ds, dt), \text{ since } |ab| \le (a^2 + b^2)/2,$$

$$= \int_{\mathbb{R}} |f(t)|^2\beta_n(dt), \quad \text{by (49).} \qquad (50)$$

It follows from (49) and (50) that

$$\left\|\int_{\mathbb{R}} f(\lambda)Z(d\lambda)\right\|_2^2 = \lim_n \int_{\mathbb{R}^2} f(\lambda)\overline{f(\lambda')}F_n(d\lambda, d\lambda')$$

$$\le \liminf_n \int_{\mathbb{R}} |f(\lambda)|^2 d\beta(\lambda).$$

This completes the proof. $\square$

To appreciate the significance of the above assertion we now indicate a general result, also of interest in this analysis, in which sequences of measures may be replaced by a single measure.

Theorem 2.2.7 *On a measurable space (Ω, Σ) consider a vector measure $\nu \colon \Sigma \to \mathcal{X}$, a Banach space. Then there exist a finite measure $\mu \colon \Sigma \to \mathbb{R}^+$, a continuous convex function $\phi \colon \mathbb{R}^+ \to \mathbb{R}^+$ growing faster than $|x|$, (i.e., $\frac{\phi(x)}{x} \uparrow \infty$ as $x \uparrow \infty$) and ν has finite ϕ-semi-variation relative to μ, meaning $(\phi(0) = 0, \phi(-x) = \phi(x))$*

$$\|\nu\|_\phi(\Omega) = \sup \left\{ \left\| \int_\Omega f(\omega)\nu(d\omega) \right\|_{\mathcal{X}} : \|f\|_{\psi,\mu} \le 1 \right\} < \infty, \quad (51)$$

where $\|f\|_{\psi,\mu} = \inf \left\{ \alpha > 0 \colon \int_\Omega \psi\left(\frac{|f(\omega)|}{\alpha}\right) \mu(d\omega) \le 1 \right\} < \infty$, and the integral in (51) is the usual vector (or Dunford-Schwartz) type, where $\psi \colon \mathbb{R}^+ \to \mathbb{R}^+$ is the complementary convex function given as usual, by $\psi(x) = \sup\{|x|y - \phi(y) \colon y \ge 0\}$, for ϕ.

The (omitted) proof depends on some standard vector integration results given in Dunford-Schwartz I (1958), and a few Orlicz space ideas. Some aspects of this will be considered below in Sections 3 and 4.

Our aim here is to present a much desired extension of Theorem 2.2.6 above so as to include most (weakly) harmonizable and certain other related generalizations of the (weakly) stationary process. Since just about the same time when Loève (1947) introduced the (strongly) harmonizable concept, K. Karhunen intoduced a general concept, still called of *Karhunen class* (as already done) which is of interest here as it conceptually includes all the harmonizable classes, and it has interest in much of our work.

Definition 2.2.8 *A process $\{X_t, t \in T\} \subset L_0^2(P)$, the centered, with covariance $r(\cdot, \cdot)$, is of Karhunen class if there exists also an auxiliary measure space $(S, \mathcal{S}, \nu)$, and a set $\{g(t, \cdot), t \in T\}$ in $L^2(S, \mathcal{S}, \nu)$ which are again complex valued it follows that $r(\cdot, \cdot)$ has the representation:*

$$r(t_1, t_2) = \int_S g(t_1, \lambda)\overline{g(t_2, \lambda)}d\nu(\lambda), \quad t_1, t_2 \in T. \quad (52)$$

Here S, T are general sets and if $T = \mathbb{R}, \mathbb{Z}$, then $S = \hat{T}(= [-\pi, \pi], \text{or } \mathbb{R}$, etc.). We now establish that the weakly harmonizable process is also of Karhunen class.

Theorem 2.2.9 *Each weakly harmonizable process $\{X_t, t \in T\}$ is also a Karhunen process, $T = \mathbb{R}$ or $[0, 2\pi]$ (or an LCA group), relative to some finite measure ν on $\hat{T}$ and for a suitable (Borel) class of functions $\{f_t, t \in T\} \subset L^2(T, \nu)$.*

Proof. It suffices to outline the argument. Since the given process is weakly harmonizable, by the preceding analysis there exists a stationary process $\{Y_t, t \in T\}$ on a larger probability space $(\Omega, \tilde{\Sigma}, \tilde{P})$ perhaps containing the given space such that $\{Y_t, t \in T\} \subset L^2(P)$, and $X_t = QY_t, t \in T$ where Q is the projection operator on $L^2(\tilde{P})$ onto $L^2(P)$, the given space, such that $X_t = QY_t, t \in T$. Here Q is the orthogonal projection onto $L^2(P)$. But by the preceding work, $Y_t = \int_{\tilde{T}} e^{it\lambda} d\tilde{Z}(\lambda), t \in T$, where $\tilde{Z}(\cdot)$ is orthogonally valued. If $\nu(A) = E(|\tilde{Z}(A)|^2)$, then $\nu(\cdot)$ is a finite positive measure on $(\hat{T}, \hat{\mathfrak{I}})$. But by a well-known theorem, due to Kolmogorov (cf. Rozannov (1959), (p. 33) or Masani (1968), Thm. 5.10), there exists an orthogonal projection Q such that $X_t = QY_t$.

$$X_t = QY_t = Q\left(\int_{\hat{T}} e^{it\lambda}\tilde{Z}(d\lambda)\right) = \int_{\hat{T}} \pi(e^{it\lambda})\tilde{Z}(d\lambda), \qquad (53)$$

for each $t \in T$ where $\tilde{Z}(\cdot)$ is orthogonally valued. Letting $f(t, \lambda) = \pi(e^{it(\lambda)}), \lambda \in \hat{T}$, we get $f_t(\cdot) \in L^2(T, v)$ and then (53) yields immediately

$$r(s, t) = E(X_s \bar{X}_t) = \int_{\hat{T}} f(s, \lambda)\overline{f(t, \lambda)}\nu(d\lambda). \qquad (54)$$

This implies that $r(\cdot, \cdot)$ is of Karhunen class on $(\hat{T}, \hat{\mathfrak{I}}, \nu)$. $\square$

Remark 10. Although we embed the given measure space in a larger one and construct the desired triple subsuming the given one, the structure of the usable (or applicable) spaces is classified by this enlargement procedure, showing the key part of geometry.

The following important extension of the above theorem, based on Bochner's Berkeley symposium paper (1956), [somewhat simplified], is of interest here:

Theorem 2.2.10 *Let $\{X_t, t \in T\}$ be a centered harmonizable process admitting an integral representation*

$$X_t = \int_{\hat{T}} e^{it\lambda} Z_x(d\lambda), \quad t \in T, \qquad (55)$$

where $\hat{T} = [-\pi, \pi]$ or $= \mathbb{R}$ if $T = \mathbb{Z}$ (integers) or $T = \mathbb{R}$ (the reals) and $Z_x(\cdot)$ is σ-additive (in the sense of mean square) set function on

Borel sets $\mathfrak{T}$ of T, $E(Z_x(A)) = 0$ and $E(Z_x(A)\overline{Z_x(B)}) = F_x(A, B)$ or $F_x(A \cap B)$, for harmonizable or stationary cases respectively $A, B \in \mathfrak{T}$. Let $\{X_t, Y_t, t \in T\}$ be processes such that $\Lambda X_t = Y_t, t \in T$, where Λ is a linear operator (or "factor") on $L_0^2(P)$ into itself, so X_t is the input and Y_t the output process. Then there exists a stationary input $\{X_t^0, t \in T\}, \Lambda X_t^0 = Y_t^0, t \in T$, where the output $\{Y_t^0, t \in T\}$ is also stationary whose spectrum coincides with that of $\{Y_t, t \in T\}$ outside of $Q = \{\lambda \colon C(\lambda) = 0\}$, the zero set for $F_Y (F_Y = C)$. The solution is representable as:

$$X_t^0 = \int_{T-Q} e^{it\lambda} C(\lambda)^{-1} Z_x(d\lambda). \tag{56}$$

Proof. Let $Q = \{\lambda \colon C(\lambda) = 0\}$, and $F_Y(\cdot)$ has the control measure of the output process $\{Y_t, t \in T\}$, and suppose that $[T = \mathbb{Z} \text{ or } \mathbb{R} \text{ so } \hat{T} = (-\pi, \pi] \text{ or } = \mathbb{R}]$, $\int_{T-Q} |C(\lambda)|^{-2} dF_Y(\lambda) < \infty$, whence X_t^0 of (56) is well-defined, since $E(|X_t^0|^2)$ is found by the above integral which is finite. That ΛX_t^0 is, in fact, the above integral, may be seen as follows:

$$\Lambda X_t^0 = \int_{\hat{T}} C(\lambda) e^{it\lambda} Z_t^0(d\lambda), \; Z_t^0 \text{ as the stochastic measure,}$$

$$= \int_{\hat{T}-Q} C(\lambda) e^{it\lambda} \frac{1}{C(\lambda)} Z_y(d\lambda), \text{ by the preceding work,}$$

$$= \int_{\hat{T}-Q} e^{it\lambda} Z_y(d\lambda) = Y_t^\epsilon, \text{ by definition.} \tag{57}$$

The spectral measures of Y_t^ϵ and Y_t processes agree on the set $\hat{T} - Q$, and it is enough to verify the integrability of Y_t on $\hat{T} - Q$.

For this, it suffices to show that the condition $\Lambda X_t = Y_t, t \in T$, holds. Now for each $\phi \in C_t(\mathbb{R})$, we have,

$$\int_T \phi(t) Y_t dt = \int_T \left(\sum_{i=0}^{k} (-1)^j \int_{\hat{T}} \phi^{(j)}(t - u) H_t(du) \right) X_t dt \tag{58}$$

where $\phi \in C^{(k)}(T)$ of compact supports and k times continuously differentiable. Since X_t is a distributional solution, by assumption, it satisfies (58) for each ϕ on T, whose Fourier transform is twice continuously differentiable, having compact support. It follows that

$$\int_{\hat{T}} C(\lambda) e^{it\lambda} l(Z_x(d\lambda)) = \int_{\hat{T}} e^{it\lambda} l(Z_y(d\lambda)),$$

since $l(\cdot)$ is a continuous linear functional and commutes with the integral. Hence by the uniqueness theorem for Fourier transforms we deduce that $l(C(\lambda)Z_x(d\lambda)) = l(Z_y(d\lambda))$. This easily implies that $C(\lambda)Z_x(dt) = Z_y(d\lambda)$.

But $C(\lambda) \neq 0$ on $\hat{T} - Q$, and hence the above implies

$$\int_{\hat{T}-Q} \frac{\psi(\lambda)}{C(\lambda)} e^{ith} Z_y(d\lambda) = \int_{\hat{T}-Q} \psi(\lambda) e^{it\lambda} Z_x(d\lambda), t \in T. \qquad (59)$$

Since the support of ψ is in $\hat{T} - Q$, this gives for the variances:

$$\int_{\hat{T}} \left| \frac{\psi(\lambda)}{C(\lambda)} \right|^2 F_y(d\lambda) = E\left(\left| \int_{\hat{T}} \psi(\lambda) e^{it\lambda} Z_y(d\lambda) \right|^2 \right), \qquad (60)$$

because $Z_y(\cdot)$ has orthogonal increments, $E\left(|Z_y(A)|^2 \right) = F_y(A)$ and both $Z_x(A), Z_y(A)$ have zero means. With the inverse Hölder inequality in (61), $E(|Z_y(A)|^2) = F_y(A), Z_x(\cdot), Z_y(\cdot)$ centered, we get on taking $\|\psi(\cdot)\|_\infty \leq 1$ arbitrarily (varying), that (60) gives

$$\int_{\hat{T}-Q} |C(\lambda)|^2 F_y(d\lambda) = \sup \left\{ \int_{\hat{T}-Q} |\psi(\lambda)|^2 |C(\lambda)|^{-2} F_y(d\lambda) : \|\psi\|_x \leq 1 \right\}$$

$$\leq \sup \left\{ \|\psi(\cdot) e^{it(\cdot)}\| \| Z_y \|^2(\hat{T}) : \|\psi\|_x \leq 1 \right\}$$

$$\leq \|Z_y\|^2(\hat{T}) < \infty,$$

where $\|Z_y\|(\hat{T})$ is the semi-variation of $Z_y(\cdot)$ which is finite. $\square$

Remark 11. A useful consequence of this result is that, if the output $\{Y_t, t \in T\}$ is stationary, and if there is a weakly harmonizable solution (in the distributional sense), then after deleting some "frequencies" from the spectrum of the Y_t-process, one can find a stationary (distributional) solution that satisfies the given filter equation. We shall include some further analysis later on (next chapter) for applications with a filter problem to understand the ramifications of this property.

The above argument (last part) contains a useful characterization of weakly harmonizable processes which is separated for reference:

Proposition 2.2.11 *Let $\{X_t, t \in T\} \subset L_0^2(P)\}$ be a mean continuous process ($T = \mathbb{R}$ or $\mathbb{Z}$). Then it is weakly harmonizable iff it is in a ball (or norm bounded) and for each integrable ϕ:*

$$E\left(\left|\int_T \phi(t)X_t dt\right|^2\right) \le c\|\hat{\phi}\|_\omega^2, \tag{61}$$

where $\hat{\phi}(\cdot)$ is the Fourier transform of ϕ and conversely.

Proof. First, note that a weakly harmonizable process having the Fourier Transform of an $L^2(P)$-valued vector measure with finite Fréchet variation is bounded, i.e., $E(|Y_t|^2) \le M < \infty, t \in T$. Regarding the inequality (61), consider with $T = \mathbb{R}$,

$$\int_{\mathbb{R}} \phi(t)X_t dt = \int_{\mathbb{R}} \phi(t)\left(\int_{\mathbb{R}} e^{it\lambda} Z(d\lambda)\right) dt$$
$$= \int_{\mathbb{R}} \hat{\phi}(\lambda) Z_x(d\lambda), \text{by interchanging integrals which is valid.}$$

Hence,

$$E\left(\left|\int_{\mathbb{R}} \phi(t)X_t dt\right|^2\right) \le \|\hat{\phi}\|_\infty^2 \|Z_x\|^2(\mathbb{R}) = C\|\hat{\phi}\|_\infty^2,$$

since $Z_x(\cdot)$ has finite Fréchet variation, and (61) holds.

Regarding the converse, it is precisely the V-boundedness condition of Bochner's, and the process is weakly continuous so that the result is a consequence of Theorem 2.2.3 above, and the proposition follows. $\square$

Remark 12. The *weak harmonizability* concept, calling it V-boundedness was introduced by Bochner in 1956 as a generalization of Loève's concept, which we now call *strongly harmonizable* for distinguishing the key differences.

Although the class of strongly harmonizable processes (or fields) is contained in the class of weakly harmonizable processes (or fields), it will be of interest to investigate their approximations. This is clarified here. The second part (on approximation) is due to Niemi (1975), and is essentially the same as in Theorem 2.2.5 above and given for comparison.

Theorem 2.2.12 *The linear space of weakly harmonizable processes* $\mathfrak{w}$ *over complex numbers forms a module of all bounded linear mappings (b.l.m) on* $\mathcal{X} = L_0^2(P)$, *so that* $(B(\mathcal{X}) \cdot \mathfrak{w}) = \mathfrak{w}$, *where* $B(\mathcal{X})$ *is the space of b.l.m on* $\mathcal{X}$.

Moreover, each weakly harmonizable process $X : \mathbb{R} \to L_0^2(P)$ *is approximable in* $L_0^2(P)$*-norm by a strongly harmonizable sequence* $X_n \colon \mathbb{R} \to L_0^2(P)$, *uniformly on compact subsets of* $\mathbb{R}$, *as* $n \to \infty$. *Here* $\mathbb{R}$ *can be replaced by an LCA group* G *where sequences are taken as nets.*

Proof. If $T \in B(\mathcal{X})$ is a bounded linear operator on $\mathcal{X}$ and $X_t \in \mathfrak{w}$ is a weakly harmonizable process, then we get:

$$Y(t) = T\left(\int_{\mathbb{R}} e^{it\lambda} Z(d\lambda)\right) = \int_{\mathbb{R}} e^{it\lambda}(T \circ Z)(d\lambda)$$

by a property of the vector (or Dunford-Schwartz) integral, and $\tilde{Z} = T \circ Z \colon \mathcal{B}(\mathbb{R}) \to \mathcal{X}$ is a stochastic (or vector) measure, since $\|\tilde{Z}\|(\mathbb{R}) \leq \|T\|\|Z\|(\mathbb{R}) < \infty$. So $Y(\cdot)$ is weakly harmonizable ($Y \in \mathfrak{w}$), and this implies the first statement, as well as $B(\mathcal{X}) \cdot \mathfrak{w} \subset \mathfrak{w}$ since the opposite inclusion is clear. The approximation is shown as in (the above) Theorem 2.2.5. The result in the case of $\mathbb{R}$ being replaced by an LCA group is also quite simple and is left to the reader. $\square$

2.3 Harmonizability and Stationary Dilations for Applications

The second concept of stationarity for second order centered processes and its crucial relation with Fourier transforms noted in early 1930s by A. Khintchine and A. Kolmogorov gave birth to an enormous growth of second moment stochastic analysis. The corresponding extensions of stationarity to strong and (later weak) harmonizability concepts by Loève and Bochner (as well as Rozanov) in early 1950s have shown the key roles played by Fourier analysis on LCA groups, with its full force on stochastic analysis and geometry of Hilbert space. We now show some deeper aspects of this interrelationship, with dilations, and some applications. We also discuss the structure of the mean functions of the harmonizable (or stationary) covariance formulas of the first moment classes, extending an early innovation of Balakrishnan's work.

Theorem 2.3.1 (Stationary Dilation) *Let* $X: G \to L_0^2(P) = \mathcal{H}$ *be a weakly harmonizable random field, indexed by an LCA group G, on a probability space (Ω, Σ, P) where as usual $L_0^2(P)$ is the centered scalar random variables forming a Hilbert space. Then there exist an extension Hilbert space $\mathcal{K} = L_0^2(\tilde{P})$ on a probability triple $(\tilde{\Omega}, \tilde{\Sigma}, \tilde{P})$ and a stationary random field $Y: G \to \mathcal{K}$ such that $X(g) = QY(g), g \in G$, where $Q: L_0^2(\tilde{P}) \to L_0^2(P)$ is the orthogonal projection with range $L_0^2(P)$. Here if $\mathcal{H}$ is determined by X (as a closed linear span) then Y determines $\mathcal{K}$, whence the latter is the minimal superspace for $\mathcal{H}$.*

Proof. Consider $X: G \to \mathcal{H}(= L_0^2(P))$ to be weakly harmonizable:

$$X(g) = \int_{\hat{G}} \langle g, s \rangle Z(ds), \quad g \in G, \tag{62}$$

by the structural representation, and there exists a finite regular Borel measure μ on the Borel σ-algebra of $\hat{G}$ such that

$$\left\| \int_{\hat{G}} f(t) Z(dt) \right\|_2^2 \leq \int_{\hat{G}} |f(t)|^2 d\mu(t), f \in C_0(\hat{G}). \tag{63}$$

Consider the mapping $v: \mathcal{B}(\hat{G} \times \hat{G}) \to \mathbb{R}^+$ given by

$$v(A, B) = \mu(A \cap B), A, B \in \mathcal{B}(\hat{G}), \tag{64}$$

the class $\mathcal{B}(\hat{G})$ being the Borel σ-ring, as is $\mathcal{B}(\hat{G} \times \hat{G})$, then $v(\cdot, \cdot)$ has finite Vitali variation, and concentrates on the diagonal of the product $\hat{G} \times \hat{G}$. Moreover

$$\int_{\hat{G}} \int_{\hat{G}} f(s, t) v(ds, dt) = \int_{\hat{G} \times \hat{G}} f(s, s) \mu(ds), f \in C_0(\hat{G} \times \hat{G}). \tag{65}$$

Let $F: \mathcal{B}(\hat{G} \times \hat{G}) \to \mathbb{C}$ be defined by $F(A, B) = (Z(A), Z(B))$, which is a bimeasure of finite semi-variation from (62). Using the D-S and MT-integration techniques, one has

$$0 \leq \left\| \int_{\hat{G}} f(s) Z(ds) \right\|_2^2 = \int_{\hat{G}} \int_{\hat{G}} f(s) \overline{f(t)} F(ds, dt). \tag{66}$$

Taking $f(s, t) = f(s) \cdot f(t)$ in (66) setting $\alpha = v - F$ one has:

$$0 \leq \int_{\hat{G}} |f(s)|^2 \mu(ds) - \left\| \int_{\hat{G}} f(s) Z(ds) \right\|_2^2$$

$$= \int_{\hat{G}} \int_{\hat{G}} f(s) \overline{f(t)} [v(ds, dt) - F(ds, dt)]$$

$$= \int_{\hat{G}} \int_{\hat{G}} f(s) \overline{f(t)} \alpha(ds, dt), \quad f \in C_0(\hat{G}) \tag{67}$$

so that $\alpha(\cdot)$ is positive (semi-) definite and $= 0$ iff $v = F$ whence F concentrates on the diagonals and so X is stationary itself. Excluding this case, $[\cdot, \cdot] \colon C_0(\hat{G}) \times C_0(\hat{G}) \to C$ gives a nontrivial inner product $[\cdot, \cdot]$ defined by

$$[f, g] = \int_{\hat{G}} \int_{\hat{G}} f(s) \bar{g}(t) \alpha(ds, dt), f, g \in C_0(\hat{G}). \tag{68}$$

If $N_0 = \left\{ f \colon [f, f] = 0, f \in C_0(\hat{G}) \right\}$, and $\mathcal{H}_1 = C_0(\hat{G})/N_0$, let $[\cdot, \cdot]$ be

$$[(f), (g)] = [f, g]', f \in (f) \in \mathcal{H}_1, g \in (g) \in \mathcal{H}_1. \tag{69}$$

This $[\cdot, \cdot]$ is an inner product on $\mathcal{H}_1$ and let $\mathcal{H}_0$ be its completion, and $\pi_0 \colon C_b(\hat{G}) \to \mathcal{H}_0$ be the canonical projection ($\mathcal{H}_0$ need not be separable!). That this is always possible is a consequence of Theorem 1.3.5 which was discussed in detail in the context of Lévy's BM (in higher dimensional time). The analogous complex case is outlined as follows. Let $\{h_i, i \in I\} \subset \mathcal{H}_0$ be a CON set, and let $(\Omega_i, \Sigma_i, P_i)$ be determined by a complex Gaussian variable, so taking $\Omega_i = \mathbb{C}, \Sigma_i = $ Borel σ-algebra of $\mathbb{C}$ and

$$P_i(A) = (2\pi)^{-1} \int_A \exp \left(\frac{-|t|^2}{2} \right) dt, t = t_1 + it_2, A \in \Sigma_i,$$

let $(\Omega', \Sigma', P') = \otimes_{i \in I} (\Omega_i, \Sigma_i, P_i)$, the product space given by the Fubini-Jessen theorem. If $X_i(\omega) = \omega(i), \omega \in \Omega' = \mathbb{C}^I$, then the set $\{X_i, i \in I\}$ forms a CON basis of $\mathcal{L} = \bar{sp}\{X_i, i \in I\} \subset L_0^2(P)$. The correspondence $\alpha \colon h_i \to X_i$ sets up a (linear) isomorphism with $\mathcal{L} = \bar{sp}\{X_i, i \in I\} \subset L_0^2(P)$ and is onto. Also

$$\|\tau(h_i)\|_2^2 = E(|X|^2) = 1 = [h_i, h_i], i \in I.$$

This implies that $\tau \colon h_i \to X_i$ is an isometric isomorphism, after polarization, of $\mathcal{H}_0$ onto $\mathcal{L} \subset L_0^2(P)$, as desired.

Let $\pi = \tau \circ \pi_0 \colon f \to \tau(\pi_0(f)) \in \mathcal{H}' \subset L_0^2(P), f \in C_0(\hat{G})$, be the composite mapping shown, and let $X(t) = \pi(e_t(\cdot)) \in \mathcal{H}', l_t \colon s \to (t, s)$ is a character of G at $t \in G$, $(e_0 = 1 \notin N_0)$ so $\pi_0(t)$ can be identified with a constant (say) $0 \in C_0(\omega)$. Thus

$$X_0(0) = \tau(1), \quad E(|\tau(1)|^2) = 1.$$

Let $\mathcal{H}'' = \bar{sp}\{X(t), t \in G\} \subset \mathcal{H}'$. Then we can find $(\Omega'', \Sigma'', P'')$, a probability space, such that $\mathcal{H}'' \subset L_0^2(P'')$, and set $\mathcal{H} = \mathcal{H}' \oplus \mathcal{H}''$, the direct sum of Hilbert space $L_0^2(P)$ and $L_0^2(P')$. If $(\tilde{\pi}, \tilde{\Sigma}, \tilde{P})$ is the direct product $(\Omega, \Sigma, P) \otimes (\Omega', \Sigma', P')$, one can identify $\mathcal{K} \subset L_0^2(\tilde{P})$. Define $Y(t) = X(t) + iX(t), t \in G$, and we can see that $\{Y(t), t \in G\} \subset \mathcal{K}$ and if $Q \colon \mathcal{K} \to \mathcal{H}(= \mathcal{H} \oplus \{0\})$ is the orthogonal projection, then $X(t) = QY(t), t \in G$. We assert that the $Y(t)$ is a stationary process to complete the argument.

For this, consider the computations (cross products vanishing):

$$r(s, t) = (Y(s), Y(t)) = (X(s), X(t)) + (X_1(s), X_1(t)) + 0$$

$$= \int_{\hat{G}} \int_{\hat{G}} \langle s, \lambda \rangle \overline{\langle t, \lambda' \rangle} F(d\lambda, d\lambda')$$

$$+ \int_{\hat{G}} \int_{\hat{G}} \langle s, \lambda \rangle \overline{\langle t, \lambda' \rangle} \alpha(ds, d\lambda)$$

$$= \int_{\hat{G}} \int_{\hat{G}} \langle s, \lambda \rangle \overline{\langle t, \lambda' \rangle} v(d\lambda, d\lambda'), (\alpha = v - F)$$

$$= \int_{\hat{G}} \langle s, \lambda \rangle \overline{\langle t, \lambda \rangle} \mu(d\lambda) = \int_{\hat{G}} \langle s - t, \lambda \rangle \mu(d\lambda),$$

by the composition of characters and this implies $r(s + h, t + h) = \tilde{r}(s - t)$ and so $\{Y(t), t \in G\}$ is stationary, and $\mathcal{H} = \bar{sp}\{X(t), t \in G\}$. $\square$

An interesting consequence of this result is the following:

Corollary 2.3.2 *If G is an LCA group, $\mathcal{B}(G)$ its Borel σ-algebra and $\mathcal{H}$ is a Hilbert space, then a vector measure $v \colon \mathcal{B}(G) \to \mathcal{H}$ has an orthogonally valued dilation.*

Proof. Let $X \colon \hat{G} \to \mathcal{H}$ be a mapping given by $X(\hat{g}) = \int_G \langle \hat{g}, \lambda \rangle v(d\lambda)$, as a vector (or D-S) integral. Then $X(\cdot)$ is weakly harmonizable as it is V-bounded. Then by the preceding theorem, there is a super Hilbert space $\mathcal{K} \supset \mathcal{H}$, an orthogonal projection $Q \colon \mathcal{K} \to \mathcal{H}$ and a stationary field Y, such that $X(\hat{g}) = QY(\hat{g}), \hat{g} \in \hat{G}$. If $Z(\cdot)$ is the stochastic measure representing Y, then we have

$$\int_G \langle \hat{g}, \lambda \rangle (\nu(d\lambda), h) = (X(\hat{g}), h) = (QY(\hat{g}), h) = \int_{\hat{G}} \langle \hat{g}, \tilde{\lambda} \rangle (Q \cdot Z(d\lambda), h).$$

These are scalar (Lebesgue) integrals. Then by the uniqueness of Fourier representation, one has

$$\langle \nu(A) - Q \circ Z(A), h \rangle = 0, A \in \mathcal{B}(G), h \in \mathcal{K}.$$

By the uniqueness of Fourier representation $\nu = Q \circ Z$. Since $Z(\cdot)$ is orthogonally valued, as Y is stationary, the result follows. $\square$

Remark 13. The preceding result shows an interesting relation between a positive definite contractive family of operators on a Hilbert space and their restriction to certain subspaces having special geometric properties observed by Sz.-Nagy and Naĭmark in early 1950s. This will be made explicit, as it implies a deep internal equivalence of these authors' works!

The following representation is of interest here:

Theorem 2.3.3 *Let* $X \colon G \to L^2(P) = \mathcal{X}, G$ *an LCA group and* (Ω, Σ, P) *a probability space, X being a centered (i.e., means zero) weakly harmonizable mapping (or field). Then there is a super (Hilbert) space* $\mathcal{K} = L_0^2(P) \supset \mathcal{X}$, *on an (possibly) enlarged probability space* $(\tilde{\Omega}, \tilde{\Sigma}, \tilde{P})$, $Y_0 \in L_0^2(\tilde{P})$, *a set of weakly continuous contractive linear operators* $\{T(g), g \in G\}$ *on* $\mathcal{K} \to \mathcal{X}$ $(T(0) = id)$ *such that the* $T(g)$-*family is positive type, and* $X(g) = T(g)Y_0$ *where* $g \in G$.

Conversely, every weakly continuous contractive family $\{T(g), g \in G\}$ *of the above type from a Hilbert space* $\mathcal{K} \supset \mathcal{X}$ *into a subspace* $\mathcal{X}$, *is weakly harmonizable (so* $X \colon G \to \mathcal{X}, X(g) = T(g)X_0$, *where* $T(0) = id$ *on* $\mathcal{X}$*).*

Proof. For the direct part, let $X \colon G \to \mathcal{X} - L_0^2(P)$ be weakly harmonizable. Then there is a $\mathcal{K} = L_0^2(\tilde{\beta}) \supset \mathcal{X}$, and a stationary $Y \colon G \to \mathcal{X}$ with $X(g) = QY(a), g \in G$, by Theorem 2.3.1 above. On the other hand $Y(g) = U(g)Y(0)$ for a strongly continuous unitary group $\{U(g), g \in G\}$ on $\mathcal{K}$. If $T(g) = QU(g), g \in G$, then it is asserted that this is the desired class. Clearly $T(0) = Q = I_{\mathcal{X}}$ and $\|T(g)\| \leq \|Q\| \|U(g)\| \leq 1$.

The continuity of the $U(k)$-family implies the weak continuity of the $U(t)$-class and its positive definiteness is seen from $\tilde{T}(-g) = (\tilde{T}(g))^*$ and $(\tilde{T}(g) = T(g)\big|_{\mathcal{X}})$ for $h_i \in \mathcal{X}$:

$$
\begin{aligned}
\left(\tilde{T}(-\theta)h_{s_1}, h_{s_2}\right) &= (U^*(g)h_{s_1}, Qh_{s_2}) \\
&= (h_{s_1}, U(g)h_{s_2}), \text{ since } Qh_s = h_s, U^{**} = U \\
&= (Qh_{s_1}, U(g)h_{s_2}), h_{s_i} \in \mathcal{X}, i = 1, 2.
\end{aligned}
$$

It follows from this that

$$
\begin{aligned}
\sum_{i=1}^{n}\sum_{j=1}^{n}\left(\tilde{T}(s_i^{-1}s_j)h_{s_i}, h_{s_j}\right) &= \sum_{i=1}^{n}\sum_{j=1}^{n}(QU(-s_j)U(s_i)h_{s_i}, h_{s_j}) \\
&= \sum_{i,j=1}^{n}((Us_j)^*U(s_i)h_{s_i}, h_{s_j}) \\
&= \left\|\sum_{i=1}^{n}U(s_i)k_{s_i}\right\|^2 \geq 0.
\end{aligned}
$$

The converse depends on Sz.-Nagy's extension of the Naĭmark result, stating that if $\tilde{T}(\cdot) = T(\cdot)|_{\mathcal{X}}$ then there is an extension Hilbert space $\mathcal{H}_1 \supset \mathcal{H}$ and a weakly continuous group of unitary operators on it such that $\tilde{T}(g) = Q_1 V(g)|_{\mathcal{H}}$ where Q_1 is an orthogonal projection of $\mathcal{H}_1$ onto $\mathcal{H}$. If $x_0 \in \mathcal{H}_1 \cap \mathcal{H}$ then

$$
T(g)x_0 = \tilde{T}(g)x_0 = QV(g)x_0 = x(g), \quad g \in G,
$$

and $\{Y(g) = V(g)x_0, g \in G\} \subset \mathcal{H}$ is stationary, and the earlier result applies. Thus $\{T(g)x_0, g \in G\}$ is weakly harmonizable. $\square$

2.4 Domination of Vector Measures and Application to Cramér and Karhunen Processes

Let (Ω, Σ, μ) be a measure space, $\mathcal{X}$ a Banach space and $\nu\colon \Sigma \to \mathcal{X}$ a vector measure (i.e., σ-additive on Σ), with p-variation ($p \geq 1$) denoted $|\nu|_p(\cdot)$, defined on $A \subset \Sigma$ as:

$$
|\nu_f|_p(A) = \sup\left\{\sum_{i=1}^{n}\|\nu(A_i)\|\|a_i\|: A_i \in \Sigma(A), \text{ disjoint}, \|f\|_{q,\mu} \leq 1\right\},
$$

$$
\tag{70}
$$

where $f = \sum_{i=1}^{n} a_i \chi_{A_i}, q = p/(p-1) \geq 1$, and $f \in L^q(\Omega, \Sigma, \mu) = L^q(\mu)$, $\Sigma(A)$ being the trace of Σ on A. If $|\nu|_p(A) < \infty$, then ν has finite p-semi-variation on A relative to μ. If $p = 1$, usually $L^\infty(\mu)$ is replaced by $B(\Omega, \Sigma)$, the space of bounded measurable scalar functions on Ω with uniform norm, (whatever μ is) and 1-variation is merely termed *variation*, and (70) becomes

$$|\nu|(A) = \sup \left\{ \sum_{i=1}^{n} \|\nu(A_i)\| : A_i \in \Sigma(A), \quad \text{disjoint} \right\}. \qquad (71)$$

Note that $|\nu|(\cdot)$ is σ-additive if ν is, but $|\nu|_p(\cdot)$ is not necessarily so if $p > 1$. We also need p-semi-variation of ν on A:

$$\|\nu\|_p(A) = \sup \left\{ \left\| \sum_{i=1}^{n} a_i \nu(A_i) \right\| : A_i \in \Sigma(A), \text{disjoint}, \|f\|_{q,\mu} \leq 1 \right\} \qquad (72)$$

where $f = \sum_{i=1}^{n} a_i \chi_{A_i}, p^{-1} + q^{-1} = 1$. If $\|\nu\|_p(A) < \infty$, then ν has *p-semi-variation finite on A relative to μ*. If $p = 1$, then

$$\|\nu\|(A) = \sup \left\{ \left\| \sum_{i=1}^{n} a_i \nu(A_i) \right\| : |a_i| \leq 1, A_i \in \Sigma(A), \text{disjoint} \right\}. \qquad (73)$$

It is clear that

$$|\nu|(A) \leq |\nu|_1(A), \quad \|\nu\|(A) \leq \|\nu\|_1(A),$$

and $\|\nu\|_p(A) \leq |\nu|_p(A)$, with (usually) a strict inequality if $\dim(\mathcal{X}) = \infty$.

Also (72) may be written as:

$$\|\nu\|_p(A) = \sup \left\{ \left\| \int_A f d\nu \right\| : \|f\|_{q,\mu} \leq 1 \right\}. \qquad (74)$$

The technical result of interest here, termed the *domination problem*, is this: Does $\nu: \Sigma \to \mathcal{X}$, σ-additive, have the finite p-semi variation for a σ-finite $\mu: \Sigma \to \bar{\mathbb{R}}^+$? A solution of this problem is of importance in our analysis. This will be answered 'yes' for a class of Banach spaces of great interest here.

It is convenient to treat the problem for *a general class of spaces* that includes the Lebesgue class $L^p(\mu), p \geq 1$, namely the Orlicz class

$L^\phi(\mu)$ where $\phi \colon \mathbb{R} \to \mathbb{R}^+$ is a symmetric convex function with $\phi(0) = 0$ termed a Young function. For each such ϕ, there exists a similar function ψ, called the *complementary function* given by $\psi(x) = \sup\{|x|y - \phi(y) : y \geq 0\}$. [If $\phi(x) = \frac{|x|^p}{p}$, then $\psi(y) = \frac{|y|^q}{q}$, $p^{-1} + q^{-1} = 1$, $p, q \geq 0$.] For each measurable $f \colon \mathbb{R} \to \mathbb{R}$ (or $\mathcal{X}$), one defines the ϕ-semi-variation of $\nu \colon \Sigma \to \mathcal{X}$, on $A \in \Sigma$ as:

$$\|\nu\|_\phi(A) = \sup\left\{\left\|\int_A f(w)\nu(dw)\right\| : \|f\|_{\psi,\mu} \leq 1\right\}, \qquad (75)$$

where ϕ, ψ satisfy, $|xy| \leq \phi(x) + \psi(y)$ and one has, by definition

$$\|f\|_\phi = \inf\left\{\alpha > 0 : \int_\Omega \phi\left(\frac{|f|}{\alpha}\right) d\mu \leq 1\right\}, \qquad (76)$$

which when $\phi(x) = |x|^p, p \geq 1$, becomes $\|f\|_\phi = \|f\|_p$, the Lebesgue norm. The problem that we want to solve is that, if $\nu \colon \Sigma \to \mathcal{X}$ is a vector measure, is it dominated by a pair (ϕ, μ) for some $\phi(x) = |v|^p, p \geq 1$? A positive solution is given by the following result:

Theorem 2.4.1 *On a measurable space (Ω, Σ), a vector measure $\nu \colon \Sigma \to \mathcal{X}$, (a Banach space) is dominated by a pair (ϕ, μ), where $\frac{\phi(r)}{r} \uparrow \infty$, $\phi \colon \mathbb{R} \to \mathbb{R}^+$ is a continuous Young function, $\|\nu\|_\phi(\Omega) < \infty$.*

Proof. Since weak and strong σ-additives of a Banach space valued measure are known to be equivalent, consider for $x^* \in \mathcal{X}^*$, $\|x^*\| \leq 1$, let $A_n, n \geq 1$, disjoint sets $A_n \in \Sigma_n$. We have

$$0 = \lim_{n \to \infty}\left\|\nu\left(\bigcup_{n \geq 1} A_n\right) - \sum_{k=1}^n \nu(A_k)\right\|$$
$$= \limsup_{n \to \infty}\left\{\left|(x^* \circ \nu)\left(\bigcup_{n \geq 1} A_n\right) - \sum_{k=1}^n (x^* \circ \nu)(A_k)\right| : \|x^*\| \leq 1\right\}. \qquad (77)$$

Thus the signed measures $\{x^* \circ \nu : x^* \in S^*\}$, $(S^*$ unit ball of $\mathcal{X}^*)$ are uniformly σ-additive on Σ. Now one invokes a result of Bartle-Dunford-Schwartz) (cf. D-S (1), IV.10.5) to conclude that there is a "control measure" $\mu \colon \Sigma \to \mathbb{R}^+$ such that $x^* \circ \nu$ is μ-continuous for each $x^* \in S^*$. By the Radon-Nikodym theorem, $g_{x^*} = d(x^k \circ \nu)/d\mu$ exists and one has by (77),

$$\lim_{\mu(A)\to 0}\int_A g_{x^*}(\omega)d\mu(\omega) = 0 \ \left(=\lim_{\mu(A)\to 0}|x^* \circ \nu(A)|\right),$$

uniformly in $x^* \in S^*$. Hence $\{g_{x^*}: x^* \in S^*\} \subset L^1(\mu)$ is bounded and uniformly μ-integrable. This gives by a classical de la Vallée Poussin's theorem, that there is a convex function $\phi: \mathbb{R} \to \mathbb{R}^*$ of the given description such that

$$\int_\Omega \phi(|g_{x^*}(\omega)|)\mu(d\omega) \le K_0 < \infty, \quad \forall x^* \in S^*. \tag{78}$$

Let ψ be the complementary function of ϕ, so $\psi: \mathbb{R} \to \mathbb{R}^+$, and

$$\|\nu\|_\phi(\Omega)$$
$$= \sup\left\{\left\|\int_\Omega f(\omega)d\nu(\omega)\right\| : \|f\|_{\psi,\mu} \le 1\right\}$$
$$= \sup\left\{\sup\left[\left|\int_\Omega f(\omega)g_{x^*}(\omega)d\mu(\omega)\right| : \|x^*\| \le 1, : \|f\|_{\psi,\mu} \le 1\right]\right\}$$
$$\le 2\sup\left\{\|g_{x^*}\|_{\phi,\mu}\|f\|_{\psi,\mu} \le |\lambda|_{x^*} \le 1\right\},$$
$$\qquad \text{by Hölder's inequality for Orlicz spaces,}$$
$$\le 2\sup\left\{\|g_{x^*}\|_{\phi,\mu}\|x^*\| \le 1\right\} \le 2K_0 < \infty, \text{by (2.4.9).} \ \Box$$

[Here we used some elementary analysis of Orlicz spaces, a reference to which is Zygmund (1959, Vol. I) or Rao and Ren [1991].]

From the general (and classical) theory of the Lebesgue and Orlicz spaces, it is known, and easy to verify, that each $L^\phi(\mu) \subset L^1(\mu)$ for $\mu(\Omega) < \infty$, $(L^\phi(\mu)$ is the Orlicz space), it is not simple to determine if a given vector measure $\nu: \Sigma \to \mathcal{Y}$ is dominated by a particular pair (ϕ, μ) to study the useful properties of μ-dominated ν when $\mathcal{Y}$ is an L^ϕ-space $(L^\phi(\mu))$. This nontrivial problem was studied and the case $\phi(x) = x^2$, was solved by Linderstrauss and Pelczyński (1968) and then some interesting (as well as important) applications for stochastic analysis followed. Here we present their specialization which will be used in the next section(s) for Cramér and Karhunen classes of second order processes. The problem and the resulting stochastic analysis are intrinsically tied to the Hilbert structure and hence to second order processes!

The applications here include most $L^p(\mu')$-spaces on a measure space (Ω', Σ', μ') as well as the abstract (M)-spaces such as $B(\Omega', \Sigma', \mu')$,

$L^\infty(\mu)$ and some others. The $L^2(\mu')$ thus includes Banach spaces iso-morphic to Hilbert spaces. The desired concept is given as follows:

Definition 2.4.2 *A Banach space* $\mathfrak{X}$ *is an* $\mathcal{L}_{p,\lambda}$-*space,* $p \geq 1, \lambda \geq 1,$ *if for each* n-*dimensional space* $F \subset \mathfrak{X}$, *there is an* n-*dimensional subspace* $F \subset \mathfrak{X}$, $(E \subset F)$ *such that* $d(F, l_\mu^n) \leq n$, $(l_p^m$ *is the usual Lebesgue sequence* p-*space) and* $d(E_1, E_2) = \inf\{\|T\|\|T^{-1}\| : T \in B(E_1, E_2)\}$, *the space of bounded linear mappings on* E_1 *to* E_2. *Then the above defined space* $\mathcal{L}_p$ *is an* $\mathcal{L}_{\mu,\lambda}$-*space for some* $\lambda > 1.$

This unmotivated concept was found by these authors (Lindenstrauss and Polczyński (1965)) and used it to solve the above problems. The motivation and discovery of this key concept are not obvious.

The above concept (and space) is used to prove the following *important result*:

Theorem 2.4.3 *Let* $B(\Omega, \Sigma)$ *be the Banach space of bounded scalar measurable functions* $f : \Omega \to \mathbb{C}$, *with uniform norm, and* $\mathcal{Y}$ *be an* $\mathcal{L}_p$-*space* $1 \leq p \leq 2$ *introduced above. Then a vector measure* $\nu : \Sigma \to \mathcal{Y}$ *is* $(2, \mu)$-*dominated, so that there is a (finite) measure* $\mu : \Sigma \to \mathbb{R}$ *satisfying*

$$\left\| \int_\Omega f(w)\nu(dw) \right\|_{\mathcal{Y}} \leq \|f\|_{2,\mu}, \quad f \in B(\Omega, \Sigma) \tag{79}$$

and ν *has 2-semi-variation finite, relative to* μ.

Proof. The detail is given in three steps for convenience. We use the standard results from Dunford-Schwartz (Chapter IV, Sec. 10). Now let $T : f \mapsto \int_\Omega f(\omega)\nu(d\omega)$, so that $T : B(\Omega, \Sigma) \to \mathcal{Y}$ is a well-defined bounded linear operator.

I. Let $\mathfrak{X} = C(S)$, the real continuous function space on a compact S, and $q : s \to l_0 \in \mathfrak{X}^*$, $l_s(f) = f(s), (f \in \mathfrak{X})$ the evaluation functional. The set $K \subset \mathfrak{X}^*$, the collection of extreme points of the unit ball of $\mathfrak{X}^*$, is, by the (Krein-) Milman theorem, closed and $K = q(S) \cup (-q(S))$ where $q : s \mapsto l_s \in \mathfrak{X}^*$, each element of which is $= \alpha l_s, |\alpha| = 1$. Thus if $\mathcal{Y}$ is an $\mathcal{L}_\mu$-space and $T \in B(C(S), \mathcal{Y}), 1 \leq p \leq 2$, then Linden-struss and Pelczyínski using a *key* Grothendieck-Pietsch inequality have deduced the existence of a regular probability measure μ_0 on K such that

$$\|Tf\|_{\mathcal{Y}}^2 \leq C_1 \int_{q(S)} |l_s(f)|^2 \mu_0(dl_s) + C_2 \int_{-q(S)} |l_s(f)|^2 \mu_0(dl_s)$$

$$\leq C_3 \int_S |f(s)|^2 \mu_0(ds), \quad f \in \mathcal{X},$$

where S and $q(S)$ are identified, $C_3 = 2\max(C_1, C_2)$. For the complex functions, replace C_3 by $2C_3$.

II. Suppose now $\mathcal{X} = B(\Omega, \Sigma)$, and $\mathcal{Y}$ is an $\mathcal{L}_p$ space, $1 \leq p \leq 2$. So $\mathcal{X}$ is a closed subalgebra of $B(\Omega)$. Then there is an isometric and algebraic isomorphism between $\mathcal{X}$ and $\mathcal{X}_0 = C(S_0)$, S_0 being a compact Hausdorff space, preserving all algebraic operations.

If $I: \mathcal{X} \to \mathcal{X}_0 = C(S_0)$, and $\tilde{T} = T \circ I^{-1}: I_0 \to \mathcal{Y}$, then $\tilde{T}: \mathcal{X}_0 \to \mathcal{Y}$ is continuous and satisfies the hypothesis of Step I. So there is a regular Borel measure μ_1 on S_0 into $\mathbb{R}^+$ such that

$$\|\tilde{T}f\|_{\mathcal{Y}} \leq \|f\|_{2,\mu_1}, \quad f \in \mathcal{X}_0.$$

Since for $f \in \mathcal{X}$, we now have $\tilde{f} = I(f) \in \mathcal{X}_0$, the following simplification obtains:

$$\|Tf\|_{\mathcal{Y}}^2 = \|\tilde{T}\tilde{f}\|_{\mathcal{Y}}^2 \leq \|\tilde{f}\|_{2,\mu}^2, \quad f \in \mathcal{X}$$

$$= (\bar{\tilde{f}}\tilde{f}, \mu_1), \text{ since } \mu_1 \in \mathcal{X}_0^*, (\cdot, \cdot) \text{ is duality pairing,}$$

$$= (I(f)I(\tilde{f}), \mu_1) = (I(f\tilde{f}), \mu_1),$$

$$\text{by the algebraic property of } I,$$

$$= (f\bar{f}, I^*(\mu_1)), I^*: \mathcal{X}_0^* \to \mathcal{X}^* \text{ is adjoint,}$$

$$= \int_\Omega |f|^2(\omega)\mu_2(dw),$$

where $0 \leq \mu_2 = I^*(\mu_1) \in \mathcal{X}^* = ba(\Omega, \Sigma)$, the bounded additive set functions on Σ with total variation norm. It is to be shown that the additive μ_2 can be replaced by a σ-additive measure.

III. Here we need to use the Carathéodory procedure to complete in replacing μ_2 with a σ-additive measure using this (Carathéodory) procedure which needs a standard but additional argument.

Let Σ_μ be the class of μ-measurable sets in this procedure. Then $\Sigma_\mu \supset \Sigma$ and $\mu(A) \leq \mu_2(A), A \in \Sigma, \mu$ being σ-additive on Σ_μ, with equality only if μ_2 is also σ-additive. Now (79) follows if (70) is valid for μ_2 when f is a step function.

We show this (for simple functions) as follows. Let $f = \sum_{i=1}^{m} a_i \chi_{A_i}$, $A_i \in \Sigma_0$ disjoint, $a_i \neq 0$, and for given $\varepsilon > 0$, choose $A_{in}^{\varepsilon} \in \Sigma$, with $A_i \subset \bigcup_{n=1}^{\infty} A_{in}^{\varepsilon}$, and

$$\mu(A_i) + \frac{\varepsilon}{m|a_i|^2} > \sum_{n=1}^{\infty} \mu_2(A_{in}^{\varepsilon}). \tag{80}$$

Replacing A_{in}^{ε} by $A_i \cap A_{in}^{\varepsilon}$ in Σ, if necessary, we take $A_i = \bigcup_{n \geq 1} A_{in}^{\varepsilon}$, in (80). Let $f_N^{\varepsilon} = \sum_{i=1}^{m} a_i \chi_{\cup_{k=1}^{N} A_{ik}^{\varepsilon}}$. Then $f_N^{\varepsilon} \in \mathcal{X}$, $f_N^{\varepsilon} \to f$ as $N \to \infty$ pointwise and boundedly. Hence Step II simplifies to:

$$\|Tf_N^{\varepsilon}\|_y^2 = \left\| \int_{\Omega} f_N^{\varepsilon}(\omega) \nu(d\omega) \right\|_y^2 \leq \int_{\Omega} |f_N^{\varepsilon}(\omega)|^2 \mu_2(d\omega)$$

$$= \sum_{i=1}^{n} |a_i|^2 \sum_{k=1}^{N} \mu_2(A_{ik}^{\varepsilon}), \text{ since } \mu_2 \text{ is additive.}$$

Letting $N \to \infty$ on both sides, and using the usual vector integration properties,

$$\|Tf\|_y^2 = \left\| \int_{\Omega} f(\omega) \nu(d\omega) \right\|_y^2$$

$$\leq \sum_{i=1}^{m} |a_i|^2 \left| \mu(A_i) + \frac{\varepsilon}{m|a_i|} \right|, \text{ by (80)},$$

$$= \int_{\Omega} |f(\omega)|^2 \mu(d\omega) + \varepsilon.$$

Since $\varepsilon > 0$ is arbitrary (79) follows for simple functions, and then as observed before, the result holds generally. $\square$

An application of the above result to *more general* classes of processes that include the (weakly/strongly) harmonizable families as introduced by H. Cramér and K. Karhunen separately and will be recalled here to see the use as well as their potential for some applications. This can complement the analysis included already.

Definition 2.4.4 *A process* $\{X_t, t \in \mathbb{R}\} \subset L_0^2(P)$, *(so it is centered) is of* Cramér *class (general) if its covariance function* $r \colon (s,t) \to \mathbb{C}$ *is expressible as the strong Morse-Transue integral*

$$r(s,t) = \int_{\mathbb{R}} \int_{\mathbb{R}} g(s,\lambda)\bar{g}(t,\lambda')\tilde{F}(d\lambda, d\lambda'), s.t \in \mathbb{R}, \qquad (81)$$

relative to a class $\{g(s,\cdot), s \in \mathbb{R}\}$ of $\tilde{F}$-integrable functions, where $\tilde{F}$ has finite Fréchet variation. If $\tilde{F}(\cdot,\cdot)$ has finite Vitali variation, then the integrals in (81) become Lebesgue-Stieltjes type, and the concept reduces to the classical (Cramér) class, as originally introduced and studied above. [This difference turns out to be significant.]

Regarding these two general classes of processes, we can present the following type of (quasi-) inclusion relations.

Theorem 2.4.5 *Let $X\colon \mathbb{R} \to L_0^2(P)$ be a process, and $\{g(t,\cdot), t \in \mathbb{R}\}$ a family of Borel functions. If X is a Karhunen process for this $g(k,\cdot)$-family and a σ-finite measure F on $\mathcal{B}(\mathbb{R})$, and $T\colon L_0^2(p) \to L_0^2(p)$ is a bounded linear mapping, then $Y(t) = TX(t), t \in \mathbb{R}$, is a Cramér process for this g-family and a covariance bimeasure. On the other hand, if $\{g(t,\cdot), t \in \mathbb{R}\}$ is a bounded Borel family and $X\colon \mathbb{R} \to L_0^2(P)$ is a Cramér process for this g-family and a covariance bimeasure, then there exists an extension space $L_0^2(\tilde{P})(\supset L_0^2(P))$ determined by the given process, a Karhunen process $Y\colon \mathbb{R} \to L_0^2(P)$ for the same g-family and a Borel measure on $\mathbb{R}$ with $X(t) = QY(t), t \in \mathbb{R}$, where $Q\colon L_0^2(\tilde{P}) \to L_0^2(P)$ is an orthogonal projection.*

Remark 14. 1. It must be observed that the g-family involved here is not of the character type $\{e^{itx}, t \in \mathbb{R}\}$, and so one *cannot conclude* that each Cramér process can be obtained as a projection of some Karhunen process. In fact, Erik Thomas has constructed an example showing that this projection property, which is so true between the stationary and harmonizable classes, need not hold for the Cramér and Karhunen classes!

2. The preceding remark indicates that the Cramér class of processes is quite large and some of its members may not be dilated to Karhunen processes. It may be concluded however that the Cramér class is closed under bounded linear mappings. Also clearly if $g(t,\lambda) = e^{it\lambda}$, a character of $\mathbb{R}$, the Cramér process reduces to a weakly harmonizable process.

The preceding discussion can be given an abstract form for applications and easy reference as follows, without further detail:

Theorem 2.4.6 *Let G be an LCA group and $X\colon G \to L_0^2(P)$ be a (centered) second order random field and $T\colon L_0^2(P) \to L_0^2(P)$ be a* *bounded linear operator. If X is weakly stationary and $Y = TX$ (or $Y(t) = (TX)(t)$, $t \in G$), then $\{Y(g), g \in G\}$ is weakly harmonizable. On the other hand, if $Y : G \to L_0^2(P)$ is weakly harmonizable, there exists an extension space $L_0^2(\tilde{P}) \supset L_0^2(P)$, a weakly stationary $X : G \to L_0^2(\tilde{P})$ such that $Y(t) = QX(t)$, $t \in G$, where $Q : L_0^2(\tilde{P}) \to L_0^2(P)$ is the orthogonal projection of $L_0^2(\tilde{P})$ onto $L_0^2(P)$.*

The point of this result is that all weakly harmonizable processes or fields are accounted for in this construction. However, the class of these super spaces may only have just $L_0^2(P)$ as a subspace. Moreover, this probabilistic result has an immediate operator theoretical consequence which may be of some interest and so we indicate it here.

Proposition 2.4.7 *A vector measure $\nu : \mathcal{B}(\mathbb{R}) \to L_0^2(P)$ is derived from a generalized spectral family.*

Proof. Let $X(t) = \int_{\mathbb{R}} e^{it\lambda} d\nu(\lambda)$, for the vector measure ν, so that $\{X(t), t \in \mathbb{R}\}$ is weakly harmonizable and by the above result there is an extension space $L_0^2(P') \supset L_0^2(P)$, a self-adjoint operator $\tilde{A}$ on it, an $\tilde{X}_0 \in L_0^2(P')$ with, for $t \in \mathbb{R}$, (cf., Thomas (1970)):

$$\int_{\mathbb{R}} e^{it\lambda} \nu(d\lambda) = X(t) = (Qg(t, \tilde{A})) \tilde{X}_0 = \int_{\mathbb{R}} e^{it\lambda} (Q \circ \tilde{E})(d\lambda) \tilde{X}_0, \quad (82)$$

where $\{\tilde{E}(t), t \in \mathbb{R}\}$ is the resolution of the identity of $\tilde{A}$ in $L_0^2(P)$. If $E(t) = Q\tilde{E}(t)$, it is the resolution of the identity of $\tilde{A}$ in $L_0^2(P')$, and (82) implies $\nu(\cdot) = E(\cdot)\tilde{X}_0$, and the result follows. $\square$

The above theorem also implies the following self-adjoint dilations of certain (abstract) operators in a Hilbert space.

Theorem 2.4.8 *Let A be a symmetric linear operator in a Hilbert space $\mathcal{H}$ with dense domain, and $\{g_t, t \in \mathbb{R}\}$ be a family of bounded Borel functions, $g_0 = 1$. Then $\{T_t = g_t(A), t \in \mathbb{R}\}$ defines a set of bounded linear operators in an extension Hilbert space $\mathcal{H}_1 \supset \mathcal{H}$, a self-adjoint operator $\tilde{A}$ on $\mathcal{H}_1$ extending A and $T_t = Qg_t(\tilde{A}), Q : \mathcal{H}_1 \to \mathcal{H}$ is an orthogonal projection.*

Conversely, every densely defined self-adjoint $\tilde{A}$ on a Hilbert space $\mathcal{H}_1$, and a Borel function family $\{g_t, t \in \mathbb{R}\}$ define closed operators $T_t = Qg_t(\tilde{A})|_{\mathcal{H}} = g_t(A)|_{\mathcal{H}}$ where $A = Q\tilde{A}, \mathcal{H} = Q(\mathcal{H}_1)$, Q is an orthogonal projection on $\mathcal{H}$.

Here $g_t(\lambda) = e^{it\lambda} \Rightarrow g_t(\tilde{A}) = e^{it\tilde{A}} = U_t$ is unitary, and so $T_t = Qg_t(\tilde{A}) = \int_{\mathbb{R}} e^{it\lambda}(Q \circ \tilde{E})(d\lambda)$ gives a weakly continuous positive definite contractive operator on $\mathcal{H} = Q\hat{\mathcal{H}}_1$. This implies the following result due to B. Sz.-Nagy ((1955), Theorem IV) which is obtained from the above result that depends only on Naĭmark's theorem, and shows that both these are *equivalent* at a deeper level. Thus Sz.-Nagy's (1955) theorem can be stated as follows:

Theorem 2.4.9 *Let* $\{T_t, t \in \mathbb{R}\}$ *be a weakly continuous positive definite contractive set of operators on a Hilbert space* $\mathcal{H}, T_0 = id$; *then there is an extension Hilbert space* $\mathcal{H}_1 \supset \mathcal{H}_0$, *a unitary group of operators* $\{U_t, t \in \mathbb{R}\}$ *with* $T_t = QU_t, t \in \mathbb{R}$. *On the other hand every weakly continuous group* $\{U_t, t \in \mathbb{R}\}$ *of unitary operators defines* $\{T_t = QU_t, t \in \mathbb{R}\}$ *a class of positive definite contractive family on* $\mathcal{H}_0 = Q\mathcal{H}_1$ *for each orthogonal projection* $Q: \mathcal{H}_1 \to \mathcal{H}_0$.

An independent proof of the last theorem was given by Sz.-Nagy who then deduced Naĭmark's theorem from his. But the above analysis shows that the opposite procedure is also valid so that both theorems are independently proved and deduced from each other implying a deep internal equivalence, of great interest.

In the sense of the preceding two results, we include a final characterization of weak harmonizability of a process $X: G \to L_0^2(P)$ comprising (and completing) the preceding work, the complete details of which are given in the author's comprehensive paper (Rao (1982), Section 7) which uses several key results from abstract harmonic analysis. The desired result is the following: [details are in the above paper].

Theorem 2.4.10 *Let G be an LCA group* $\mathcal{X} = L_0^2(p)$ *be separable. For a weakly continuous mapping* $X: G \to \mathcal{X}$, *the following statements (i)–(v) are equivalent:*

  (i) X is weakly harmonizable,
 (ii) X is V-bounded,
(iii) X is the Fourier transform of a regular measure on the Borel sets of $\hat{G}$ (the dual group of G) into $\mathcal{X}$,
 (iv) for each $p \in \hat{L}^1(\hat{G})$, the process $X_p(= p \cdot X): G \to L_0^2(P)$ is weakly harmonizable and bounded.
    Moreover, the above equivalent conditions are implied by the following assertion:

(v) If $\mathfrak{X}_0 = \bar{sp}\{X(g), g \in G\} \subset \mathfrak{X}$, then there exists a weakly continuous positive contractive set of operators $\{T(g), g \in G\} \subset B(\mathfrak{X})$ with $T(0) = $ identity, and $X(g) = T(g)X(\sigma), g \in G$.

It will be of interest to restate the work on weakly harmonizable class, in an abbreviated way, for a quick reference with $\mathfrak{X} = L_0^2(P)$ as a separable Hilbert space on a probability space (Ω, Σ, P):

$\mathcal{V}$ = weakly continuous V-bounded random fields on $G \to \mathfrak{X}$,

$\mathfrak{w}$ = weakly harmonizable random fields on $G \to \mathfrak{X}$,

$\mathcal{F}$ = the random fields which are Fourier transforms of regular vector measures on $\hat{G} \to \mathfrak{X}$,

$\mathcal{M}$ = the module over $\hat{L}^1(\hat{G})$ of functions on $G \to \mathfrak{X}$, that are in $\hat{\mathcal{M}}_\mathfrak{X}(\hat{G})$, i.e., $\mathcal{M} = \{X : G \to \mathfrak{X} | X.\hat{L}^1(G) \subset \hat{\mathcal{M}}_\mathfrak{X}(\hat{G})\}$,

$\mathcal{P}$ = the random fields $X : G \to \mathfrak{X}$, which are projections of stationary fields on $G \to \mathfrak{K}$, $\mathfrak{K} \supset \mathfrak{X}$ is some extension (or super) Hilbert space of $\mathfrak{X}$.

With the above abbreviations, the following comprehensive result holds:

Theorem 2.4.11 *We have: $\mathcal{F} = \mathcal{M} = \mathcal{P} = \mathcal{V} = \mathfrak{w}$.*

This result and the preceding one contain essentially all the known results on the structure of weakly harmonizable random fields (and process). Some applications and adjuncts of these results will be indicated in exercises below which should be of interest for other applications.

2.5 Multiple Generalized Random Fields

A random field is a mapping observed at $(t, x_1, \ldots, x_n)$, say it denotes the time and an n-space, denoted $\mathfrak{K}$, with outcome $X(t, x_1, \ldots, x_n)$. The outcome typically depends on chance fluctuations and is abbreviated as X_τ. Thus $\mathfrak{K}$ is an observation set at $(t, x_1, \ldots, x_n)$ with the random outcome at points of $\mathfrak{K}$ into $L^2(P)$, so that one has $X : \mathfrak{K} \to L_0^2(P)$ if the variables are centered. Then the problem of interest is thus an analysis of the "data" $\{X(f), f \in \mathfrak{K}\} \subset L_0^2(p)$, if the set of $X(f)$'s, termed random fields are assumed to have a natural structure for the index set $\mathfrak{K}$. This problem was abstracted and considered by K. Ito (1954) and I. M. Gel'fand (1955) independently and by several others thereafter. A natural space $\mathfrak{K}$ now is the one introduced by L. Schwartz (1957), and the subject thereafter greatly developed, both in theory and applications.

We use a few key facts of Schwartz's theory in our applications, recalling the definitions, but referring the reader to L. Schwartz's own book (1950), or to that of Gel'fand and Vilenkin (Translation 1964). We now can introduce a *generalized random field* (g.r.f.) in three ways:

Definition 2.5.1 *(a)Let $\mathcal{K}$ be the Schwartz space of infinitely differentiable (complex) functions on $\mathbb{R}$, with compact supports (with each function). Then a linear mapping $F\colon \mathcal{K} \to \mathbb{C}$ is a generalized random field (grf) if $f_n \in \mathcal{K}, f_n \to 0$ in $\mathcal{K}$ (and they all vanish off a bounded set) along with all of their derivatives, implies $F(f_n) \to 0$ in probability.*
(b)A linear mapping $F : \mathcal{K} \to L^2(P)$, is a grf if $f_n \in \mathcal{K}, f_n \to 0$ in $\mathcal{K} \Rightarrow F(f_n) \to 0$ in $L^2(P)$.
(c)A map $F\colon (\Omega, \Sigma) \to (\mathcal{K}^*, \mathbb{R})$ is a grf if it is $\mathcal{B}$-measurable i.e., if $\mathcal{B}$ is the σ-algebra determined by $\{l \in \mathcal{K}^*\colon \mathrm{Re}\,(l(f)) < c, \mathrm{Im}\,(l(f)) < c_2, \text{where } c_1, c_2 \in \mathbb{R}, \text{and } f \in \mathcal{K}\}$.

It can be verified that a grf F with two moments, and defined as in (a), (b) or (c) above, *all agree* when compared so any one of these definitions can be used below as needed and there will be no conflicts. These ideas will be applied to the *Cramér class*, which is the most general as seen above in this work.

Definition 2.5.2 *Let $\mathcal{K}$ be the Schwartz space on $\mathbb{R}^n$ and $F\colon \mathcal{K} \to C$ be a second order grf, centered and $B(\cdot, \cdot) = \rho(\cdot, \cdot)$ as its covariance functional. Then $F(\cdot, \cdot)$ is of class (C) relative to ρ and a Borel function $g\colon \mathbb{R}^n \times \mathbb{R}^n \to \mathbb{C}$, if $\rho(\cdot, \cdot)$ determines a tempered measure $\tilde{\rho}$ on the Borel sets of $\mathbb{R}^n \times \mathbb{R}^n$, in that ($|x| = $ euclidean length)*

$$\int_{\mathbb{R}^n} \int_{\mathbb{R}^n} \frac{|d^2\tilde{\rho}(x,y)|}{[(1+|x|^2)(1+|y|^2)]^{\frac{k}{2}}} < \infty, \tag{83}$$

for some $k \geq 0$ and $\int_{\mathbb{R}^n} \int_{\mathbb{R}^n} g(t,x)\bar{g}(s,x)d^2\rho(t,s) = b(x)$ exists, $x \in \mathbb{R}^n, b(\cdot)$ is bounded on bounded sets of $\mathbb{R}^n$, so that

$$B(u,v) = \int_{\mathbb{R}^n} \int_{\mathbb{R}^n} \tilde{u}(t)\bar{\tilde{v}}(s)d^2\rho(t,s), \quad u, v \in \mathcal{K}, \tag{84}$$

with $\tilde{u}(t) = \int_{\mathbb{R}} g(t,x)u(x)dx$ as the g-transform of u, and similarly $\tilde{v}$. Then the g.r.f. F is called a generalized harmonizable random field, *and $B(\cdot, \cdot)$ is of class (C) covariance (and harmonizable when $g(t,x) =$*

e^{itx}, *the complex exponential). The corresponding tempered measure* $\rho(\cdot, \cdot)$ *is called the spectral measure of F.*

We now present an integral representation of the random field F of class (C) which includes the harmonizable class, and this will be of real interest in many applications as well as extensions.

Theorem 2.5.3 *Let Φ be a test space on $\mathbb{R}^n$ as introduced above so that it contains $\mathcal{K}$ as a dense set, and $F: \Phi \to L^2(P)$ be a generalized random functional with mean zero and covariance $B(\cdot, \cdot)$ where F is of class (C) relative a $g(\cdot, \cdot)$ and tempered covariance $\rho(\cdot, \cdot)$. Suppose the g-transforms of members of Φ exist (so $\tilde{f}(t) = \int_{\mathbb{R}^n} g(t, x) f(x) dx$ exists, $f \in \Phi$). Then there is a random measure $Z(\cdot)$ relative to ρ and that one has:*

$$F(f) = \int_{\mathbb{R}^n} \tilde{f}(t) dZ(t), \quad f \in \Phi \tag{85}$$

where $\tilde{f}$ is the $g(\cdot)$-transform of f. Moreover, $B(\cdot, \cdot)$ is given as:

$$B(u, v) = \int_{\mathbb{R}^n} \int_{\mathbb{R}^n} \tilde{u}(t) \bar{\tilde{v}}(s) d^2 \beta(t, s), \quad u, v \in \Phi.$$

The above stochastic integral is defined in the mean-square sense. Conversely, if $g(\cdot, \cdot)$ and $\rho(\cdot, \cdot)$ are given with the properties of Definition 2.5.2 above, and $Z(\cdot)$ is a random measure on the Borel algebra of $\mathbb{R}^n$, relative to ρ, and $\mathcal{K} \subset \Phi$ is dense, then $F(\cdot)$ given by (85) is a grf of class (C) on $\mathcal{K}$ and has a continuous extension to Φ.

Proof. I. We sketch the essential argument in steps for convenience, and take $E(F(f)) = 0$ and assume $B(\cdot, \cdot)$ is strictly positive definite so that $B(f, f) = 0$ only for $f = 0$. Thus the inner product in Φ is now

$$(f_1, f_2) = B(f_1, f_2) = \int_{\mathbb{R}^n} \int_{\mathbb{R}^n} f_1(t) \bar{f}_2(s) d^2 \rho(t, s), f_1 \in \Phi. \tag{86}$$

II. Let $\mathcal{K}_0 = \text{sp}\{F(f): f \in \Phi\}$ and $\mathcal{K}$ be its completion in $L^2(P)$, where $(X, Y) = E(X\bar{Y})$ for $X, Y \in L^2(P)$. Then from the equation

$$(F(f_1), F(f_2)) = B(f_1, f_2) = (f_1, f_2),$$

the mapping $f \mapsto F(f)$ defines an isometry of Φ onto $\mathcal{K}$, and the extension onto $\mathcal{K}_0$ is unique and the map $f \mapsto F(f)$ gives an isometry from $L^2(P)$ onto $\mathcal{K}$. If $Z(A)$ and $\chi_A, A \in \mathcal{B}$, (bounded) $Z(A) \in$

$\mathcal{H}, \chi_A \in L^2(\rho)$, then $Z : \mathcal{B} \to \mathcal{H}$ defines a random measure relative to ρ. Since $Z(\cdot)$ is clearly additive, its σ-additivity is verified as follows. Let $A_i \in \mathcal{B}$, disjoint, $A = \bigcup_{i=1}^n A_i$, and be bounded. Then with the standard notation, we have:

$$\left\| Z(A) - \sum_{i=1}^n Z(A_i) \right\|_2 = \left\| \chi_A - \sum_{i=1}^n \chi_{A_i} \right\|_2$$

$$= \left\| \chi_{\bigcup_{i \geq n} A_i} \right\|_2$$

$$= \tilde{\rho}(\cup_{i \geq n}(A_i \times A_i))$$

$$\leq \rho(A \times A) < \infty,$$

$\tilde{\rho}$ being the measure determined by ρ. It follows from this that $Z(\cdot)$ is σ-additive, and $Z(A) = \sum_{i=1}^\infty Z(A_i)$ holds in $L^2(\rho)$. For general bounded S_1, S_2 in $\mathcal{B}$, we get

$$E(Z(S_1)\bar{Z}(S_2)) = \int_{S_1} \int_{S_2} d^2\rho(x, y),$$

so that $Z(\cdot)$ is a random measure relative to $\rho(\cdot)$.

III. If $\tilde{\Phi} = \{\tilde{f} : f \in \Phi\}$, the g-transforms of f, then $\tilde{\Phi} \subset L^2(\rho)$, and this is a dense subspace, since by (86), $(\tilde{f}_1, \tilde{f}_2) = B(f_1, f_2) = (f_1, f_2)$. It follows that $\Phi \subset L^2(\rho) \Rightarrow f \to F(f)$ is also an isometry, and $F(f_n) = \sum_{i=1}^m a_{m_i} Z(A_i) = \int_{\mathbb{R}^n} \tilde{f}(t)dZ(t)$, $\|f - f_m\| \to 0$ shows $\{F(f_m), m \geq 1\} \subset \mathcal{H}$ is Cauchy and so the above is well-defined, and

$$F(f)_n = \int_{\mathbb{R}^n} \tilde{f}(t)dZ(t), \quad f \in \Phi$$

is uniquely defined. This gives (85).

IV. It is to be shown that the random field $F(\cdot)$ of (85) is of class (C). The linearity of F being clear, only its continuity is to be shown. Now $B(f, g) = \int_{\mathbb{R}^n} \int_{\mathbb{R}^n} \tilde{f}(t)\bar{\tilde{g}}(s)d^2(t, s)$, $f, g \in \Phi$, so $F(\cdot)$ is of class (C), if it is shown to be continuous. But $f_n \in \mathcal{K} \subset \Phi$, $f_n \to 0$ in $\mathcal{K} \Rightarrow$ the set $\{f_n, n \geq 1\}$ is compactly based and tends to zero uniformly there. Since $g(\cdot, \cdot)$ is locally bounded, it follows that $f_n \to 0$ a.e. and boundedly. It then is seen to follow that $B(f_n, f_n) \to 0$ so $F(f_n) \to 0$ in $\mathcal{H}$. Thus $F(\cdot)$ is continuous on $\mathcal{H}$. Since $\mathcal{K} \subset \Phi$, with a stronger topology, $F(\cdot)$ is continuous on $\mathcal{K}$ in the topology of Φ also. Now $\mathcal{K}$ is dense in Φ, F has a unique extension to Φ, and so is a g.r.f. of class (C). $\square$

The following consequences of the preceding result are useful.

Corollary 2.5.4 *Let* $F\colon \Phi(= \mathcal{K}) \to L^2(P)$ *be a g.r.f. of class* (C) *for a tempered covariance* ρ *and* $g(\cdot, \cdot)$. *Then* F *admits the representation relative to a random measure* $Z\colon \mathcal{B} \to L^2(P)$ *so that*

$$F(f) = \int_{\mathbb{R}^n} \tilde{f}(t)dZ(t) = \int_{\mathbb{R}^n}\int_{\mathbb{R}^n} f(x)g(t,x)dxdZ(t), f \in \mathcal{K}, \quad (87)$$

uniquely. Conversely, a functional $F\colon \mathcal{K} \to L^2(p)$ *given by* (87) *with* $Z(\cdot), g(\cdot, \cdot)$ *and* $\rho(\cdot)$ *is a g.r.f. of class* (C) *relative to* $g(\cdot, \cdot)$ *and* ρ.

This specializes to the harmonizable class which is given as follows:

Corollary 2.5.5 *Let the test space* $\mathcal{K}$ *satisfy* $\mathcal{K} \subset \Phi \subset \mathcal{L}$, *the inclusions being both algebraic and topological. If* $F\colon \Phi \to L^2(P)$ *is a grf which is harmonizable relative to a tempered covariance* ρ, *then there is a random measure* $Z\colon \mathcal{B} \to L^2(P)$ *relative to* ρ *satisfying*

$$F(f) = \int_{\mathbb{R}^n} \hat{f}(t)dZ(t) = \int_{\mathbb{R}^n}\int_{\mathbb{R}^n} e^{itx}f(x)dxdZ(t), f \in \Phi, \quad (88)$$

uniquely. Conversely, F *given on* Φ *by* (88) *with* $Z(\cdot)$ *and* $\rho(\cdot)$ *is a generalized harmonizable random field relative to* ρ.

With the above discussion, we can present a multivariate Cramér representation of the corresponding class for reference.

Theorem 2.5.6 *Let* Φ *be the (test) space as in Theorem 2.5.3 above. If now* $F = (F_1, \ldots, F_k)$ *is a* k-*dimensional grf on* Φ *of class* (C) *with respect to a* $g(\cdot, \cdot)$ *and* $\rho = (\rho_{ij}, 1 \leq i, j \leq k)$ *of tempered covariances forming a positive definite matrix, then there exists a random vector measure* $Z = (Z_1, \ldots, Z_k)$ *on the Borel field of* $\mathbb{R}^n$ *such that* $F(\cdot)$ *is uniquely representable as*

$$F(f) = \int_{\mathbb{R}^n} \tilde{f}(t)dZ(t), \quad f \in \Phi, \quad (89)$$

where $\tilde{f}$ *is the g-transform of* f, *defined before. Conversely, if* $g(\cdot, \cdot), \rho,$ *and a random vector measure* Z *relative to* ρ *are given, satisfying* (89), *then* $F(\cdot)$ *defines a* k-*dimensional class* (C) *random field on* Φ, *relative to* $g(\cdot, \cdot)$ *and* $\rho(\cdot)$.

We next present the corresponding harmonizable form which gives an immediate (extended) version of K. Ito (1954) and A. M. Yaglom (1957).

Theorem 2.5.7 *Let Φ be a complete countably normed space (such as $\mathcal{K}(M_r)$) so that $f_n \in \Phi, f_n \to 0$ in Φ, then $f_n(x) \to 0, x \in \mathbb{R}^n$, in the topology of Φ, and let $F \colon \Phi \mapsto L^2(P)$ be a grf, centered, with covariance $B(\cdot, \cdot)$. If F is of class (C) relative to $g(\cdot, \cdot)$ and a tempered covariance ρ, let $\tilde{f}(t) = \int_{\mathbb{R}^n} g(t, x) f(x) dx$ exist for all $f \in \Phi$. Then there is a random measure $Z(\cdot)$, relative to ρ, such that the following unique representation holds:*

$$F(f) = \int_{\mathbb{R}^n} \tilde{f}(t) dZ(t), \quad f \in \Phi, \tag{90}$$

where $\tilde{f}$ is the g-transform of f and we have

$$B(u, v) = \int_{\mathbb{R}^n} \int_{\mathbb{R}^n} \tilde{u}(t) \bar{\tilde{v}}(s) d\rho(t, s), \quad u, v \in \Phi. \tag{91}$$

The stochastic integral in (90) is in the mean square sense.

Conversely, if g and ρ are as in Definition 2.5.2, and $Z(\cdot)$ is a random measure, on the Borel sets of $\mathbb{R}^n$, relative to ρ, and $\mathcal{K} \subset \Phi$ is dense, then F given by (90) is a g.r.f. of class (C) on $\mathcal{K}$, with unique extension to Φ.

Proof. Let F be a grf of class (C) with covariance $B(\cdot, \cdot)$, so that we have $(E(F(f)) = 0$ is assumed, and)

$$(u, v) = B(u, v) = \int_{\mathbb{R}^n} \int_{\mathbb{R}^n} \tilde{u}(t) \tilde{v}(s) d^2 \rho(t, s), \quad u, v \in \Phi.$$

The hypothesis on Φ implies by the earlier work, $B(u, v) = G(u\bar{v})$ for some $G \in \Phi_2$, where $\Phi_2 = \Phi \times \Phi$. Since F has orthogonal values, and $(X, Y) = E(X\bar{Y})$ for $X, Y \in L_0^2(P)$, one has

$$(F(f_1), F(f_2)) = B(f_1, f_2) = (f_1, f_2) \tag{92}$$

so that the relation $f_1 \mapsto F(f_1)$ defines an isometry of Φ onto $\mathcal{H}_0 = \bar{sp}\{F(f) : f \in \Phi\} \subset L^2(P)$. If $A = \cup_{i=1}^\infty A_i, A_i \in \mathcal{B}$, disjoint, then

$$\left\| Z(A) - \sum_{i=1}^n Z(A_i) \right\|^2 = \left\| \chi_A - \sum_{i=1}^n \chi_{A_i} \right\|^2 = \left\| \chi_{\cup_{i \geq n} A_i} \right\|^2$$

$$= \rho \left(\cup_{i \geq n} A_i \times A_i \right) \leq \rho(A \times A) < \infty$$

where $\rho(\cdot)$ is the measure determined by "$\rho(s, t)$". Since $Z(\cdot)$ is a vector measure (σ-additive!) and so $Z(\cdot)$ is σ-additive, we get

$$E(Z(s_1)\overline{Z(s_2)}) = (\chi_{s_1}, \chi_{s_2}) = \int_{s_1} \int_{s_2} d^2\rho(x, y),$$

implying that $Z(\cdot) : \mathcal{B} \to \mathcal{X}$ is a random measure relative to ρ.

Consider the class $\tilde{\Phi} \subseteq \{\tilde{f} : f \in \Phi\}$ of the g-transforms, so that $\tilde{\Phi} \subset L^2(\rho)$ as a dense subset. Indeed we have by (92):

$$(\tilde{f}_1, \tilde{f}_2) = \int_{\mathbb{R}^n} \int_{\mathbb{R}^n} \tilde{f}_1(t_1)\tilde{f}_2(t_2)d^2\rho(t_1, t_2) = B(f_1, f_2) = (f_1, f_2)$$

and $f \mapsto \tilde{f}$ is an isometry, so $\tilde{f} \in L^2(\rho)$. Since $\Phi \subset L^2(\rho)$ is dense so is $\tilde{\Phi}$. Thus $\tilde{f} \in \tilde{\Phi}$ is approximable in $L^2(P)$-norm and so $\tilde{\Phi}$ is dense in $L^2(\rho)$. Hence

$$\tilde{f}_m = \sum_{i=1}^{m} a_{m_i}\chi_{A_i} \in L^2(\rho), \quad \|\tilde{f} - \tilde{f}_m\| \to 0, \text{as } n \to \infty.$$

We can conclude that $f \mapsto F(f)$ is an isometry and if $f_n \in L^2(\rho)$ corresponds to $\tilde{f}_m$, where $F(f_n) = \sum_{i=1}^{m} a_{m_i}Z(A_i) = \int_{\mathbb{R}^n} \tilde{f}(t)dZ(t)$. Then $\|f - f_n\| \to 0$ so that $\{F(f_m), m \geq 1\} \subset \mathcal{H}$ is Cauchy, and hence

$$F(f_n) = \int_{\mathbb{R}^n} \tilde{f}_n(t)dZ(t), \quad \tilde{f}_n \in L^2(\rho), \tag{93}$$

is well-defined for step functions, and then for all $f \in \Phi$, by the isometry seen above. Hence $F(f) = \int_{\mathbb{R}^n} \tilde{f}(t)dZ(t), f \in \Phi$, is defined uniquely. This establishes (90). It remains to verify that $F(\cdot)$ is of class (C) in order to complete the argument.

Thus let $Z(\cdot), \rho$ and $g(\cdot, \cdot)$ be as in the (converse) hypothesis. Since $F(\cdot)$ defined above is clearly linear, only its continuity has to be shown. It is clear that $B(f, g) = \int_{\mathbb{R}^n} \int_{\mathbb{R}^n} \tilde{f}(t)\tilde{g}(s)d^2\rho(t, s)$, for $f, g \in \Phi$ so that it is of class (C) if it is shown continuous. Now $f_n \in \mathcal{K} \subset \Phi, f_n \to 0$ in $\mathcal{K}$ implies f_n's are compactly based on a fixed set and converge uniformly there to zero, and are bounded. Interchanging the limit and integral which is permissible, we get $B(f_m, f_n) \to 0$ so $F(f_n) \to 0$ in $\mathcal{K}$, and $F(\cdot)$ is continuous on $\mathcal{K}$. Since the topology of $\mathcal{K}(\subset \Phi)$ is stronger than that of Φ, we see that $F(\cdot)$ is continuous on $\mathcal{K}$ in the topology of Φ, and by the density of the former, F has a unique continuous extension to Φ, and hence is a g.r.f. of class (C) on Φ as desired. $\square$

The following specialization is of interest in applications, comparing harmonizable and more general Cramér classes. Also, as noted before, class (C) is the most general class Φ and it includes all the known applications of interest.

Corollary 2.5.8 *Let* $\mathcal{K} \subset \Phi \subset \mathcal{S}(= \mathcal{K}(M_2))$, *the Schwartz space, [i.e., the infinitely differentiable functions on* $\mathbb{R}^n$, *which are fast decreasing in that* $|x|^n |D^\alpha f(x)| \to 0$ *as* $|x| \to \infty, n \geq 1$] *these inclusions being algebraic as well as topological. If* $F : \Phi \to L^2(P)$ *is a g.r.f. which is harmonizable relative to a tempered* ρ, *then there is a random measure* $Z : \mathcal{B} \to L^2(P)$, *relative to* ρ *such that we have the unique integral representation for* $f \in \Phi$ *as:*

$$F(f) = \int_{\mathbb{R}^n} \hat{f}(t) dZ(t) = \int_{\mathbb{R}^n} \int_{\mathbb{R}^n} e^{itx} f(x) dx dZ(x). \tag{94}$$

Conversely, F *on* Φ *defined by* (94) *with* $Z(\cdot)$ *and* $\rho(\cdot)$, *is a generalized harmonizable random field relative to* ρ.

The result follows from the theorem since $\Phi \subset \mathcal{S}$ and the Fourier transform on $\mathcal{S}$ is an onto isomorphism.

We now present a general (multidimensional) representation of a set of random fields that are related to the Cramér class which clarifies the structure of the problem, indicating some extensions.

We next consider a somewhat more general class of test spaces $\mathcal{K}(M_r)$, of Gel'fand-Shilov-Vilenkein type in which the weights $\{M_r(x), r \geq 1\}$ satisfy a growth condition called *nuclearity*, also denoted $(\mathcal{N})$, and is defined as: for each $r \geq 1$, there is $r' > r$ so that $m_{r,r'}(x) = M_r(x)/M_{r'}(x)$, is Lebesgue integrable on $\mathbb{R}^n$ and $m_{r,r'}(x) \to 0$ as $|x| \to \infty$. [It can be shown that $m_{r,r'}(x) \to 0$ as $|x| \to \infty$, and $m_{r,r'}(\cdot)$ is then Lebesgue integrable on $\mathbb{R}^n$.] This gives a general form of the *class* (C) *g.r.f.'s* of use in our study.

Proposition 2.5.9 *Let* Φ *be a test space which is a* $\mathcal{K}(M_r)$ *space where the sequence* $\{M_r, r \geq 1\}$ *satisfies the condition* $(\mathcal{N})$ *above. If* $F : \Phi \mapsto L^2(P)$ *is a grf of class* (C) *relative to a* $g(\cdot, \cdot)$ *and a tempered covariance* ρ *and* $F(\cdot)$ *has orthogonal values so that* $F(f_1) \perp F(f_2)$ *if* $f_1 \cdot f_2 = 0$, *then* ρ *concentrates on* $x = y$ *and has a tempered measure* $\sigma : \mathcal{B}(\mathbb{R}^n) \to \mathbb{R}^+$ *and a random measure* $Z : \mathcal{B}(\mathbb{R}^n) \to L^2(P)$ *relative to* σ, *with orthogonal values, such that we have:*

$$F(f) = \int_{\mathbb{R}^n} \int_{\mathbb{R}^n} f(x) g(t, x) dx dZ(t) = \int_{\mathbb{R}^n} \tilde{f}(t) dZ(t), f \in \Phi, \tag{95}$$

uniquely $\tilde{f}$ being the g-transform. Conversely, F on Φ defined by (95) *relative to g, Z, and σ is a grf of class (C) with orthogonal values.*

Proof. For a grf of class (C) with covariance $B(\cdot, \cdot)$, we have

$$B(u, v) = \int_{\mathbb{R}^n} \int_{\mathbb{R}^n} \tilde{u}(t)\tilde{v}(s)d^2\rho(t, s), \quad u, v \in \Phi,$$

and by hypothesis on Φ, $B(u, v) = G(u\bar{v})$ for some $G \in \Phi^*$. Since F is orthogonally valued and the weights $\{M_r, n \geq 1\}$ satisfy the hypothesis $(\mathcal{N})$, it follows from Gel'fand and Vilenkin ((1964), p. 287), that $f \in \Phi \times \Phi$ vanishing in a neighbourhood of $x = y$, implies $G(f) = 0$. Thus G is only concentrating on the diagonal $x = y$ and since $G_m = G|\mathcal{K}_2^m(M_r)$, has the same property, one has

$$\int_{\mathbb{R}^n} \int_{\mathbb{R}^n} u(x)\tilde{v}(y) \int_{\mathbb{R}^n} \int_{\mathbb{R}^n} g(t, x)\overline{g(s, y)}d^2\rho(t, s)dxdy$$

$$= \sum_{|\alpha|+|\beta|\leq p_m} \int_{\mathbb{R}^n} \int_{\mathbb{R}^m} M_{p_m}(x)M_{p_n}(y)h_{\alpha,\beta}(x, y)D^\alpha u(x)\overline{D^\beta u(y)}dxdy,$$

for $u, v \in \mathcal{K}^m(M_r)$. But G_m concentrates on $x = y$ implying that $h_{\alpha,\beta}$ again vanishes away from $x = y$, so that $\rho(\cdot, \cdot)$ in the above concentrates on $x = y$ also as the contrary hypothesis easily leads to a contradiction. Thus

$$\tilde{\rho}(A \times A_2) = \int_{A_1 \cap A_2} d\rho(x, x) = \rho(A_1 \cap A_2), \tag{96}$$

and the temperedness of $\tilde{\rho}$ implies that of ρ.

Thus we are back to the argument of Theorem 2.5.7, and the existence of $Z(\cdot)$ as stated in (95) obtained. The converse follows immediately from the condition (95) as in that theorem, and since Φ is the inductive limit of the spaces $\mathcal{K}^m(M_r)$ with orthogonal values, it follows from known properties of such spaces that F is a grf on Φ itself. $\square$

As a simple multidimensional extension of Theorem 2.5.7 and the above proposition, we present the following result for ready use.

Theorem 2.5.10 *Let Φ be the test space as in Theorem 2.5.7 above, and let $F = (F_1, \ldots, F_k)$ be a k-vector g.r.f. on Φ of class (C) relative to some $g(\cdot, \cdot)$ and a positive definite matrix $\rho = (\rho_{ij}, i, j = 1, \ldots, k)$ of tempered covariances also relative to $g(\cdot, \cdot)$. Then there is a random vector $Z = (Z_1, \ldots, Z_k)$ on the Borel field of $\mathbb{R}^n$ relative to ρ such that*

$$F(f) = \int_{\mathbb{R}^n} \tilde{f}(t) dZ(t), \quad f \in \Phi, \tag{97}$$

uniquely where $\tilde{f}(\cdot)$ is the g-transform of f.

Conversely, if $g(\cdot, \cdot), \rho(\cdot)$ and a random vector Z relative to ρ, are given, then (97) defines a k-vector g.r.f. on Φ of class (C), relative to $g(\cdot, \cdot)$ and $\rho(\cdot)$.

The result is established by considering the scalar product $a \cdot F$ ($= \sum_{i=1}^{n} a_i F_i$) for an arbitrary real vector 'a', and following the familiar arguments. Note that if each $\rho_{ij}(\cdot, \cdot)$ concentrates on the diagonal $x = y$, then the above result reduces to the stationary g.r.f., and the converse can be obtained that includes both K. Ito (1956) and A. M. Yaglom (1957) theorems (cf. next result), but we are concentrating on Cramér's extensions.

Theorem 2.5.11 *Let $\mathcal{K}$ be the Schwartz space of infinitely differentiable scalar functions on $\mathbb{R}^n$ with compact supports and $\tilde{C}$ be the space of complex random variables with means zero and finite variances. If $F : \mathcal{K} \to \tilde{C}$ is with mean zero and covariance functional $B(\cdot, \cdot)$ of compact support, then it can be represented as:*

$$B(f, g) = \int_{\mathbb{R}^n} \int_{\mathbb{R}^n} f(x) \overline{g(y)} h(x, y) dx dy, \quad f, g \in \mathcal{K}, \tag{98}$$

where $h(\cdot, \cdot)$ is a continuous (ordinary) covariance function on $\mathbb{R}^n \times \mathbb{R}^n$ based on a compact set, (so $h(\cdot, \cdot)$ is determined by B).

Proof. By definition of the functional $B \colon \mathcal{K} \times \mathcal{K} \to \tilde{C}$, being a covariance, and the defining property of F, it is continuous and hermitian as well as bilinear. By the famous Kernel theorem of L. Schwartz, (see e.g., Gel'fand and Vilenkin (1964), p. 74) there is a continuous bilinear form $G : \mathcal{K}_2(\mathbb{R}^n \times \mathbb{R}^n) \to \mathbb{C}$ with compact support and if $(f \cdot g)(x, y) = f(x)g(y)$ so $f, g \in \mathcal{K} \mapsto f \cdot g$ is in $\mathcal{K}_2$ that $B(f, g) = G(f \cdot g)$ then one has the integral representation as

$$B(f, g) = G(f \cdot \bar{g}) = \int_{\mathbb{R}^n} \int_{\mathbb{R}^n} f(x) \overline{g(y)} h(x, y) dx dy, f, g \in \mathcal{K},$$

for a unique $h(\cdot, \cdot)$ with compact support. But $B(\cdot, \cdot)$ is a positive definite Hermitian functional so that the $f \cdot g \in \mathcal{K}$ form a dense set, and so it is easily seen that h satisfies (98) as desired. $\square$

Recall that a random field $F : \mathcal{K} \to \mathbb{C}$ has orthogonal values if $f, g \in \mathcal{K}$ are disjointly supported then $F(f) \perp F(g)$, i.e. are orthogonal. The following result extends the earlier representations of stationary generalized fields of K. Ito (1954) and A. M. Yaglom (1957) to those of Cramér class (C) which is of interest in our study.

Theorem 2.5.12 *Let $F : \mathcal{K} \to \mathbb{C}$ be a generalized random field of class (C) having orthogonal values. Then there exists a random measure $Z(\cdot)$ on the Borel field of $\mathbb{R}^n$ with orthogonal increments such that*

$$F(f) = \int_{\mathbb{R}^n} \tilde{f}(t) dZ(t), \quad f \in \mathcal{K}, \tag{99}$$

where $\tilde{f}(t) = \int_{\mathbb{R}^n} f(x) g(x) dx$, and the covariance B of F is given by $B(u, v) = \int_{\mathbb{R}^n} \tilde{u}(t) \bar{\tilde{v}}(t) d\sigma(t)$, the tempered measure σ on $\mathbb{R}^n$ being the spectral measure of F, relative to the $g(\cdot)$ of class (C). Conversely, if $Z(\cdot)$ is the random measure with orthogonal values, relative to σ, then F given by (99) is a generalized random field on $\mathcal{K}$ of class (C) with orthogonal values.

Proof. Let $F : \mathcal{K} \to \tilde{C}$ be of class (C) and suppose it is strictly positive definite, for simplicity, so $B(f, f) = 0$ only if $f = 0$. Let $L^2(\rho) \supset \mathcal{K}$ be the completion of $\mathcal{K}$ for the inner product

$$(f, g) = B(f, g) = \int_{\mathbb{R}^n} \int_{\mathbb{R}^n} \tilde{f}(x) \bar{\tilde{g}}(y) d^2 \rho(x, y), f, g \in \mathcal{K}, \tag{100}$$

where $\tilde{f}, \tilde{g}$ are the g-transforms (i.e., $\tilde{f}(x) = \int_{\mathbb{R}^n} f(t) g(t, x) dt$ the $g(\cdot, \cdot)$ being the defining "g-element" of (C)). If F has, on $\mathcal{K}$, orthogonal values, then B of (100) is concentrated on the set $x = y$. The proof of this statement is essentially the same as in Gel'fand and Vilenkin ((1964), p. 287). Since F is of class (C), $\tilde{\rho}$ also must concentrate on $x = y$. If the resulting measure is denoted by σ, then the result is seen to be a simple consequence of Theorem 2.5.11, and further details can now be left to the reader. $\square$

In relation to A. V. Balakrishnan's characterization of the mean functions of a second order process with a given (stationary) covariance function, detailed in Chapter 1, it is of interest to know the structure of the class of all mean functions of a process with a given covariance function. Not surprisingly, it is not a linear set in general as the following example due to Skorokhod (1970) shows. Let (Ω, Σ, P) be a canonically represented probability space for $(X_t, t \in T)$ where $\Omega = \mathbb{R}^T$, Σ is the σ-algebra of Ω relative to which each $X_t(\cdot) : \Omega \to \mathbb{R}$, is measurable $t \in T$. If $Y(t) = X(t) + f(t)$, a translate of the X-process, then $f(\cdot)$ is an *admissible mean (or translate)* of $X(\cdot)$ if $P_f(\cdot) = P \circ \tau_f^{-1}(\cdot)$, where $\tau_f X_t = X_t + f(t)$ is P-continuous or $P_f \ll P$, (where $\Omega = \mathbb{R}^T$ as noted above).

Let $f_0 \in \Omega - M_\rho$ where M_p denotes the set of all admissible mean values of P so that $f \in M_p$ implies $P_f \ll P$ where P_f is the measure of the translated process $Y_t = X_t + f(t)$, defined above. Assume that the X-process is Gaussian with mean zero, and $r(\cdot, \cdot)$ as covariance. It is now known that for $f_9 \in \Omega - M_p$, the Gaussian measures P_0 and P_{f_0} are mutually singular ($P_0 \perp P_{f_0}$) by the well-known Hájek-Feldman theorem. Define the mixture measure on Σ as:

$$Q = \sum_{k=-\infty}^{\infty} \alpha_k P_{kf_0}, \quad \alpha_k > 0, \quad \sum_{k=-\infty}^{\infty} \alpha_k = 1. \qquad (101)$$

Then $f_0 \in M_Q$ but $tf \notin M_Q, 0 < t < 1$. Thus M_Q is not convex, although it is a semi-group under addition.

The following *positive* result is due to Pitcher (1963), and is stated for comparison and information:

Proposition 2.5.13 *Let $\{X_t, t \in T\} \subset L^2(P), E(X_t) = 0, r(s, t) = E(X_s \bar{X}_t), r(\cdot, \cdot)$ being continuous on the compact interval $[a, b]$. Then the operator $R : L^2(T, dt) \to L^2(T, dt)$ defined by $(Rg)(s) = \int_T r(s, t)g(t)dt$, is positive definite, compact, and $M_p \subset R^{1/2}(L^2(T, dt))$, so $f \in M_p \Rightarrow f = R^{1/2}h$, for some $h \in L^2(T, dt)$.*

Let $\{X_t, t \in T\}$ be a second order process on (Ω, Σ, P), a probability space, with $E(X_t) = 0$ and $K(s, t) = E(X_s \bar{X}_t)$, the covariance. Let M_P be the set of all admissible means of the X-process, so that the probability measures P and $P_f, f \in M_P$, are such that $P_f \ll P$. The structure of M_P is clearly useful and one can ask for $\frac{dP_f}{dP}$ for such a class of f and so the space M_P will be of interest in applications. If P is Gaussian then this space of "admissible means" M_P can be characterized now. The space M_P may be seen to be a vector space, but its topological characterization will be useful for applications. In fact an inner product can be introduced and an interesting analysis of it may be given. We thus introduce for each covariance K, a space $\mathcal{H}_K = \{f : f = \sum_{i=1}^{n} c_i K(s_{i,\cdot}), n \geq 1 \text{ and } c_i \in \mathbb{C}\}$. If $f, g \in \mathcal{H}_K$ define a complex number (f, g):

$$(f, g) = \sum_{i=1}^{n}\sum_{j=1}^{m} K(s_i, t_j) c_i \bar{d}_j = \sum_{i=1}^{n} c_i \bar{g}_i(s_i) = \sum_{j=1}^{m} d_j f(t_j), \quad (102)$$

where $f = \sum_{i=1}^{n} c_i K(s_i, \cdot)$ and $g = \sum_{j=1}^{m} d_j K(\cdot, t_j)$. We then have:

Proposition 2.5.14 *The space of admissible means* $(\mathcal{H}_k, (\cdot, \cdot))$ *of a Gaussian process is a Hilbert space. Moreover for each* $f \in \mathcal{H}_k$, *there is a* Z *such that* $E(Z\bar{X}_t) = f(t)$ *and the likelihood ratio is given by*

$$\frac{dP_i}{dP} = \exp\{Z - \frac{1}{2}E(|Z|^2)\}, a.e. \ [P]. \quad (103)$$

Proof. The demonstration depends on a basic result that two Gaussian measures in $\mathbb{R}^n$ are either mutually absolutely continuous or singular; known as the Hájek-Feldman theorem [proved in the comparison volume (2nd ed. 2014, p. 226)], and the reader is referred to the book for details which we omit here. Also as a byproduct of the demonstration there, the useful formula (103) emerges, which is also in the above book on p. 230, and we omit the detail referring the reader to it. $\square$

The general problem of characterizing admissible means for process with a given covariance function is not easy, in contrast to Balakrishnan's work noted earlier. We give a solution in Theorem 2.5.15 below which is indicative of the nature of the problem.

Using the methods and results of harmonic analysis we can present a general characterization of the set of admissible means of a second order process of a given covariance, supplementing Proposition 2.5.13 above, motivated by an idea of Hida and Ikeda (1967), for second order (not necessarily Gaussian) processes of interest in the general study. So let $\Omega = \mathbb{R}^T$ and $D \subset \Omega$ be the class of all $R^\alpha, \alpha \subset T$, finite subsets of T. If $\alpha, \beta \in D, \alpha < \beta$ (to mean $\alpha \subset \beta$), let $\Pi_\alpha : \Omega \to \mathbb{R}^\alpha$, and $\Pi_{\alpha\beta} : \mathbb{R}^\beta \to \mathbb{R}^\alpha, (\alpha \subset \beta)$ so $\Pi_\alpha = \Pi_{\alpha\beta} \circ \Pi_\beta$, and $\Pi_{\alpha\alpha} = $ identity, $\Pi_{\alpha\beta} \circ \Pi_{\beta\gamma} = \Pi_{\alpha\gamma}$ for $\alpha < \beta < \gamma$. Then $\Omega = \lim_{\leftarrow}(\mathbb{R}^\alpha, \Pi_\alpha)$ the projective limit. We take $\sum = \sigma(\cup_\alpha \Pi_\alpha^{-1}\mathcal{B}_\alpha)$ and $P \circ \Pi_\alpha^{-1} = P_\alpha$ on $\mathcal{B}_\alpha$. [The detailed presentation is given in the paper (Rao (1995), p. 546).] Thus $\tau : L^2(P) \to \mathcal{H}_C$, is an isometric isomorphism onto, where $\mathcal{H}_C$ is the Aronszajn space of continuous functions in Ω^*, the dual of the space

$\Omega = \lim_{\leftarrow} (\mathbb{R}^\alpha, \Pi_\alpha)$ where $\Omega = \mathbb{R}^T$ noted above. We use some simple properties of $\mathcal{H}_C$ as recalled in Chapter 1 and above.

Theorem 2.5.15 *Let $L^2(P)$ be as above on (Ω, Σ, P) and $\mathcal{H}_C$ and $\tau : L^2(P) \to \mathcal{H}_C$ be defined by*

$$(\tau\phi)(l) = \int_\Omega e^{il(\phi)} \phi(\omega) dP, \phi \in L^2(P), l \in \Omega^* \qquad (104)$$

is an onto isometric isomorphism, and M_P is the set of admissible means for P. Then τ induces a linear onto isometry between $L^2(P)$ and $\mathcal{H}_C$. Let $\tilde{M}_P = \{f \in M_P : \rho_{kf} \in L^1(P_{kf}), k \geq 1\}$ where $\rho_f = \frac{dP_f}{dP}$ so that ρ_{kf} exists a.e. Then $f \in \tilde{M}_P$ if and only if $\exp\{i(\cdot, nf)\} \in \mathcal{H}_c$ for $n \geq 1$, and then $\tilde{M}_P$ is a positive cone (i.e., $f \in \tilde{M}_P \Rightarrow \alpha f \in \tilde{M}_P, \alpha \geq 0$) iff we have

$$\sum_{n \geq 0} \binom{\alpha}{n} (J - I)^m \rho_f \qquad (*)$$

exists in $L^2(P)$ where $J^k \rho_f = \rho_{kf}, k \geq 0$ and I is the identity. If (*) holds for all $\alpha \in \mathbb{R}$ and $f \in \tilde{M}$, then $sp(\tilde{M}_P) \subset M_P$. In particular if for each $f \in M_P$, there is a $0 < K_f < \infty$ such that $P_f(A - f) \leq K_f P(A), A \in \Sigma$, then $\tilde{M}_P = M_P$. In particular if for each $f \in M_P$, there is a $0 < k_f < \infty$ such that $P_f(A - f) \leq k_f P(A), A \in \Sigma$, then $\tilde{M}_P = M_P$ in the above representation.

Proof. We include the argument for convenience and real feeling. The previous discussion shows that τ of (104) is an onto isometric isomorphism. The fact that M_P is a semi-group implies $k_f \in M_P, k \geq 1$, and so ρ_{kf} exists. Thus $f \in \tilde{M}_P \Rightarrow k_f \in \tilde{M}_P$ for $k \geq 0$. So $\rho_{kf} \in L^2(P)$ and $C_{nf} = C \exp(i(\cdot, nf)) \subset \mathcal{H}_C$. In the opposite direction, if $C_{nf} \in \mathcal{H}, n \geq 1$, then $\rho_{n,f} = \tau^{-1}(C_{nf}) \in L^2(P)$ and

$$\tau(\rho_{nf})(\ell) = C_{nf}(\ell) = \int_\Omega e^{i(\ell,\omega)} dP_{nf}(\omega) = \int_\Omega e^{i(\ell,\omega)} \rho_{nf}(\omega) dP(\omega).$$

$$(105)$$

The uniqueness of the Fourier transforms gives $\rho_{nf} = \frac{dP_{nl}}{dP}$, a.e., and then $\rho_{nf} \in L^1(P_{nf}), n \geq 1$, so that $f \in \tilde{M}_P$.

Next let $\alpha \geq 0$ and $f \in \tilde{M}_P$, so $\rho_{kf} \in L^2(P)$, and $\tau(\rho_{kf}) = C_{kf}$. Suppose that the condition (*) holds, so the series converges in $L^2(P)$. Since τ is a bounded linear map on $L^2(P)$, we can interchange it with the summation of (*) and the series converges in $L^2(P)$. Since τ is a bounded linear mapping on $L^2(P)$, and $\mathcal{H}_C$ is complete, we get,

$$\sum_{k=0}^{\infty} \binom{\alpha}{k} \tau((J-I)^k \rho_f) \in \mathcal{H}_C. \tag{106}$$

Now from $J^k(\rho_f) = \rho_{kf} = \tau^{-1}(C_k)$, one can define

$$\tau((J-T)^n \rho_f) = \sum_{k=0}^{n} \binom{n}{k} (-1)^k C_{kf}(\cdot) = \sum_{k=0}^{n} \binom{n}{k} (-1)^k e^{ik(\cdot, f)}(\cdot). \tag{107}$$

If $\chi_f = \exp(\cdot, f)$ (a character) so $\chi_{tf} = (\chi_f)^t$, and (106) and (107) $\Rightarrow$

$$\sum_{n=0}^{\infty} \binom{\alpha}{n} \sum_{k=0}^{n} (-1)^k \chi_{kf} C(\cdot) = \sum_{n=0}^{\infty} g^{(n)}(1)(\chi - 1)^n C(\cdot)$$
$$= g(\chi_f) C(\cdot) \in \mathcal{H}_C, \tag{108}$$

where $g(\chi) = \chi^\alpha$ and $g^{(n)}(1) = \left(\frac{d^n y}{dx^n}\right)(1)$. It then follows that $\chi_{\alpha f} \in \mathcal{H}_C \Rightarrow \rho_{kf} = \tau^{-1}(C_{\alpha f}) \in L^2(P)$. Thus with α for n there, shows $\rho_{\alpha f} \in L^2(P)$ and $\alpha f \in \tilde{M}_P \subset M_P$.

For the converse, let $\alpha \in \tilde{M}_P$, then $C_{\alpha, f} \in M_P$, and so (108) holds. Apply τ^{-1} to both sides of this result. It follows that the series converges in $L^2(P)$ and (*) holds and $\rho_k f \leq kf$ a.e. so ρ_{kf} is bounded, and it is seen that $\tilde{M}_P = M_P$. If (*) also holds for all $\alpha \in \mathbb{R}$ and $(f, g) \in \tilde{M}_f$ then the pair $(\alpha f, \beta g)$ also is in it. By the semi-group property of M_P, it follows that $sp(\tilde{M}_P) = M_P$ as desired. $\square$

Remark 15. In the Gaussian case, the condition (*) above is automatic and $\tilde{M}_P = M_P$.

The following result, due to T. S. Pitcher (1963), does not use the Aronszajn space methods and is based on Karhunen's expansion for second order processes (not restricted to Gaussian) we state the interesting result, referring to the original paper for the proof.

Theorem 2.5.16 *Let* $X = \{X_t, -\infty < a \leq t \leq b < \infty\} \subset L_0^2(P)$ *be a process with a continuous covariance* $r : (s, t) \mapsto E(X_s \bar{X}_t)$. *If* $\{\lambda_n\}$ *and* $\{\phi_n\}$ *are the eigenvalues and the corresponding eigenfunctions of* r, *define* X_n *and* f_n *by the integrals* $(T = [a, b])$

$$X_n = \lambda_n^{-\frac{1}{2}} \int_T X(t)\phi_n(t)dt, f_n = \lambda_n^{-\frac{1}{2}} \int_T f(t)\phi_n(t)dt, n \geq 1, \quad (109)$$

so $\{X_n, n \geq 1\}$ are orthogonal and $\{f_n, n \geq 1\} \in l^2$. Let P_n, p_n be the distribution and density of $\{X_i, 1 \leq i \leq n\}$ relative to the Lebesgue measure. Suppose the $\{p_n, n \geq 1\}$ satisfies the conditions: (i) $p_n > 0$, a.e., (ii) $\lim_{t_i \to \infty} p_n(t_1, \ldots, t_n) = 0, 1 \leq i \leq n$, for all t_i, and (iii) $\frac{\partial p_n}{\partial t_i}, 1 \leq j \leq n$ exists and $\sum_{j=1}^{n} \int_{\mathbb{R}^n} \left(\frac{\partial \log p_n}{\partial t_i} \right)^2 dP \leq K_0 < \infty, n \geq 1$.

Then the set $M_1 = \{f \in M_p : f = \{f_n, n \geq 1\} \in l^1\}$ is a positive cone. Further, $M_1 = M_P$ holds if also $\sum_{n \geq 1} \lambda^{-\frac{1}{2}} < \infty$. If each p_m is symmetric about the origin of $\mathbb{R}^n$, then M_1 is also linear. In general M_P need not even be a convex set.

We omit the proof, referring it to Pitcher (1963). The point of this result is to emphasize that the set of means is not generally linear for non Gaussian processes. The proof of the result itself depends on an approximation result of semi-groups of operators, due to Trotter (1958), in Banach spaces. The point of the result is to emphasize that the structure of the set of means of (second order) processes is involved and exemplifies the earlier discussion of nonlinearity! We next turn to studying some useful functionals based on certain *local* properties.

2.6 Local Functionals in Probability; Their Integral Representations and Applications

The classical probability problems involving sums of independent random variables have been generalized by the Russian mathematicians I. M. Gel'fand, N. Ya. Vilenkin and their associates to the class of (continuous) linear functions on smooth function spaces (e.g., infinitely differentiable functions on $\mathbb{R}^n$) with independent values. The 'smooth' space considered here is denoted by $\mathcal{K}$, the L. Schwartz space of infinitely differentiable real functions (on $\mathbb{R}$) vanishing off compact sets and $l : \mathcal{K} \to \mathcal{F}$ where $\mathcal{F}$ is a topological vector space of (scalar) random variables on (Ω, Σ, P), with $l(f_1 + f_2) = l(f_1) + l(f_2)$ for $f_1 \cdot f_2 = 0, (f_i \in \mathcal{K}), \ell(f_i), i = 1, 2$ independent, such a fundamental $l(\cdot)$ is called *local*, and its characterization under suitable conditions is desired. It is a vast generalization of classical limit laws for sums of independent random variables. Here is a characterization, with some applications, handled by Gel'fand and were obtained by the author.

Theorem 2.6.1 *Let $C_c(\Omega)$ be the space of continuous compactly supported real functions on a locally compact space Ω, and $\Lambda : C_c(\Omega) \to \mathbb{R}$ be a mapping (called a **local functional**) satisfying*

(i) *(Sequential continuity) If $\{f_n, n \geq 1\} \subset C_c(\Omega)$ is a pointwise convergent bounded set, then $\{\Lambda(f_n), n \geq 1\} \subset \mathbb{R}$ is Cauchy.*

(ii) *(Additivity) $\Lambda(f_1 + f_2) = \Lambda(f_1) + \Lambda(f_2)$ if $f_1 \cdot f_2 = 0$ for which the quantities are defined.*

(iii) *(Bounded Uniform Continuity) For each $\varepsilon > 0, \gamma > 0$ there is a $\delta (= \delta_{\varepsilon,\gamma}) > 0$ with $\|f_i\| = \gamma, f_i \in C_c(\Omega), i = 1, 2, \|f_1 - f_2\| \leq \delta$ implies $|\Lambda(f_1) - \Lambda(f_2)| < \varepsilon$, where $\| \cdot \|$ is the uniform norm.*

Under these conditions, $\Lambda(\cdot)$ admits an integral representation as:

$$\Lambda(f) = \int_\Omega \Phi(f(\omega), \omega) \mu(d\omega), f \in C_c(\Omega) \qquad (110)$$

where μ is a finite regular Borel measure on Ω, and $\Phi : \mathbb{R} \times \Omega \to \mathbb{R}$ satisfies the following three conditions:

(a)$\Phi(0, \omega) = 0$, *and $\Phi(\cdot, \omega)$ is continuous for $a \cdot a \cdot \omega \in \Omega, (\mu)$,*

(b)$\Phi(x, \cdot)$ *is μ-measurable for all $x \in \mathbb{R}$,*

(c)*for $f \in C_c(\Omega), \Phi(f(\omega), \omega)$ is bounded for $a \cdot a \cdot \omega \in \Omega$, and for sequences $\{f_m : n \geq 1\}$ as in (i) $\{\Phi(f_n, \cdot), n \geq 1\}$ is Cauchy in $L^1(\mu)$.*

Conversely, if the pair (Φ, μ) satisfies conditions (a)–(c) above, then it is a local functional on $C_c(\Omega)$ for which the statements (i)–(iii) hold.

Before establishing the theorem we first show, as a consequence, that the classical Riesz–Markov theorem follows immediately. This gives a motivation for applications and other uses later on.

Theorem 2.6.2 (Riesz-Markov) *Let Ω be a locally compact space and $C_c(\Omega)$ be the space of real continuous functions with compact supports. If $l : C_c(\Omega) \to \mathbb{R}$ is a positive linear functional, then there exists a unique regular Borel measure μ on the Borel σ-ring of Ω such that*

$$l(f) = \int_\Omega f(\omega) d\mu(\omega), f \in C_c(\Omega). \qquad (111)$$

Proof (of Theorem 2.6.2). This result is a significant generalization of the classical F. Riesz's representation stating that a positive linear functional $l(\cdot)$ on $C_c(\Omega)$, the space of continuous scalar (in real or complex) functions on a locally compact space Ω, vanishing off compact sets, corresponds to a finite Borel measure $\mu \geq 0$, such that

$$l(f) = \int_\Omega f(w)d\mu(w), \quad f \in C_c(\Omega), \tag{112}$$

and we present a quick proof of the general formula (111), containing (112).

Since $l(\cdot)$ is a continuous linear functional, $|l(f)| \leq K_0\|f\|$ for some $0 \leq K_0 < \infty$, and $f \in C_c(\Omega)$. In particular $l(\cdot)$ is local. So by Theorem 2.6.1, there is a finite signed Borel measure verifying (110).

Consider the set

$$\mathcal{C} = \{A \subset \Omega : \int_A \Phi(f+g, \cdot)d\mu = \int_A \Phi(f, \cdot)d\mu + \int_A \Phi(g, \cdot)d\mu\}. \tag{113}$$

If $l_S = l|C(S)$, for each compact $S \subset \Omega$, then l_S is continuous and linear satisfying $l_{S_1} = l_{S_2} = l_{s_1 \cap s_2}$ on $C(S_1 \cap S_2)$ for any compact subsets S_1, S_2 of S, and $\mu_S(\cdot) = \mu(S \cap \cdot)$ represents l_S in (110). Thus $\mathcal{T}$ of (113) contains compact, and then B is a Baire subset of Ω_1 and $l_B(\cdot)$ is a local functional, satisfying (i)–(iii) of Theorem 2.6.1.

It follows from this that $\mathcal{T}$ contains each Borel set, and $l_B(\cdot)$ is again a local functional satisfying the conditions of Theorem 2.6.1. Hence $\mathcal{C} \supset \mathcal{B}$, the Borel σ-algebra, and for $a \cdot a \cdot (w)$.

$$\Phi(f + g)(w) = \Phi(f)(w) + \Phi(g)(w), \quad w \in \Omega. \tag{114}$$

However, one can take $\Phi(\cdot, w)$ to be continuous for each w. Since f, g are arbitrary (114) can be identified with the classical Cauchy functional equation. Then the well-known (and familiar) solution is $\Phi(f(w), w) = \beta(w)f(w), w \in \Omega$ for some Borel function $\beta : \Omega \to \mathbb{R}$, which must be also integrable from the hypothesis of Φ, (we use the fact that $\int_\Omega fgd\mu \in \mathbb{R}$ for each $g \in L^\infty(\mu) \Rightarrow f \in L^1(\mu)$). Let $d\nu = \beta d\mu$ and then the (local but now) linear functional $l(\cdot)$ satisfies

$$l(f) = \int_\Omega f(w)d\nu(w), \quad f \in C_c(\Omega). \tag{115}$$

But this is the same as (111), and it also implies as a byproduct that a bounded linear $l : C_c(\omega) \to \mathbb{R}$ is representable as $l = l_1 - l_2$, l_i positive. $\square$

With the preceding important and motivational application, we now outline the main representation theorem for local functionals whose quite long proof is only presented in outline here for convenience. [Complete details are in the author's Measure Theory book (Rao (1987), pp. 456–486 or second Ed. (2004), pp. 676–684).]

Proof sketch of Theorem 2.6.1. Here we use a series of steps in establishing the result.

1. Initial simplification: From $C_c(\Omega)$, of compactly based scalar continuous functions, consider all bounded pointwise limits from $C_c(\Omega)$, denoted by $B_0(\Omega)$ which is a vector space of bounded Baire functions, vanishing at infinity, so that each $\{\omega : |f(w)| \geq \varepsilon > 0\}$ is compact. Then $C_0(G)$ is uniformly dense in $B_0(G)$. If A is compact and U open with $A \subset U(\subset G)$, then let for $h > 0$, $p_{A,U}^h \in C_\infty(G)$ with $p_{A,U}^h = h$ on $A = 0$ off U, (possible to find by Urysohn's lemma) to be called a *peak function* of height h and base A; and $p_{A,U}^h \downarrow h\chi_A$. So $\chi_A \in B_0(G)$. If A, B are compact Baire sets then $\chi_{A-B}, \chi_{A\cup B}$ are in $B_0(G)$, and $\mathcal{T} = \{A : \chi_A \in B_0(G)\}$ is a ring containing all compact Baire sets. Also, $\mathcal{T}$ includes the ring $\mathcal{R}$ of compact Baire sets. This reduction is needed for the ensuing analysis.

2. Next, we extend the (local) functional $l(\cdot)$ to $B_c(G)$ from $C_c(G)$, the compactly supported continuous function space, on to the locally compact G. If $f_n \downarrow \chi_A$, so that $\{l(f_n), n \geq 1\}$ is Cauchy in $\mathbb{R}$, its limit $\tilde{M}(\chi_A) = \lim_n M(f_n)$, and extends. The generalized functional $\tilde{M}(\cdot)$ is additive on simple functions on $\mathcal{R}$.

3. For each $A \in \mathcal{C}, h \in \mathbb{R}$, let $\mu_h(A) = \tilde{M}(h\chi_A)$. Then $\mu_h(\cdot)$ is additive and can be extended, with some standard work, to be a regular content and with some further work, to the Borel σ-algebra $\mathcal{R}$ of G. It is necessary to verify that $\mu_h(\cdot)$ depends on h continuously, and $\mu_0(\cdot) = 0$.

4. If $h_1, h_2, \ldots$ is some enumeration of rationals, define $\mu(\cdot)$ as

$$\mu(\cdot) = \sum_{n=1}^{\infty} \frac{1}{2^n} \frac{|\mu_0^{h_n}|(\cdot)}{1 + |\mu_0^{h_n}|(G)}$$

and show that $\mu(\cdot)$ is a measure and after some further work, we get

$$\tilde{M}\mid_{C_n(G)} (\cdot) = M(\cdot)$$

which is well-defined, and one gets after some further work that

$$\tilde{M}(h\chi_A) = \lambda_0^h(A) = \int_A a(h)d\lambda = \int_G \Phi(h\chi_A)d\lambda,$$

where $\Phi(h\chi_A(t)) = a(h), t \in A$, and $= 0$ if $t \notin A$. Finally,

$$M_1(f) = \int_G \Phi(f(t))d\lambda(t), \quad f \in C_\infty(G),$$

and $M_1 \mid_{C_\infty(G)} = M$ in all cases. The many details (omitted) should be obtained from my paper (Rao (1980), pp. 25–29).

5. The converse is simpler. Under the given representation one sees that if $\{f_n, n \geq 1\} \subset C_\infty(G)$, then $\{\psi(f_n), n \geq 1\}$ is Cauchy in $L^1(\mu)$, and one can verify the remaining conditions. [The reader can see all the details from the easily accessible paper (Rao (1980), Theorem 2 and also from the author's *Measure Theory book* (2004), second Edition, pp. 676–684).] This completes the outline of the proof. $\square$

We now briefly indicate a key probabilistic application of the above theorem as well as its place in extending the Lévy–Khintchine representation in this context as an important motivation here. For this, it is useful to recall a concept, termed a *Sazanov topology*, on a locally convex vector space. Thus Sazanov's topology $\mathcal{S}$ on a locally convex linear space $\mathcal{F}$ is a locally convex one defined by the set of continuous seminorms generated by all the quadratic forms $Q(\geq 0)$ of finite trace on $\mathcal{F}$ as follows:

If Q and H on $\mathcal{F}$ are such forms, and $\{e_i, i \in I\}$ is an orthonormal set in $\mathcal{F}$, and $\sup_n \sum_{i=1}^n Q(e_i) = \text{trace}(Q/H) < \infty$; let us consider all such pairs in $\mathcal{F}$ for H, with $\sup_n \sum_{i=1}^n Q(e_i) = \text{trace}(Q/H) < \infty$. All these pairs define a topology termed *Sazanov or $(\mathcal{S})$ topology* which is so named and detailed by N. Bourbaki (Livre VI, Integration, Chapitre III). It is locally convex and the neighbourhood system at $f_0 \in \mathcal{F}$ is

$$N(f_0 : \varepsilon_1, \dots, \varepsilon_n) = \{f : Q_j(f - f_0)^2 < \varepsilon_j^2, j = 1, \dots, n\}. \quad (116)$$

This is verified to be a locally convex topology and is coarser than the given one, which coincides with it iff the given one is nuclear. Its importance is exemplified by the following useful result.

Theorem 2.6.3 *As before let $C_\infty(G)$ be the vector space of real continuous compactly supported functions on a locally compact Hausdorff space G, and $L : C_\infty(G) \to \mathbb{C}$ be a mapping. If it is the ch. f of a g.r.f., then one has (i) $L(0) = 1$, (ii) $L(\cdot)$ is positive definite, and (iii) $L(\cdot)$ is continuous in the topology of $C_\infty(G)$. In the opposite direction let $L(\cdot)$ satisfy (i), (ii) and (iii'); namely given $\varepsilon > 0$ and $k > 0$, there is a $\delta(= \delta_\varepsilon > 0)$ such that for $\|g_i\| \leq k, g_i \in C_\infty(G), i = 1, 2, \|g_1 - g_2\| < \delta$*

$\Rightarrow |L(g_1) - L(g_2)| < \varepsilon$, *so that $L(\cdot)$ is continuous now in the 8-topology of $C_\infty(G)$. On the other hand, a mapping L satisfying (i)–(ii) and is continuous in the 8-topology on $C_\infty(G)$ is the ch.f. of a g.r.f. $F : C_\infty(G) \to L^0(P)$, on some probability space (Ω, Σ, P).*

A proof of this result uses some properties of the space $C_\infty(G)$ which is a "bornological space", and the details are in (Rao (1980), Theorem 7). An interesting consequence of it is a generalized Lévy–Khintchine formula: Since this result has such a distinguished place in the modern studies of Probability Theory, we present it here.

As before, let G be a locally compact Hausdorff space and $C_\infty(G)$ as a real continuous compactly supported function space, $\psi : \mathbb{R} \times G \to \mathbb{R}$ satisfies $\psi(0,t) = 0, \psi(\cdot, t)$ is continuous, $\psi(x, \cdot)$ is measurable, and $\psi(f(t), t)$ is bounded for $a \cdot a \cdot (t)$, and for all $f \in C_c(\mathbb{R})$ with $\int_{\mathbb{R}} f(x)dx = 0$, then we have

$$\int_A \int_{\mathbb{R}} \psi(x,t)(f * \bar{f})(x)dx d\mu(t) \geq 0, \qquad (117)$$

where $\bar{f}(x) = \overline{f(-x)}$, any Borel sets $A \subset G$ with '*' as convolution. Since $\int_{\mathbb{R}} f'(x)dx = 0$, (117) can be written also as:

$$\int_{\mathbb{R}} \psi(x,t)(f' * \bar{f}')(x)dx \geq 0, \qquad a \cdot a \cdot (t), f \in \mathcal{K} \subset C_c(\mathbb{R}).$$

With this set up the Lévy–Khintchine analog is now given by:

Theorem 2.6.4 *Let $L : C_c(G) \to \mathbb{C}$ be the characteristic function of a generalized random field with independent values at each point, and obey conditions (i)–(iii) of the above theorem.*

Then the functional $L(\cdot)$ can be represented as

$$L(f) = \exp\left(\int_G \psi(f(t), t)d\mu(t)\right), \qquad f \in C_\infty(G) \qquad (118)$$

where μ is a Radon measure on $(G, \mathcal{G})$, Borel couple, $\psi(x, \cdot)$ is μ-measurable, $x \in \mathbb{R}$, and $\psi(x, \cdot)$ is given by

$$\psi(x,t) = \int_{[|y|>0]} [e^{iyx} - \alpha(y)(1 + ixy)]\sigma(dy, t) + a_0(t)$$

$$+ ia_1(t)x - a_2(t)\frac{x^2}{2}, \qquad (119)$$

where $a_2(t) \geq 0, a_1 : G \to \mathbb{C}$ Borel measurable, such that;

(i) $\alpha : \mathbb{R} \to \mathbb{C}$ is an analytic function of exponential type and $\alpha(y) - 1$ has a zero of order three at $y = 0$,

(ii) $\sigma(\cdot, t)$ is a Radon measure on $\mathbb{R}$ and $\sigma(A, \cdot)$ is Borel for each $t \in G$ and Borel set $A \subset \mathbb{R}$,

(iii) for $a \cdot a \cdot (t)$ one has

$$\int_{0 < |y| \leq 1} y^2 \sigma(dy, t) + \int_{|y| > 1} \sigma(dy, t) < \infty; \int_{|y| > 0} (1 - \alpha(y)) d\sigma(y, t) = -a_0(t).$$

$$(120)$$

Conversely, if μ is Radon and ψ satisfies (i)–(iii), then $L(\cdot)$ of (118) with (i)–(iii) defines $L(\cdot)$ of (118) and is the ch.f. of a g.r.f. on $C_\infty(G)$ having independent values at each point, and $L(\cdot)$ verifies also conditions of Theorem 2.6.3 above.

This general result as well as the preceding ones are detailed in the author's paper (Rao, (1980), *Functional Analysis* Journal 39, 23–41) and will be useful for readers to study it carefully and in detail. The above theorem is related to (and is an extension of) a key result due to A. M. Yaglom and N. Ya. Vilennkin, described in the book by Gel'fand and Vilenkin (English translation, 1964, Sec II. 4). The work is included in this section for the purpose of inviting the reader to proceed with this important extended analysis.

An interesting question is to characterize functionals $L(\cdot)$ on $\mathcal{K}$ of order m that are exponentials, i.e.

$$L(f) = e^{M(f)}, \quad f \in \mathcal{K} \tag{121}$$

where $M(\cdot)$ is a local functional of order m, finite. The next result has a characterization of the $L(\cdot)$ which is an extension of the fundamental Lévy–Khintchine representation, and hence will be of special interest for an advanced analysis in this area:

Theorem 2.6.5 *Let $L : \mathcal{K} \to \mathbb{C}$ be a functional given by*

$$L(f) = \exp\{M(f)\}, \quad f \in \mathcal{K}, \tag{122}$$

where $M(\cdot)$ is local of order $m < \infty$. Then $L(\cdot)$ is the characteristic functional of a g.r.f. on $\mathcal{K}$, with independent values if and only if it is representable as, for $f \in \mathcal{K}$,

$$L(f) = \exp\left\{ \int_{\mathbb{R}^n} \Phi(f(t), (Df)(t), \dots, (D^\alpha f)(t), t) d\mu(t) \right\}, \tag{123}$$

where $|\alpha| \leq m$, μ *is a Radon measure on* $\mathbb{R}^n$ *(and nonatomic for the 'only if' part) and* $\Phi : \mathbb{R}^v \times \mathbb{R}^n \to \mathbb{R}$, *is given by*

$$\Phi(x,t) = \int_{|y|>0} [e^{i(x,y)} - \alpha(ty)(1 + i(x,y))]\sigma(dy;t)+$$

$$a_0(t) + \sum_{|k|=1}^{2} a_k(t)\frac{(ix)^k}{k!}. \tag{124}$$

Here $\sigma(\cdot, t)$ *is a positive tempered measure on* $\mathbb{R}^v$ *for* $t \in \mathbb{R}^n$, $\sigma(A, \cdot)$ *is Borel on* $\mathbb{R}^n$ *for each Borel set* $A \subset \mathbb{R}^v$, (x,y) *is the scalar product in* $\tilde{R}$, $\alpha(\cdot)$ *is an entire analytic function of exponential type such that* $\alpha(y) - 1$ *has a zero of order 3 at* $y = 0$, *and the* $a_i(\cdot)$ *are Borel functions in* $\mathbb{R}^n$, *determined by* $L(\cdot)$ *(or* Φ),

$$\int_{0<|y|<1} |y|^2 \sigma(dy, t) + \int_{|y|>1} \sigma(dy, t) < \infty$$
$$\int_{|y|>0} (1 - \alpha(y))\sigma(dy, t) + a_0(t) = 0, \sum_{|r|=|s|=1} a_{r+s}(t)\xi_r\xi_s \geq 0, \tag{125}$$

where the ξ's *are complex numbers* $(|k| = k_1 + \cdots + k_n, |y|^2 = (y, y), k! = k_1! \ldots k_v!$ *and* $x^k = x_1^k \ldots x_r^k$).

Finally, if $L(\cdot)$ *is also translation invariant and* Φ *of (124) is independent of* t, *then* $\sigma(\cdot)$ *and* $a_i(\cdot)$ *also do not depend on* t.

A proof of this result which generalizes the classical Lévy–Khintchine representation, is detailed in the author's paper (1971) in the well-known Russian probability journal, given in English, which is available in the U.S.A., and most other places, and so the repetition will be skipped. Several other results in this paper, which originally emerged from the Russian authors' contributions, have been carefully revised there, and so we invite the readers to read and extend it further. It also has some more analysis on *strongly local functionals* which are local and the continuity of $M : \mathcal{K} \to \mathbb{C}$ is strengthened to uniform continuity, and there the corresponding characterization was also detailed.

2.7 A Probabilistic Proof of Riemann's Hypothesis

It is first necessary to recall and review some works that considered this famous classical problem, called the *Riemann hypothesis* on the zeroes of the zeta function of this author, and some other approaches without using probability although the subjects are closely related and viewed from one point of the subject. This may show the affinity of the subjects even though they started from different perspectives as well as advantage points.

(a) *The harmonic method.* To explain this method we recall:

$$G(x, s) = \frac{1}{\Gamma(s)} \int_0^x e^{-t} t^{s-1} dt, \quad s > 0, x > 0, \tag{126}$$

and

$$\tilde{G}(x, \alpha) = \frac{1}{\sqrt{2\pi\alpha^2}} \int_{-\infty}^x e^{-\frac{t^2}{2\alpha^2}} dt, \quad -\infty < x < \infty, \alpha > 0. \tag{127}$$

These two distributions $G(\cdot, s), \tilde{G}(\cdot, \alpha)$, called the *gamma* and the *Gaussian*, are both basic in probability theory. They lead to the Riemann problem (or hypothesis) quickly as we show now.

It is trivial that $G(+\infty, s) = 1$ so that $1 = \frac{n^s}{\Gamma(s)} \int_0^\infty e^{-nt} t^{s-1} dt$, and

$$\Gamma(s) \sum_{n=1}^\infty \frac{1}{n^s} = \int_0^\infty t^{s-1} \left(\sum_{n=1}^\infty e^{-nt} \right) dt = \int_0^\infty \frac{t^{s-1}}{e^t - 1} dt. \tag{128}$$

If the series on the left is denoted by $\zeta(s)$, so $\zeta(1) = +\infty$, or the function $\zeta : (0, \infty) \to \mathbb{R}, \zeta(1) = \infty$, and that 1 is a simple pole of ζ. Using complex analysis, $\zeta(\cdot)$ can be extended to all of the complex plane $\mathbb{C}$, with $(-s)^s = \exp[s \log(-s)]$ where $\log Z$ is defined for complex Z, and using the contour integral from $+\infty$ on going to $\delta > 0$ and back to $+\infty$, using $(-t)^s = \exp[s \log(-t)]$ with $\log Z$ for complex Z defined as usual. Thus one obtains

$$\oint_{+\infty}^{+\infty} \frac{(-t)^s}{e^t - 1} \frac{dt}{t} = \lim_{\delta \downarrow 0} \left(\int_{+\infty}^\delta + \int_{|t|=\delta} + \int_\delta^{+\infty} \right) \frac{(-t)^s}{e^t - 1} \frac{dt}{t}$$

$$= \lim_{t \downarrow 0} \int_{+\infty}^\delta \frac{e^{s \log t - \pi i}}{e^t - 1} \frac{dt}{t} + \int_\delta^{+\infty} \frac{e^{(s \log t + \pi i)}}{e^t - 1} \frac{dt}{t} + 0$$

$$= \left(e^{i\pi s} - e^{-i\pi s} \right) \int_0^\infty \frac{t^s - 1}{e^t - 1} dt + 0$$

$$= 2i \sin(\pi s) \Gamma(s) \zeta(s), \quad \text{using (128).} \tag{129}$$

Now with the classical result $\frac{\pi}{\sin \pi s} = \Gamma(s)\Gamma(1 - s)$, and $\Gamma(\cdot)$ extended to $\mathbb{R}^-$ using the classical Complex Function Theory (cf. Ahlfors (1975), p. 98) the integral for $\zeta(\cdot)$ gives

$$\zeta(s) = \frac{\Gamma(-s+1)}{2\pi i} \oint_{+\infty}^{+\infty} \frac{(-t)^s}{e^t - 1} \frac{dt}{t}. \tag{130}$$

But by the usual Taylor expansion of $\frac{t}{e^t-1}, 0 < t < 2\pi$, in (128),

$$\frac{t}{e^t - 1} = \sum_{n\geq 0} \frac{B_n t^n}{n!}, \tag{131}$$

where the B_n are the Bernoulli numbers $B_0 = 1, B_1 = -\frac{1}{2}, B_{2n+1} = 0$, $n \geq 1$, but $B_{2n} \neq 0$, so $\zeta(s) = 0, s = -2n, n \geq 1$, termed the trivial zeros of $\zeta(\cdot)$ and $\zeta(-n) = (-1)^n \frac{B_{n+1}}{n+1}, \zeta(2n) = (2\pi)^n \frac{(-1)^{n+1}B_x}{2(2n)!}, n > 1$. The basic question is to find the nontrivial zero of $\zeta(\cdot)$, and study its analytical properties. And then Riemann conjectured that the nontrivial zeros of $\zeta(\cdot)$ all lie on the complex line Re $(s) = \frac{1}{2}$ (with the obvious simple pole of $Z(\cdot)$ at $s = 1$). This is the *Riemann hypothesis*. In 1975, Norman Levinson has shown that almost all roots of $\zeta(s) = 0$ are near the critical line $Re(s) = \frac{1}{2}, s = \frac{1}{2} + it, t \in \mathbb{R}$, and the same is true of the roots of $\zeta(s) = a$. We strengthen this assertion below. To explain the problem further, we now consider $\tilde{G}$ the Gaussian distribution (127) and get some new insights.

To understand the key aspects of the problem let us consider the zeta function, derived from the classical Euler product representation of prime numbers as given by (p denoting primes) the Euler product:

$$\zeta(s) = \Pi_p (1 - p^{-s})^{-1} \tag{132}$$

so that $\zeta_1 = +\infty$ (i.e., there are infinitely many prime numbers), and that $s \mapsto [\zeta(s)]^{-1}$ is an entire function on $\mathbb{C}$. The classical work by Khintchine (1923) shows that $\phi_\sigma : t \to \zeta(\sigma + it)$ is a positive definite and non vanishing function for all $\sigma > 1$. It is given, more precisely, by the following simple but useful result.

Proposition 2.7.1 *The mapping* $\phi_\sigma : t \mapsto \frac{\zeta(\sigma+it)}{\zeta(\sigma)}, \sigma > 1$, *is an infinitely divisible characteristic function and so never vanishes. Thus* $s \mapsto \zeta(s)$ *is a nonzero entire function in the right half-plane determined by* $\sigma > 1$ *of* $\mathbb{C}$ *where* $s = \sigma + it$, *and where* $\zeta(\cdot)$ *is defined by* (132).

Proof. In the Euler product (132) above, set $s = \sigma + it$ and let $\log \phi_\sigma(t)$ denote the principal branch (for definiteness), and we get

$$\log \phi_\sigma(t) = \sum_p \left[\log(1 - p^{-\sigma}) - \log(1 - e^{-\sigma - it}) \right]$$

$$= \sum_p \sum_{m=1}^\infty p^{-m\sigma}(p^{-imt} - 1)/m$$

$$= \sum_p \sum_{m=1}^\infty p^{-m\sigma}(e^{-imt\log p} - 1)/m,$$

and each of the terms inside of the display above is the logarithm of the characteristic function of a Poisson distribution with the $-\log p > 0$, as its parameter, and hence is infinitely divisible. Therefore $\phi_\sigma(\cdot)$ is also an indefinitely divisible ch.f., continuous at $t = 0$, and never vanishes by the Lévy-Khintchine representation theorem. Thus $\zeta(\cdot)$ is an entire function in the right plane. $\square$

Remark 16. It was noted by Gram (1903) that there are exactly 15 solutions of $\zeta(\frac{1}{2} + it) = 0$ for $0 < t < 50$, and evaluated each for a few decimal places, substantiating Riemann's conjucture. Now a computer expert Odlysko, found over 10^{22} zeros of $\zeta(\frac{1}{2} + it)$, reported in Derbyshire (2003).

For a different approach towards the possible solution of RH, consider the reciprocal of $\zeta(\cdot)$,

$$\zeta^*(s) = \frac{1}{\zeta(s)} = \Pi_p(1 - p^{-s}) = \sum_{n=1}^\infty \frac{\mu(n)}{n^s},$$

where $\mu(\cdot)$ is the Móbius function defined as $\mu(n) = +1, 0, -1$, according as 'n' is a product of odd number of distinct primes divisible by a square integer, $(+1)$, or a product of an even number of primes (-1) respectively, or zero otherwise. Although $\zeta^s(1) = 0$ was considered by Euler in 1750, the convergence properties were discussed by von Mangoldt (1897) much later. The Merters' function $M(\cdot)$ changing values at integers where $M(x) = \sum_{n \le x} \mu(n)$, and that

$$\frac{M(n - \varepsilon) + M(n + \varepsilon)}{2} = \sum_{k=1}^{n-1} \mu(k) + \frac{\mu(n)}{2},$$

whence $|M(\cdot)| \nearrow$, and $|M(x)| \le x, x > 0$, so that one has

$$\xi^*(s) = \int_0^\infty x^{-s} dM(x) = \int_0^\infty M(x) x^{-s-1} dx,$$

since $|M(x)| \leq |x|^\alpha$ and $\alpha < 1$. Then $x^{-s} M(x) \to 0$ as $x \to \infty, \sigma \geq 1$ with $s = \sigma + it$. The following result of Littlewood's is useful here.

Theorem 2.7.2 (Littlewood (1912)) *With the Martens' function $M(\cdot)$ above, for each $\varepsilon > 0, M(x) x^{-\frac{1}{2}-\varepsilon} \to 0$ as $x \to \infty$ if and only if the RH holds.*

The details of proof of this theorem are in several books, e.g., see H. M. Edwards (1974). Here the introduction of $M(\cdot)$ brings us back to using the probability method with the occurrence of prime numbers whose appearance is sometimes called the 'game of chance', by M. Kac, (1959), and their behavior (in the probabilistic sense) was already used by Denjoy (1931). A usable discussion of the behavior will be included and utilized in our solution.

Let $\Omega = \{1, 2, \ldots\}$, the set of positive integers and consider its subset $\tilde{\Omega}$ with square free divisors n, so $\mu(n) \neq 0$. This set is about $1 - \frac{1}{\zeta(2)}$ in proposition, so that by Euler's formula its value is $\frac{6}{\pi^2}$. Let $\tilde{\Omega}_i (\subset \tilde{\Omega})$ be the products of even and odd primes $i = 1, 2$ on which $\mu(n) = 1$, or -1 respectively. They have the corresponding proportions $= \frac{3}{\pi^2}$. Thus if the set Ω of integers is given the 'volume'$= 1$, then the corresponding volumes of $\tilde{\Omega}_i, i = 1, 2$ will have values $\frac{3}{\pi^2}, i = 1, 2$ and $\tilde{\Omega}$ will have the size $= 1 - \frac{6}{\pi^2}$. Thus one may define a probability measure on Ω so that $\mu(n) = 0, \pm 1$, will become independent random variables with $P[\mu(n) = 1] = P[\mu(n) = -1] = \alpha$ and $P[\mu(n) = 0] = 1 - 2\alpha$ where $\alpha = \frac{3}{\pi^2}$. This describes a probability function on the power set 2^Ω.

The existence of such a (σ-additive) probability model is to be verified. This can be done nontrivially with the following (standard) probability method. One starts here by using the crucial result that the density of prime numbers is approximately given by $(\log x)^{-1}, x > 1$. Then the existence of such a P is needed. For this one has to refer to the work of Jensen and his collaborators with extensions (e.g., as detailed in the book by Laurinčikas (1996)).

(The existence of such a probability measure was assumed and used by Denjoy (1931).) This fact is made explicit here for convenience. With this explanation, we can proceed using the *pairwise independence* (in the sense of Probability Theory). We can now complete the argument by using the classical (Probability Theory) central limit theorem and establish the RH as follows *with just the pairwise independence property* of the $\mu(n)$'s, and hence for $\{M(n), n \geq 1\}$.

It may be restated as $P[\mu(n) = 1] = P[\mu(n) = -1] = \frac{3}{\pi^2}$ and $P[\mu(n) = 0] = 1 - \frac{6}{\pi^2}$, where the $\mu(n)$ being *pairwise independent* random variables. Hence $E(\mu(n)) = 0$, $Var(\mu(n)) = \frac{6}{\pi^2}$. Now $M(\mu) = \sum_{i=1}^{n} \mu(i)$, the Mertan's function, satisfies $E(M(\mu)) = 0, \sigma^2(\mu(n)) = \frac{6}{n^2}$ so that $M(\mu) = \sum_{i=1}^{n} \mu(i)$, and then one verifies. $E(M(w)) = 0, \sigma(M(u)) = \frac{3n}{\pi^2}$. Note that $\frac{\sigma^2(M(n))}{n} \to \infty$ which is the appropriate condition for the central limit theorem of Probability Theory for pairwise independent random variables to hold. Hence by such a known central limit theorem for pairwise independent variables we have (cf., Rao (1984), p. 399) the limit relation:

$$\lim_{n \to \infty} P\left[\frac{M_n - 0}{\sqrt{\sigma^2(M(n))}} \leq x \right] = \frac{1}{\sqrt{2\pi}} \int_{-\infty}^{x} e^{-u^2/2} du,$$

so that the sequence $\{M_n/\sqrt{n^\alpha}, n \geq 1\}, \alpha = \frac{6}{\pi^2}$, is *bounded in probability*. Hence $|M(n)|n^{-\frac{1}{2}-\varepsilon} \to 0$ with probability 1, for each $\varepsilon > 0$. This result and Theorem 2.7.2 above imply:

Theorem 2.7.3 *Under the preceding conditions leading to pairwise independence of $\mu(n)$, it follows that $M(\mu)n^{-\frac{1}{2}-\varepsilon} \to 0$ with probability one for each $\varepsilon > 0$ as $n \to \infty$, so that the Riemann Hypothesis holds with probability one, as a consequence of the central limit theorem of probability theory.*

Remark 17. 1. This is the best conclusion that we can present on the classical RH problem. The only question is to know if the null set is actually the empty set. This problem is not simple to answer and there is no general method to analyze the structure of the sets of measure zero. The details will be omitted again, and referring the reader to the readily available reference to the author [Rao (2012)].

2. A philosophical discussion about the sets of measure zero (on which the RH can be false) should ideally be empty for a conclusion from the above result for making an absolute assertion. But there is no method available to assert that a Lebesgue null set is empty. On this point, the mathematician J. E. Littlewood (1912), who spent a considerable effort on RH, thinks that all roots of $J(\frac{1}{2} + it)$ are on the

line $t \in \mathbb{R}$ to be false, and this seems to be based in part on the following: If $L(x) = \int_2^x \frac{dt}{\log t}+$ constant, $x > 2$, then the great Gauss's conjecture states that $\pi(x) = L(x), x > 2$ where $\pi(x)$ denotes the number of prime numbers less than x and Littlewood showed that this is true for almost all x (Lebesgue measure). But a South African mathematician by the name of Skews showed that the conjecture fails for many x starting at $10(x) = 10^{10_{k-1}(x)}$, $k > 1$ where $10_1(x) = 10^x$ (and $k > 1 \Rightarrow 10_k(x) = 10^{10_{k-1}(x)}$). See the article by Littlewood (1912), 'The Riemann Hypothesis', (ed. I. J. Good et al.), Basic Books, New York.

2.8 Admissible Means of Second Order Processes

This chapter will be completed with an analysis of possible mean functions of a given second order process whose covariance functions are subject to the types of conditions considered in the above sections in which the mean functions should obey certain restrictions that will be acceptable for a given covariance. The first steps have been detailed in Section 1.2 in the early result by A. V. Balakrishnan, detailed there. We now include some key facts that enhance and complement the earlier analysis, with the Gaussian as well as some general cases to conclude this chapter.

It is convenient to recall the Aronszajn space concept, already used in Chapter 1, for a quick reference here. Thus let $K(\cdot, \cdot)$ be a Hermitian positive definite function on $T \times T \to \mathbb{C}$ and let $\mathcal{H}_r(T) = \{f : f = \sum_{i=0}^n a_i r(s_i, \cdot), n \geq 1$, and $a_i \in \mathbb{C}\}$. Define an inner product on $\mathcal{H}_1(T)$ by the equation:

$$(f, g) = \sum_{i,j=1}^{n,m} r(s_i, t_j) c_i \bar{d}_j = \sum_{i=1}^{n} c_i \bar{g}(s_i) = \sum_{j=1}^{m} \bar{d}_j f(t_j), \qquad (133)$$

where $f = \sum_{i=1}^n c_i r(s_i, \cdot), g = \sum_{i=1}^m d_j \bar{r}(\cdot, t_j)$, are in $\mathcal{H}_r(T)$. With the positive definiteness of the covariance $r(\cdot, \cdot)$, one has the inner product $(\cdot, \cdot)$ and let $\| \cdot \|$ be the norm derived from it and $\mathcal{H}_r$ be the space completed for $(\cdot, \cdot)$. Here we note that $f \in \mathcal{H}_r(T) \Rightarrow f(t) = (f, r(t, \cdot))$ and $(r(s, \cdot), r(\cdot, t)) = r(s, t)$ as well as, $\{r(s, \cdot), s \in T\} \subset \mathcal{H}_r$ is dense, $|f(t)| \leq \|f\|\|r(t, \cdot)\|, t \subset T$. The space $\{\mathcal{H}_r$, with inner product $(\cdot, \cdot)\}$ is usually called the *Aronszajn space* (in honor of its author N. Aronszajn (1962).

We now record a useful property of certain admissible means as:

Proposition 2.8.1 *The space of admissible means of a Gaussian process, with covariance r, is a Hilbert space.*

Proof. We use the Aronszajn space technique and the key result that the mapping $f \in M_P$ to $Y = \frac{dP_f}{dP} > 0$ a.e., by the classical Hájet-Feldman dichotomy theorem on the equivalence of Gaussian measure P_f and P respectively with mean f, and mean zero (cf., e.g., Rao ((2014), 226)):

$$f(t) = \int_\Omega \bar{X}_t dP_f = \int_\Omega Y_f \bar{X}_t dP = E(Y_f \bar{X}_t) = L(\bar{X}_t). \tag{134}$$

It is seen that $L : \mathcal{L} \to \mathbb{C}$ is a continuous linear mapping, and hence by the Riesz representation, (134) implies that $f = u(Y)$ of (133) and so $f \in \mathcal{H}_r$. Thus $M_P \subset \mathcal{H}_r$. We need to show the converse inclusion.

For the converse, let $L(\cdot)$ be continuous and $f \in M_P$ so $P_f \ll P$. Let $\{\mathcal{F}_i, i \in I\}$ be a right order continuous family of σ-algebras from $\Sigma = \sigma(X_t), t \in T$ generating it, so $\Sigma = \sigma(\cup_{i \in T}\mathcal{F}_i)$ and let $P_{if} = P_f|\mathcal{F}_i$, and let $H_j(P, P_f) = \int_\Omega (dPdP_f)^{\frac{1}{2}}$, the *Hallinger distance* $(= \int_\Omega (dP_{jf}/dP_j)^{\frac{1}{2}} dP_j)$, then it follows from the theorem of E. J. Brady (1971) that $\lim_i H_i(P_f, P_i) = c$ exists and $c = 0$ if and only if $P_f \perp P$. Also $H_i \downarrow c, (0 \le c \le 1)$, and $c = 0$ iff $P_f \perp P$. Since by hypothesis this is not the case, $c > 0$, because $H_i \ge c > 0$ now, choose $\mathcal{F}_0 = \sigma(X_t)(= \mathcal{F}_{t_0})$ for an arbitrarily fixed $t_0 \in T$ and then H_{t_0} using the distributions of X_t and $X_t + f(t)$ to get

$$c \le H_{i_0} = \int_\Omega (dP_0 dP_{0f})^{\frac{1}{2}} = \int_\mathbb{R} \left(e^{-\frac{u^2}{2}} - e^{\frac{-(u-f(t))^2}{2a}} \right)^{\frac{1}{2}} \frac{dx}{\sqrt{2\pi a}}$$

$$= \exp\left[-\frac{1}{8a} f^2(t) \right], \text{ of (134).} \tag{135}$$

If $c_0^2 = -8 \log c > 0$, then (135) implies $f(t)^2 \le c^2 a$, or $f(t) = |L(X_t)|$ implies that $L(\cdot)$ is continuous, so $f \in M_P$, as desired. $\square$

The preceding analysis has the following consequence.

Corollary 2.8.2 *For a Gaussian process $\{X_t, t \in T\}$ on (Ω, Σ, P) and $f \in M_p$, the admissible mean, there is a $Z = u(f) \in \mathcal{L}, f(t) = E(2\bar{X}_t)$, and the likelihood ratio is:*

$$\frac{dP_f}{dP} = \exp\{Z - \frac{1}{2}E(|Z|^2)\}, \quad a.e. \ [P]. \tag{136}$$

Further analysis in the Gaussian case is clearly possible. Here we present a key result on a regression problem if the process is also a martingale (1975). This includes the Brownian motion case considered by Liptser and Shiryayev (1971). We thus have:

Proposition 2.8.3 *Let* $Y = \{Y_t, \mathcal{F}_t, t \geq 0\}$ *be a real continuous Gaussian martingale and* X *be a random variable depending on* Y *in that* $(X, Y_{t_1}, \ldots, Y_{t_n})$ *is a Gaussian vector for any* $(t_1, \ldots, t_n)$ *finite index of times. Then the regression function* $E(X|Y_s, s \leq t)$ *is representable as*

$$E(X|Y_s, 0 \leq s \leq t) = E(X|Y_0) + \int_{0+}^{t} K(t, s) dY_s, \qquad (137)$$

where $K(\cdot, \cdot)$ *is a nonstochastic (jointly) measurable and locally integrable function, so that the (stochastic) integral in (137) is well-defined (by the Bochner* $L^{2,2}$*-boundedness principle).*

Proof. We present the argument in steps for convenience.

I. Let $E(X) = \mu_x, E(Y) = \mu_y$ and the covariance matrices of (X, Y) and (Y, Y) be R_{xy} and R_{yy} and for simplicity let the vectors be linearly independent so that R_{yy}^{-1} exists. Then by the Gram-Schmidt orthogonalization we can find a matrix A so that $(Y - \mu y) \perp Z = ([X - \mu x] - A[Y - \mu_y])$ so that $E(Z(Y - \mu_y)^*) = 0$, i.e., $Z \perp (Y - \mu_x)$. A simplification of this equation gives:

$$0 = E(Z(Y - \mu_y)^*) = R_{xy} - AR_{yy} \Rightarrow A = R_{xy}R_{yy}^{-1}. \qquad (138)$$

Thus with this A, $Y - \mu_y$ and Z are (being Gaussian) independent, so that $E(Z|Y) = E(Z) = 0$, and $E(X|Y) = \mu_x + A(Y - \mu_y)$. We can now obtain the conditional covariance of ZZ^*, given Y:

$$E(ZZ^*|Y) = E(ZZ^*) \text{ since } Y, Z \text{ are independent, being Gaussian}$$
$$= R_{xx} - R_{xy}R_{yy}^{-1}R_{xy}^*, \qquad (139)$$

which is nonstochastic. [In case R_{xy} is singular, if R_{xy}^{-1} is taken as a generalized inverse (in the Moore–Penrose sense), the above statements are still valid, but this is not needed here.]

II. The argument is a nontrivial extension of that of Lipster and Shiryayev (1977) for the general Gaussian case from their work on Brownian motion. Thus consider the dyadic partition of $[0, t], t > 0$, as $0 = t_0^n < t_1^n < \ldots < t_{2^n}^n = t$, where $t_k^n = \frac{kt}{2^n}, k = 0, 1, \ldots, 2^n$.

Let $\mathcal{F}_t^n = \sigma(Y_{t_k^n}, 0 \leq k \leq 2^n)$, so by the path continuity of the process, $\mathcal{F}_t^n \uparrow \mathcal{F}_t$ as $n \to \infty$. If now $X_t^n = E(X|\mathcal{F}_k^n)$, then

$$E(X_t^{n+1}|\mathcal{F}_l^m) = E[E(X|\mathcal{F}_t^{n+1})|\mathcal{F}_t^n] = E(X|\mathcal{F}_t^n) = X_t^n, a.e. \quad (140)$$

and by applying the conditional Jensen inequality

$$E((X_t^n)^2) \leq E(X^2) < \infty, n \geq 1. \quad (141)$$

Hence the uniformly integrable martingale $\{X_t^n, n \geq 1\}$ converges pointwise and in $L^2(P)$-mean, $X_t^n \to E(X|\mathcal{F}_t), a.e.$
    III. Since $\mathcal{F}_t^n = \sigma(Y_{t_k^n} - Y_{t_{k-1}}, k = 1, \ldots, 2^n - 1)$, we can use the analysis of step II, and get

$$E(X|\mathcal{F}_t^n) = E(X|Y) + \sum_{j=0}^{2^n-1} L_n(t, t^n)(Y_{t_{j-1}^n} - Y_{t_j^n}), \quad (142)$$

where $L_n(\cdot, \cdot)$ is a product moment function jointly continuous by the continuity of the Y-process. To see that (142) converges to (137), as $n \to \infty$, consider $K_n(\cdot, \cdot)$ as (the nonstochastic function)

$$K_n(t, s) = \sum_{j=1}^{2^n-1} L_n(t_i t_j^n)\chi_{[t_j^n, t_{j*}^n]}(s).$$

It is (jointly) measurable, integrable on compact sets, so that (142) becomes

$$E(X|\mathcal{F}_t^n) = E(X|Y_0) + \int_{0+}^{t} K_n(t, s)dY_t, a.e.. \quad (143)$$

But for a simple function $f = \sum_{j=1}^n a_j \chi_{[t_j, t_{j+1}) \times A_j}, A_j \in \mathcal{F}_{t_j}$, and a square integrable martingale $\{Y_s, \mathcal{F}_s, s \geq 0\}$ one has by using the Doleans–Dade measure (cf. Rao (1991, p. 66) in the following, which works for general martingale integrals, extending Brownian process):

$$E\left(\left|\int_0^t f(s)dY_s\right|^2\right) = \sum_{j=1}^n a_j^2 E\left[\chi_{A_j}(Y_{t_{j+1}} - Y_{t_j})^2\right] + 0$$

$$= \sum_{j=1}^n a_j^2 \mu[(t_j, t_{j+1}] \times A_j] = \int_{[0,t]\vee\Omega} |f|^2 d\mu, \quad (144)$$

where μ is the above noted Doleans–Dade measure determined by the $L^1(P)$-bounded submartingale $\{Y_t^2, \mathcal{F}_t, t \geq 0\}$. By a standard argument (144) can be used to obtain the result for all locally bounded measurable f. Then one can have

$$E(X_t^2 - X_t^m)^2 = E\left[\int_{0+}^t [K_n(t,s) - K_m(t,s)]\, dY_s\right]^2$$

$$= \int_{[0,t] \times \Omega} (K_n(t,s) - K_m(t,s)]^2 d\mu(s), \qquad (145)$$

where $\mu(\cdot)$ is the Dolean–Dade's measure of (144). If now, we set $\tilde{\mu}(0,s) = \mu((0,s] \times \Omega)$, then (145) becomes for $t > 0$

$$E(X_t^n - X_t^m)^2 = \int_{0+}^t [K_n(t,s) - K_m(t,s)]^2 d\tilde{\mu}(s). \qquad (146)$$

Then by the early work, $\{X_t^n, n \geq 1\}$ is Cauchy, so $K_n(t,\cdot) \in L^2(R^+, \tilde{\mu})$, $n \geq 1$ and being Cauchy it tends to some $K(t,\cdot) \in L^2(\tilde{\mu})$ for each t. This implies that (143) converges to (137) and completes the argument. $\square$

We end this section by presenting a quite general result on the structure of the admissible mean values of a second order process with continuous covariances, to indicate its complex structure and to understand the problem raised by Balakrishnan in its generality. The following comprehensive result on the linearity of the set of admissible means is due to Pitcher (1963) which is based essentially on the Karhunen expansion of a second order process and amplifies the problem showing the nontrivial structure involved.

Theorem 2.8.4 *Let $X = \{X_t, t \in T\}$ be a centered second order process with a continuous covariance r where $T \subset \mathbb{R}$ is a bounded closed interval. Let $\{\lambda_n, n \geq 1\}$ and $\{\phi_n, n \geq 1\}$ be the eigenvalues and the corresponding eigenfunctions of r which exist. Let X_n, f_n be given by*

$$X_n = \lambda_n^{-\frac{1}{2}} \int_T X(t)\phi_n(t)dt, \qquad f_n = \lambda_n^{-\frac{1}{2}} \int_T f(t)\phi_n(t)dt, f \in M_p,$$

where M_p is the set of admissible mean values of the process X. [Thus $\{X_n, n \geq 1\}$ is a sequence of orthonormal random variables and $\{f_n, n \geq 1\} = f \in l^2$, so $\sum_{n=1}^{\infty} |f_n|^2 < \infty$.] Let P_n be the n-dimensional distribution of $\{X_i, 1 \leq i \leq n\}$ with densities p_n relative *to the Lebesgue measure, which exist. Suppose that the p_n satisfies the following three conditions:*

(i) $p_n > 0$ a.e, $n \geq 1$, (ii) $\lim_{|t_i| \to \infty} p_n(t_1, \ldots, t_i, \ldots t_n) = 0, 1 \leq i \leq n$, for almost all $t_i, 1 \leq i \leq n, i \neq j$: and (iii) $\frac{\partial p_n}{\partial t_j}, 1 \leq j \leq n$ exist, $n \geq 1$, and $\sum_{j=1}^{n} \int_{\mathbb{R}^n} \left(\frac{\partial \log p_x}{\partial t_j} \right)^2 dP_n$ is bounded by $K < \infty$. Then $M_1 = \{ f \in M_p : f = \{ f_n, n \geq 1 \} \in l^1 \}$ is a positive cone, and $M_1 = M_p$ holds if further $\sum_{n=1}^{\infty} \lambda_n^{-\frac{1}{2}} < \infty$.

If also each p_n is symmetric about the origin of $\mathbb{R}^n$, then M_p is linear. In general $M_1 \subset M_p$ only, and M_p need not even be a convex set.

The complete detail is to be found in Pitcher's paper, and we refer the reader to that work to understand the intricacies of the assertion.

It may be of interest to end this chapter with a simple note on the relations between weakly harmonizable class of processes and that of the Karhunen's. This gives a general relationship between them.

Theorem 2.8.5 *Let $\{ X_t, t \in \mathbb{R} \}$ be a weakly harmonizable process on a probability space (Ω, Σ, P). Then it is also a Karhunen process relative to a finite positive measure ν on $\hat{\mathbb{R}}$ and a suitable Borel set $\{ f_t, t \in \mathbb{R} \} \subset L^2(\hat{\mathbb{R}}, \mathcal{B}, \nu)$.*

Proof. Since $\{ X_t, t \in \mathbb{R} \}$ is weakly harmonizable, it has a stationary dilation, denoted $\{ Y_t, t \in \mathbb{R} \} \subset L^2(\tilde{\Omega}, \tilde{\Sigma}, \tilde{P})$, on a larger space as shown in Section 2.3 above (cf. Theorem 2.3.1). If $Q : L^2(\tilde{P}) \to L^2(P)$ is the orthogonal projection from $L^2(\tilde{P})$ onto $L^2(P)$, then we have

$$X_t = QY_t = Q \left(\int_{\tilde{T}} e^{it\lambda} d\tilde{Z}(\lambda) \right) = \int_{\tilde{T}} \pi(e^{it(\cdot)})(\lambda) \tilde{Z}(d\lambda),$$

and if $\tilde{f}(t, \lambda) = \pi(e^{it(\cdot)})(\lambda)$, then $\{ \tilde{f}(t, \cdot), t \in T \} \subset L^2(T, \nu)$, where $\nu(\cdot)$ is a measure determined by $\tilde{Z}(\cdot)$, an orthogonally valued random set function. Thus,

$$r(s, t) = E(X_s \bar{X}_t) = \int_{\tilde{T}} \tilde{f}(s, \lambda) \bar{\tilde{f}}(t, \lambda) d\nu(\lambda).$$

Hence $r(\cdot, \cdot)$ is of Karhunen class, relative to $\{ \tilde{f}(t, \cdot), t \in T \}$. $\square$

To complete this chapter, it is useful to present a multivariate version of a (weakly) harmonizable process which indicates how much of the preceding analysis can be given for this extension. Thus let $X_t = (X_t^1, \ldots, X_t^n), t \in T$, centered with an $n \times n$ matrix covariance r which is representable as

$$r(s,t) = \int_{\hat{T}} \int_{\hat{T}} e^{ius - iu't} F(du, du'), \quad s, t \in T$$

relative to an $n \times n$ matrix of (complex) bimeasures F, as in the scalar case which is extended to n-dimensions, making $r(\cdot, \cdot)$ a positive definite matrix function. The desired analog then can be established and is presented as follows.

Theorem 2.8.6 *Let $\{X_t, t \in T\}$ be a multivariate harmonizable process $(T = \mathbb{R}$ or $[0, 2\pi))$, with $F(\cdot, \cdot)$ as its spectral matrix and*

$$L^2(F) = \left\{ f : \hat{T} \to M_P \Big| \int_{\hat{T}} \int_{\hat{T}} f(\lambda) f^*(\lambda') F(d\lambda, d\lambda') = (f, f), \|f\|_F < \infty \right\}$$

where $\|f\|_F^2 = trace(f, f)$, of the positive definite matrix (f, f). Then $L^2(F)$ is a Hilbert space of equivalence classes of matrices with inner product $((f, g)) = trace\ (f, g)$, where the linear space is considered with constant matrix coefficients.

Although the result is more or less a direct extension of the scalar case of harmonizable processes, some care is needed because of the non-commutative problems with matrix operations.

In this connection, we should also mention some early work by J. Kampé de Feriet and F. N. Frenkiel (1954) (called class KF) which is continued by them for several years later by extending (weak) stationarity to include some (particularly strong) harmonizability of processes with applications. These results are motivated by the summability methods of classical analysis. They were also related to some work by E. Parzan (1962) and Yu. A. Rozanov (1959) and have interest in several applications. We state this concept and later include an example to show that weakly harmonizable class is not included. (More is in Chapter 3.)

Definition 2.8.7 *A centered second order process $\{X_t, t \in T\}(\subset L_0^2(P))$ with a continuous covariance r (here $T = \mathbb{R}$ or $\mathbb{Z}$), is of class (KF) if the following limits exist:*

$$\tilde{r}(h) = \begin{cases} \lim_{a \to \infty} \frac{1}{\alpha} \int_0^{a - |h|} r(s, s + |h|) ds, & \text{if } T = \mathbb{R}, \\ \lim_{n \to \infty} \frac{1}{n} \sum_{k=0}^{n - |h| - 1} r(k, k + |h|), & \text{if } T = \mathbb{Z}. \end{cases}$$

The point of this extension is to bring in the very useful and powerful summability methods of the general analysis into the popular second order stochastic analysis. It may be verified that $\tilde{r}(\cdot)$ is positive definite (and measurable) so that by the well-known Herglotz-Bochner-Riesz theorem $\tilde{r}$ will have a representation relative to a positive measure $H(\cdot)$ as;

$$\tilde{r}(h) = \int_{\tilde{T}} e^{iht} d\tilde{H}(t), h \in T, (T = \mathbb{R} \text{ or } \mathbb{Z}).$$

It can be verified that stationary, as well as strongly harmonizable classes, are included in class (KF). However, we shall include an example (due to H. Niemi) below to show that weak harmonizability is not always in class (KF). [Cf. Exercise 2 below.]

2.9 Complements and Exercises

1. A weakly harmonizable centered process is norm bounded, but the converse implication is not true. For example, let $f \in L^1(\mathbb{R}, dt)$, and $\mu(\cdot) = \int_{(\cdot)} f(t) dt \in M(\mathbb{R})$, the space of regular signed measures on $\mathbb{R}$. If $\mathcal{Y} = \{\hat{\mu} : \mu \in M(\mathbb{R})\}$, let $f \in C(\mathbb{R}) - \mathcal{Y}$, e.g., $f(x) = sgn(x)[(\log|x|)^{-1}\chi_{[|x|\geq 0]} + \frac{|x|}{e}\chi_{[|x|<e]}]$, for $x \in \mathbb{R}$ so that this f satisfies the requirements. If we take $\phi \in L_0^2(P), \|\phi\|_2 = 1$ and $l(\phi) = 1$ for an $l \in [L_0^2(P)]^*$, then the process $X : t \to f(t)\phi$, is bounded and continuous on $L^2(P)$ but not weakly harmonizable. So there exist non harmonizable weakly continuous second order processes that have other properties.

2. Consider the intervals $C_n = [2^{2n}, 2^{2n+1})$ and $D_n = [2^{2n+1}, 2^{2n+2})$ and $a_k = \sum_{n=0}^{\infty}[\chi_{C_n} + 2\chi_{D_n}](k)$ and $a_n = a_{-n}, a_0 = 1, k > 0$ as above. The sets C, D_n are disjoint, $1 \leq a_k \leq 2$ and the series is finite.

The covariances $r(k, l) = 0$ if $k \neq l$ and

$$r_n(l) = \frac{1}{n}\sum_{k=0}^{n-l-1} r(k, k+h) = \frac{1}{n}\sum_{k=0}^{n-1} a_1^2 \text{ for } h = 0, \text{ and } = 0, \text{ for } h \neq 0.$$

Verify that for $h \neq 0, \lim_{n\to\infty} r_n(h) = 0$, and

$$r_n(0) = \begin{cases} \frac{5}{3} - \frac{1}{3 \cdot 2^{2m-1}}, & \text{if } n = 2^{2m} - 1 \\ \frac{4}{3} - \frac{1}{3 \cdot 2^{2m}}, & \text{if } n = 2^{2n+1} - 1, \end{cases}$$

and verify that $\lim_{m\to\infty} r_{2m}(0) = \frac{5}{3}, \lim_{n\to\infty} r_{2^{2m+1}}(0) = \frac{4}{3}$. Conclude that $\lim_{m\to\infty} r_n(0)$ does not exist. So $\{X_n, n \in \mathbb{Z}\}$ is not in class (KF). Thus class (KF) and weakly harmonizable classes do not include each other although both have weakly stationary and strongly harmonizable classes in their fold. [This is Niemi's example.]

3. If $X : \mathbb{R} \to L_0^2(P)$ is strongly harmonizable so that it is representable as $X(t) = \int_{\mathbb{R}} e^{it\lambda} Z(d\lambda)$, for a stochastic measure $Z : \mathcal{B}(\mathbb{R}) \to L_0^2(P)$, show that $X \in$ class (KF), the class considered above. The same result holds if the weakly harmonizable class is restricted, e.g., if the tensor product $Z \otimes Z$ is also a stochastic measure. [The preceding exercise implies that this need not hold for all stochastic measures. See the author's paper (1982), p. 338, the proof there works in this case.]

4. This result deals with a pointwise approximation of weakly harmonizable processes by the strongly harmonizable ones due to H. Niemi (1975). If $X : \mathbb{R} \to L_0^2(P)$ is a weakly harmonizable process then there exists strongly harmonizable $X_n : \mathbb{R} \to L_0^2(P)$ such that $X_n(t)\chi_A(t) \to X(t)\chi_A(t)$ as $n \to \infty$, for all compact $A \subset \mathbb{R}$, uniformly converging. [Here $\mathbb{R}$ can be replaced by an LCA group G.] (In all cases, the convergence is in $L^2(P)$-norm.)

5. This problem shows that the class of weakly harmonizable processes is a proper subspace of bounded continuous processes in $L^2(P)$. Identify the Lebesgue space $L^1(\mathbb{R})$ as a subspace of $M(\mathbb{R})$, the class of signed measures ($f \in L^1(\mathbb{R}) \to \int_{(\cdot)} f(t)dt(\in M(\mathbb{R}))$), as a signed measure). If $\mathcal{Y}_1 = \{\hat{\mu} : \mu \in M(\mathbb{R})\}$ for $f \in C_0(\mathbb{R}) - \mathcal{Y}_1$, e.g., $f(x) = sgn(x)\{(\log|x|)^{-1}\chi_{[|x|\ge e]} + \frac{|x|}{e}\chi_{[|x|<e]}\}$ is known to be such a function. Let $\phi \in L_0^2(P), \|\phi\|_2 = 1, l(\phi) = 1$ where $l \in (L_0^2(P))^*$. Verify that $X_0 : t \to f(t)\phi$ is bounded and continuous but not weakly harmonizable!

6. The V-boundedness concept can be used for strict cases also namely if $X : G \to L^\alpha(P), 1 \le \alpha \le 2$ is a mapping call it *strictly harmonizable*, with G as a compact group, if for some $0 < C < \infty$,

$$\left\| \sum_{k=1}^n a_k X g_k \right\|_\alpha \le C \sup \left\{ \left| \sum_{k=1}^n a_k \chi_{g_k} \right| : \chi_{g_k} \in \tilde{G}, a_k \in \mathbb{C} \right\}$$

where the χ_{g_k} are characters of the dual object $\tilde{G}$ of G. Then Y. Hosoya (1952) obtained for $G = \mathbb{Z}$ (so $\tilde{G} = (-\pi, \pi]$), and

$\chi_t = e^{it(\cdot)}$), the following representation:

$$X_n = \int_G e^{int} dZ(t), n \in \mathbb{Z} = G,$$

and conversely this representation of $\{X_n, n \in Z\}$ is V-bounded.

7. This problem describes an abstract result from "point process" applications. Thus if G is a locally compact group and $H \subset G$ a closed subgroup, let μ be a regular measure on $\mathcal{B}_0(G)$, λ a (left) Haar measure on G, then there is a unique Borel measure $\tilde{\mu}$ on G/H, the quotient space, such that for all compactly supported scalar f

$$\int_{G/H} \left[\int_H f(xy) d\lambda(\mu) \right] d\tilde{\mu}(\tilde{x}) = \int_G f(x) d\mu(x),$$

if and only if, with Δ_H as a left modular function on H, we have:

$$\int_G f(xy^{-1}) d\mu(x) = \Delta_H(y) \int_G f(x) d\mu(x)$$

(cf. H. J. Raiter (1965), p. 167). If G is abelian or H is compact, then $\Delta_H = 1$. We can specialize this result for point-processes. On $(\mathbb{R}^n, \mathcal{B}_i(\mathbb{R}^n))$, with $\mathcal{B}_i(\mathbb{R}^n)$ denoting the bounded Borel sets of $\mathbb{R}^n$, let $Z : \mathcal{B}_i(\mathbb{R}^n) \to L^p(P), 2 \leq k \leq p < \infty$, be a random measure defining β by the relation:

$$\beta(A_1, \ldots, A_k) = E\left(\prod_{i=1}^k Z(A_i) \right), A_i \in \mathcal{B}_0(\mathbb{R}^n),$$

and suppose β determines a (signed) measure $\tilde{\beta}$, and that Z is (k-fold) translation invariant, implying

$$\tilde{\beta}(\tau_1 A_1 \times \cdots \times \tau_k A_k) = \tilde{B}(A_1 \times \cdots \times A_k).$$

Then $\tilde{B}$ admits a factorization as $d\tilde{B} = a\beta(\lambda)d\lambda$, where $d\lambda$ is the Lebesgue measure and $\tilde{\beta}(d\lambda)$ is a Radon measure on $\mathbb{R}^{k-1}$. If this translation property of $\tilde{\beta}$ holds for $2 \leq k \leq p$, then $Z(\cdot)$ is representable as

$$Z(A) = \int_{\mathbb{R}^{k-1}} \int_{A_1} e^{it(x_1 + \cdots + x_{k-1})} d\tilde{Z}(x_1, \ldots, x_{k-1}) d\lambda(x_k)$$

where $A = \mathbb{R}^{k-1} \times A_1, (A_1 \subset \mathbb{R})$ and $\tilde{Z}$ is a random measure with $\tilde{\beta}$ as a signed measure on $\mathcal{B}_i(\mathbb{R}^{k-1})$.

8. Let $M : \mathcal{K} \to \mathbb{C}$ be a local functional of order $m(c, \infty)$ and $L(f) = \exp M(f), f \in \mathcal{K}$, the L. Schwartz space of infinitely differentiable compactly supported scalar functions on $\mathbb{R}^n$. Then the functional $L(\cdot)$ is a characteristic functional of a generalized random field with independent values if and only if

$$\exp \left(\int_A \Phi(x_1, \ldots, x_n, t) \right) d\mu(t), x_i \in \mathbb{R},$$

is positive definite as a function of $x_1, \ldots, x_n$ for all compact $A \subset \mathbb{R}^n$, (Φ, μ) being determined by M which is representable as

$$M(f) = \int_{\mathbb{R}^n} \Phi(f(t), Df(t), \ldots, D^\alpha f(t), t) d\mu(t), f \in \mathcal{K},$$

and $D^\alpha = \frac{\partial^\alpha}{\partial x_1^{\alpha_1} \ldots \partial x_n^{\alpha_n}}, |\alpha| = \alpha_1 + \cdots + \alpha_n \leq m, \mu$ a Borel measure on $\mathbb{R}^n$, $\Phi : \mathbb{R}^\nu \times \mathbb{R}^n \to$ scalars, ($M(\cdot)$ is a local functional).
[This representation is of importance in the analysis of generalized random fields, and was stated in Gel'fand–Vilenkin volume without complete details and description. It was detailed in the author's (1971) *Russian Probability Journal* paper, and the readers will find it useful (read for details and also for a further study).]

9. This problem sketches an extension of random measures. Let $(\mathbb{R}^n, \mathcal{B}(\mathbb{R}^n))$ be a Borel measurable pair with $\mathcal{B}_0(\mathbb{R}^n)$ as the δ-ring of bounded Borel sets and $Z : \mathcal{B}_0(\mathbb{R}^n) \to L^p(P)$ a random measure with k-moments, $2 \leq k \leq p < \infty$. Let β be determined as a multimeasure

$$\beta(A_1, \ldots, A_n) = E \left(\prod_{i=1}^n Z(A_i) \right), \quad A_i \in \mathcal{B}_0(\mathbb{R}^n).$$

Suppose β determines a (multi)measure $\tilde{\beta}$ and $Z(\cdot)$ is k^{th} order translation $(\tau-)$ invariant so that $\tilde{\beta}(\times_{i=1}^k \tau_i A_i) = \tilde{B} \left(\times_{i=1}^k A_i \right)$ and that $\tilde{\beta}$ can be factorized as $d\tilde{\beta}(\lambda) = \tilde{a}\tilde{\beta}_1(d\lambda)$, where $\lambda(\cdot)$ is the Lebesgue measure and $\tilde{\beta}_1(\cdot)$ is Radon on $\mathbb{R}^{k-1}$. If this condition holds for $2 \leq k \leq p$, then show that the random measure $\tilde{Z}$ can be written as:

$$Z(A) = \int_{\mathbb{R}^{k-1}} \int_{A_1} e^{it(x_1, \ldots, x_{k-1})} d\tilde{Z}(x_1, \ldots, x_{k-1}) d\lambda(x_k)$$

where $A = \mathbb{R}^{k-1} \times A_1, A_1 \subset \mathbb{R}$ and $\tilde{Z}$ is a random measure and $\tilde{\beta}$, as a signed measure on $\mathcal{B}_0(\mathbb{R}^{k-1})$.

[This is due to Brillinger (1992), Berkeley Symp. See also Vere–Jones (1971) in an appendix to a paper by D. J. Daley (JRSS, B.33, 425–428).]

10. This problem presents a general stochastic integration that includes several others in the literature, and is motivated by Bochner's $L^{2,2}$ and $L^{p,p}$ boundedness principles as follows, which are originated from the Wiener integral for Brownian motion and includes it. If (Ω, Σ, P) is a probability space, let $\mathcal{F} \subset \Sigma$ be an increasing class of σ-algebras $\mathcal{F}_s, a \leq s < t \leq b, s$ and let $\mathcal{F}_t = \cap_{s>t} \mathcal{F}_s$, where for the σ-algebras, $a \leq s < t \leq b \Rightarrow \mathcal{F}_s \subset \mathcal{F}_t$ and let $\mathcal{F}_t = \cap_{s>t} \mathcal{F}_s$, and all the σ-algebras are P-completed. Let $E_t = B(\Omega, \mathcal{F}_t)$, the space of bounded (scalar) $\mathcal{F}_t$-measurable functions on Ω, and set $\xi = \{E_t, t \in I\}, \Omega' = I \times \Omega$ and $\mathcal{L}(\Omega', \xi)$ of the above type simple functions, so that each element is of the form $f = \sum_{i=1}^n f_i \chi_{A_i}, f_i \in E_{\inf(A_i)}, E_t = B(\Omega, \mathcal{F}_t)$, bounded $\mathcal{F}_t$ measurable scalar functions. If $|||f|||_\infty = \sup_{t_i} \|f(t_i)\|_\infty$, with $\| \cdot \|_\infty$ as the uniform norm on E_t, then $\{\mathcal{L}(\Omega', \xi), ||| \cdot |||_\infty\}$ becomes a normed linear space of vector functions f (or vector fields). If $A = (t_i, t_{i+1}]$ here in the representation of f, we can set

$$\tau(f) = \int_I f(t) dX(t) = \sum_{i=1}^n f_i(X_{t_{i+1}} - X_{t_i}), f \in \mathcal{L}(\Omega', \xi).$$

All this set up is to define a generalized Bochner $L^{2,2}$-(and $L^{p,q}$) type boundedness conditions that include most of the known stochastic integrals due to P. A. Meyer, K. Itô, and others, the extension being on Bochner's ideas. Here we give the generalization based on Orlicz spaces inclusively. So let ϕ be a Young function (i.e., a symmetric convex $\phi : \mathbb{R} \to \mathbb{R}^+, \phi(-x) = \phi(x), \phi(0) = 0$), and $f \in L^\phi(P)$ on (Ω, Σ, P) to scalars, and $\|f\|_\phi = \inf\{\varepsilon > 0 : E\left(\phi\left(\frac{|f|}{\varepsilon}\right)\right) \leq 1\} < \infty$. Then $(L^\phi(P), \| \cdot \|_\phi)$ is a complete metric space including the $L^p(P)$-spaces for $\phi(t) = |t|^p, p > 0$. Then $X : I \to L^\phi(P)$ is termed a *stochastic integrator* if (a) $\{\tau(f) : f \in L(\Omega, \xi), \|f\|_\phi \leq 1\}$ is bounded in $\| \cdot \|_\phi$-metrics, and (ii) $f_n \in L(\Omega, \xi), |f_n| \downarrow 0 \Rightarrow \lim_n \tau(f_n) = 0$ in $L^\phi(P)$. [If $\phi(x) = x^2$, this was Bochner's original condition which motivated the extension.] We have the following:

Let ϕ_1, ϕ_2 be Young functions, (Ω, Σ, P) be a probability space and $X = \{X(t), \mathcal{F}_t, t \in I \subset \mathbb{R}\} \subset L^{\phi_2}(P)$, be a process which is L^{ϕ_1, ϕ_2}-bounded relative to a σ-finite measure $\alpha : \mathcal{B}(I) \times \Sigma \to \mathbb{R}^+$, as defined above. Then X is a stochastic integrator and the integral $\tau(f)$ extends from the simple functions of $\mathcal{L}(\Omega', \xi)$ to the class $M^{\Phi}(\mathcal{L}) = \bar{sp}\{\mathcal{L}(\Omega', \xi), \|\cdot\|_\phi\}$ into $L^\phi(P)$ for which the dominated convergence theorem holds. [The converse holds in a slightly restricted form. An outline of the detail of this result, in its general form, is included in the author's C. R. Acad. Sci., Paris (tome 314 (1992), pp. 629–633) paper, which should be studied by the probability (functional analysis) students. This leads to interesting general analysis and possible extension of the work given in M. Metivier and J. Pellamail (1980) on the stochastic integrals used at that time with restrictions.]

2.10 Bibliographical Notes

The work of this chapter gives a view and substantial enlargement of stochastic analysis and integration, starting with the Khintchine stationarity and going much further into the harmonizable classes introduced from a general point of view by Bochner, with more insights based on the contemporary works of Karhunan and Cramér. Staying in the powerful Fourier analysis mold, generalized by Bochner to the class called V-boundedness (V stands for variations), presents a great and (by keeping close to Fourier or harmonic analysis) powerful methods of stochastic theory which leads to numerous (new) applications, using the MT-integration crucially.

The weak and strong harmonizable classes are the two basic classes, with applications of the sharp Fourier methods pioneered by Bochner, takes a prominent place here and in the theory and some forthcoming applications further some deeper aspects of noncommutative extensions of harmonizable classes will be considered later.

The study leads to generalized random processes and fields of Gel'fand and the related deep analysis of Lévy classes, as well as processes with independent increments in the generalized case. There was an open problem on the characterization of real functions for that study. The author had the opportunity of solving the open characterization problem, and the work was handled by Gel'fand himself or his associates, published in the *Russian Probability Journal* and another aspect in the (American) *Functional Analysis journal* also handled by Gel'fand, have been of special interest and included in Section 2.5. All of this work is dependent on the key concept of positive definiteness of certain function(al)s in multidimensional definitions. But there were serious obstacles. An elaboration of the attending problems as well as their solutions giving rise to a healthy growth of the subject. It may be interesting to briefly describe this here.

The key concept of positive definiteness of a function (of two variables) in probability arose in the classical work on Brownian motion, by the botanist Robert Brown, whose studies and experiments gave rise to the analysis, which later got N. Wiener to establish it rigourously hence termed also the Wiener process. This leads to a deeper analysis and extension by P. Lévy who wanted to study $\{X_t, t \in T\}$ if $T \subset \mathbb{R}^n, n \geq 2$, the random field if T is an 'interval', $n > 2$, and found that its covariance must verify (*) $\mathrm{cov}\,(s, t) = \frac{1}{2}[\|s\| + \|t\| - \|s - t\|]$, and its positive definiteness is not easy to establish for a further analysis. Lévy with much difficulty and work was able to prove the result if $n = 2$, but could not proceed further. According to legend (detailed in P. Cartier (1971)) that P. Lévy announced in his class that he would offer a great prize for anyone who could prove the result in the general case $n > 2$. Namely, he would offer his daughter's hand to a successful future mathematician. Thereafter, in a short while, L. Schwartz came up with a complete proof of the positive definitions of (*), for any $n > 2$, and the promise was fulfilled indeed. [Later I. J. Schoenberg also proved the same result differently.] This appeared to be strange, but I sent two papers in late 1960s to P. Lévy to be communicated for publication in the Paris based C. R. Acad. Sci., and he asked me to spell out my name fully (it is mandatory there, unlike in the U.S. Acad. Sci., where I published some papers as usual with initials only). So I asked P. A. Meyer about Cartier's account, while spending my sabatical in Strasbourg, France and in all seriousness. Meyer told me that it is a true story! This emphasizes the importance of the property which is also crucial for harmonizable random fields analysis, that occupies a key position in this and coming chapters.

The work related to local functional characterization was an open problem at that time and its analysis with solution in Section 2.5 above was handled by I. M. Gel'fand and he was pleased with it and when I met him at the Advanced Mathematical Institute in Princeton, he also wanted to have a reprint of it for his files, which was provided.

The next problem is on Riemann's Hypothesis (RH) and the earlier attempts were detailed in H. M. Edward's book (now Dover) entitled Rieman's Zeta function gives a readable account along with a short probabilistic attack by A. Denjoy (1931). The current account based on the probabilistic argument with Mertin's function was recently detailed in my paper (Rao (2012)), and, as noted in that paper, the validity of RH with probability one is proved. The question of the emptiness of the null set cannot be determined with our probability argument. An earlier (different) method by J. E. Littlewood also concludes that RH is valid a.e. who even thinks and asserts that the emptiness of the null set cannot be proven. A similar question about a problem of Gauss was proved by him to be true a.e. but the exceptional null set was shown later by others (as noted in the above work also) to be nonempty. Littlewood (1962) thinks of the same fate for the RH as well. This is somewhat analogous to the following question, attributed to H. Weyl, namely prove or disprove the statement or assertion or conjecture: If the irrational number π is expanded in decimals the natural numbers $0, 1, 2, \ldots, 7, 8, 9$ occur somewhere in the natural order. Many computer experiments early were unsuccessful. But the Canadian number theorist, M. Borwein (1998), reports that two Japanese computer scientists at the University of Tokyo in 1997 using "massively parallel" Hitachi machine with 2^{10} processors have discovered the first occurrence of the above sequence after 17 billion decimals at 17,387,594,880th digit after the decimal point. The next five occurrences are found at 26,852,899,245; 30,243,957,439; 34, 549,153,953; 41,952,536,161; and 43,289,964,000 places. Thus the RH and Littlewood's feeling on the emptiness of the null set may be on the same level as H. Weyl's. Also, the admissible means problem was treated here in some detail, as it was bypassed.

We now proceed with harmonizable applications and extensions in the next chapter.

3

Applications and Extensions of Harmonizable Processes

In the preceding chapter, we considered several aspects and applications of harmonizable processes, and found that in some (serious) applications we need to study further nontrivial but useful consequences of bimeasures. Their properties allow extensions of random set functions based on the (Morse-Transue or) MT-type analysis and related theory. This work also leads to key applications of filtering and signal extraction problems. Some properties of the MT-integrals were discussed in Sections 1, 2 (of Chapter 1) and they will be used here. [These are weaker than the usual Lebesgue integrals, and such (resulting) extensions will be needed and employed with details below.]

3.1 Special Classes of Weak Harmonizability

It is of interest in applications to consider second order processes with covariance $r(\cdot, \cdot)$ having special properties, such as: (a) periodically correlated, (b) oscillatory, or (c) some related ones.

Definition 3.1.1 *A centered second order process* $\{X_t, t \in \mathbb{R}\} \subset L_0^2(P)$ *is* periodically correlated *if its covariance* $r(\cdot, \cdot)$ *satisfies* $r(s+\alpha, t+\alpha) = r(s, t)$ *for a fixed* $\alpha > 0$ *for all* $s, t \in \mathbb{R}$ *(here* α *is termed a* period*). A centered second order process* $\{X_t, t \in \mathbb{R}\}$ *is oscillatory if* $r(\cdot, \cdot)$ *is representable as*

$$r(s, t) = \int_{\mathbb{R}} e^{i(s-t)\lambda} a_s(\lambda) \bar{a}_t(\lambda) F(d\lambda), \tag{1}$$

for a positive nondecreasing $F(\cdot)$ *defining a* σ*-finite measure and* $\{a_s, s \in \mathbb{R}\} \subset L^2(F)$. *[If* $e^{is\lambda} a_s(\lambda) = \tilde{a}_s(\lambda)$ *defines some other function,*

131

more general, then we go to the Karhunan *family!] In all cases* $F(\cdot)$ *is termed a* spectral measure *of the process. If* $F(\cdot, \cdot)$ *is restricted to* $F(s,t) = M(s)\bar{M}(t)$, *then it is termed a* factorizable spectral measure *(f.s.m.). [The positive definiteness of F implies that* $F(s,t) = M(s)\bar{N}(t)$ *giving* $N(t) = a \cdot M(t)$ *and so there is no generality gained by taking* M, N *'different'.]*

We present some structural properties of these special classes and their (abstract) representations, to understand the harmonizabilities better.

Proposition 3.1.2 *A weakly harmonizable process* $\{X_t, t \in \mathbb{R}\} \subset L_0^2(P)$ *is of f.s.m. class iff its covariance* $r(\cdot, \cdot)$ *is factorizable. As a consequence, such a process (of f.s.m. class) is strongly harmonizable.*

Proof. Let the process be f.s.m. so that $F(A, B) = M(A)\overline{M(B)}$. The bimeasure condition of $F(\cdot, \cdot)$ implies that $M(\cdot)$ is σ-additive on the Borel sets of $\mathbb{R}$ and hence is bounded. It follows that

$$
\begin{aligned}
r(s,t) &= \int_{\mathbb{R}} \int_{\mathbb{R}} e^{is\lambda - it\lambda'} M(d\lambda)\bar{M}(d\lambda') \\
&= \int_{\mathbb{R}} e^{is\lambda} M(d\lambda) \cdot \int_{\mathbb{R}} e^{-it\lambda'} \bar{M}(d\lambda') \\
&= f(s) \cdot \overline{f(t)} (= \hat{M}(s)\overline{\hat{M}(t)}).
\end{aligned}
\tag{2}
$$

Suppose conversely that $r(\cdot, \cdot)$ satisfies (2) for some f. Since the process is weakly harmonizable, r is expressible as:

$$
r(s,t) = \int_{\mathbb{R}} \int_{\mathbb{R}} e^{is\lambda - it\lambda'} F(d\lambda, d\lambda'),
\tag{3}
$$

as a Morse-Transue integral. This is a well-defined integral. Using the properties of this (nonabsolute) integral discussed in Chapter 1 (see Sec. 1 there), we get with $A = (\lambda_1, \lambda_2), B = (\lambda_1', \lambda_2')$:

$$
F(A, B) = \lim_{0 \leq T_1, T_2 \to \infty} \int_{-T_1}^{T_1} \int_{-T_2}^{T_2} \frac{e^{i\lambda_2 s} - e^{-i\lambda_1 t}}{-is}, \frac{e^{i\lambda_2' s} - e^{-i\lambda_1' t}}{it} r(s,t) ds dt,
$$

where A, B are noted above, such that $F(\{\lambda_1\}, \{\lambda_1'\}) = 0 = F(\{\lambda_2\}, \{\lambda_2'\})$. Putting $r(s,t) = f(s) \cdot \overline{f(t)}$ here so that (λ_1, λ_2) and (λ_1', λ_2') are continuity points of F, this shows on simplification, that $F(A, B) =$

$M(A)\overline{M(B)}$ for a suitable $M(\cdot)$. But $F(\cdot, \cdot)$ is a bimeasure so that $M(\cdot)$ is σ-additive, and has a unique extension to be a scalar measure on $\mathcal{B}$. Thus the process is f.s.m. From this we can easily conclude, with a standard argument, that the process is strongly harmonizable. Thus the f.s.m. class is contained in the strongly harmonizable class. When $F(A, B) = M(A) \times \overline{M(B)}$, the bounded Borel measure F has a unique extension to be a bounded Borel measure on $\mathbb{R} \times \mathbb{R}$. Thus the integrals in (1) are in the Lebesgue sense so that the process is strongly harmonizable. Thus the integrals in (2) are also in the Lebesgue sense. In all cases, then the f.s.m. class is a proper subclass of the strongly harmonizable class. $\square$

We have just seen the periodically correlated class whose covariance $r(s, t)$ is given by (1) which forms a special case of the Karhunen family representable as:

$$X_t = \int_{\mathbb{R}} g(t - u)dZ(u), \quad t \in \mathbb{R}, \tag{4}$$

with $Z(\cdot)$ determining F by $F(A, B) = E(Z(A)\overline{Z(B)})$, and $g(t, \cdot) \in L^2(F), t \in \mathbb{R}, E(Z(A)) = 0$. This special class of the Karhunen family is said to have a *moving average representation*, motivated by some applications. Its relation to the harmonizable class itself has the following useful property which is separately stated for some applications as well as reference. It should also be noted here and in this area that H. L. Hurd (cf. e.g., (1992), and in much of his work referred to there and later) has done important work with real applications, and it should be consulted in relation to this area.

The following structural analysis is often used in applications.

Proposition 3.1.3 *Let $\{X_t, t \in \mathbb{R}\}$ be a second order process having a moving average representation given by (4) above. Then it is strongly harmonizable and with its spectral measure to be absolutely continuous relative to the planar Lebesgue measure.*

Proof. Since $g = \hat{f}$ in (4), hence bounded, it follows that (4) defined as in the Dunford and Schwartz sense can be simplified:

$$X_t = \int_{\mathbb{R}} \hat{f}(t-u)Z(du) = \int_{\mathbb{R}} Z(du) \left(\int_{\mathbb{R}} e^{i(t-u)\lambda} f(\lambda)d\lambda \right)$$

$$= \int_{\mathbb{R}} e^{it\lambda} f(\lambda) \int_{\mathbb{R}} e^{-iu\lambda} Z(du)d\lambda, \text{ using a form}$$

of Fubini's theorem,

$$= \int_{\mathbb{R}} e^{it\lambda} f(\lambda)\bar{Y}(\lambda)d\lambda \tag{5}$$

where $\{Y(\lambda), \lambda \in \mathbb{R}\}$ is weakly harmonizable, and the symbol becomes Bochner's integral. Let $\tilde{Z} : A \mapsto \int_A \bar{Y}(t)f(t)dt, A \in \mathcal{B}$ so that $\tilde{Z}(\cdot)$ is a stochastic measure on the Borel field of $\mathbb{R}$, and one gets:

$$E(\tilde{Z}(A)\bar{\tilde{Z}}(B)) = \int_A \int_B E(\bar{Y}(s)Y(s)) \cdot \bar{f}(s)f(t)dsdt = \nu(A, B)$$

where the covariance $r_y(s,t) = E(Y(s)\bar{Y}(t))$ is bounded and that $\nu(\cdot, \cdot)$ has finite Vitali variation. Hence (5) can be given as

$$X_t = \int_{\mathbb{R}} e^{it\lambda} \tilde{Z}(d\lambda), \quad t \in \mathbb{R},$$

and so $\{X_t, t \in \mathbb{R}\}$ is strongly harmonizable. Its covariance $r(\cdot, \cdot)$ can be calculated as:

$$r(s,t) = E(X_s\bar{X}_t) = \int_{\mathbb{R}} \int_{\mathbb{R}} e^{is\lambda - it\lambda'} \nu(d\lambda, d\lambda')$$

$$= \int_{\mathbb{R}} \int_{\mathbb{R}} e^{is\lambda - it\lambda'} r_y(\lambda, \lambda')\bar{f}(\lambda)f(\lambda')d\lambda d\lambda', \tag{6}$$

so that $\nu(\cdot, \cdot)$ is absolutely continuous relative to the Lebesgue measure. $\square$

The following consequence of the preceding discussion is useful:

Proposition 3.1.4 *The class of oscillatory processes* $\{X_t, t \in T\} \subset L_0^2(P)$ *(cf. (1)) and the class of Karhunen processes induced by* $T = \mathbb{R}$ *or* $\mathbb{Z}$, *coincide.*

We now present an operator representation of a centered mean continuous process, as a major result also covering several classes studied earlier.

Theorem 3.1.5 *Let $\{X_t, t \in T\} \subset L_0^2(P)$ be a mean continuous process that is centered. Then it is representable as:*

$$X_t = A_t U_t Y_0, t \in T(T = \mathbb{R} \text{ or } \mathbb{Z}), \tag{7}$$

for some $Y_0 \in \mathcal{H}(X) = \overline{sp}\{X_t \, t \in T\} \subset L_0^2(P)$, where A_t is a densely defined closed linear operator on $\mathcal{H}(X)$ and $\{U_t, t \in T\}$ is a weakly continuous unitary group of operators on $\mathcal{H}(X)$, which commute with each $A_t, t \in T$. If further the process is weakly harmonizable, then there exists a (bigger) Hilbert space $\mathcal{K} \supset H(X)$, operators $A_t : \mathcal{K} \to \mathcal{H}(X)$, and a weakly continuous unitary group $\{U_t, t \in T\}$ which forms a weakly continuous positive definite contractive class satisfying $A_0 = id$ on $\mathcal{H}(X)$. On the other hand, a process given by (7) is always weakly harmonizable whenever $\{A_t, t \in T\}$ is a weakly continuous positive definite family of operators on $L_0^2(P)$, with $A_0 = identity$. [Here again positive definite means $\sum_{i=1}^n \sum_{j=1}^n (A_{t_i - t_j} h_i, h_j) \geq 0, n \geq 1$, as usual.]

Proof. The argument below depends on some well-known (but deep) facts and we include the essential details for use as well as completeness.

Now, the process being continuous in mean, $\mathcal{H}(X)$ is separable, and $\{X_t, t \in T\}$ is of Karhunen class relative to a set $\{g'_t, t \in T\}$ and μ, a Borel measure. Replacing g'_t with $e^{it(\cdot)} g_t$ we can express the process as

$$X_t = \int_{\hat{T}} e^{it\lambda} g_t(\lambda) Z(d\lambda), \qquad t \in T, \tag{8}$$

where $Z(\cdot)$ is orthogonally valued measure in $\mathcal{H}(X)$ from $\mathcal{B}(\hat{T})$, the Borel σ-algebra so that $E(Z(A)\bar{Z}(B)) = \mu(A \cap B)$, $\int_{\hat{T}} |g_t(\lambda)|^2 \mu(d\lambda) < \infty$, and hence $Y_t = \int_{\hat{T}} e^{it\lambda} Z(d\lambda)$ defines a stationary field. So there is a weakly continuous group of unitary operators $\{U_t, t \in T\}$ acting on $\mathcal{H}(X)$ satisfying $Y_t = U_t Y_0$. Also by the spectral theorem $U_t = \int_{\hat{T}} e^{it\lambda} E(d\lambda), t \in T$, the $\{E(\cdot), \mathcal{B}\}$ being the resolution of I, and $\mathcal{B}$ the Borel σ-algebra of $\hat{T}$. Note that $Z(A) = E(A)Y_0, A \in \mathcal{B}$ here. Now let $A_t = \int_{\hat{T}} g_t(\lambda) E(d\lambda)$. Since $\mu : B \mapsto (E(B)Y_0, E(B)Y_0)$ and $\int_{\hat{T}} |g_t(\lambda)|^2 \mu(d\lambda) < \infty$, it follows that A_t is a closed densely defined operator in $\mathcal{H}(X)$, whose domain contains $\{Y_s, s \in T\}$ for each t (from the standard functional analysis results), and A_t and $\{E(B), B \in \mathcal{B}\}$ commute. Then A_t and $\{U_s, s \in T\}$ also commute so that

$$A_t U_t Y_0 = A_t \left(\int_{\hat{T}} e^{itv} E(dv) Y_0 \right)$$

$$= \int_{\hat{T}} g_t(\lambda) E(d\lambda) \left(\int_{\hat{T}} e^{itv} E(dv) Y_0 \right),$$

$$= \int_{\hat{T}} e^{it\lambda} g_t(\lambda) E(d\lambda) Y_0,$$

(by a property of spectral integrals),

$$= \int_{\hat{T}} e^{it\lambda} g_t(\lambda) Z(d\lambda) = X_t, \text{ by (8).} \qquad (9)$$

This shows that the representation (7) holds.

The converse direction depends on a deep result due to von Neuman and F. Riesz (cf. Riesz-Nagy (1955), p. 35). It follows that

$$A_t = \phi_t(U_t) = \int_{\hat{T}} \phi_t(\lambda) E(d\lambda). \qquad (10)$$

This representation implies

$$X_t = A_t(U_t Y_0) = \int_{\hat{T}} \phi_t(\lambda) E(d\lambda) \int_{\hat{T}} e^{itv} E(dv) Y_0,$$

$$= \int_{\hat{T}} e^{it\lambda} \phi_t(\lambda) E(d\lambda) Y_0, \text{ (property of spectral integral),}$$

$$= \int_{\hat{T}} e^{it\lambda} \phi_t(\lambda) Z(d\lambda).$$

If now $g'_t(\lambda) = e^{it\lambda} \phi_t(\lambda), \mu(B) = (E(B)Y_0, E(B)Y_0)$, then this implies that $g'_t \in L^2(\hat{T}, \mathcal{B}, \mu)$ so that $\{X_t, t \in T\}$ is of Karhunen class.

For the last part, let $Q_t = A_t U_t$ then it can be verified that $\{Q_t, t \in T\}$ must be positive definite, weakly continuous and contractive, by known results (cf., e.g. Rao (1982), p. 330). Here the U_t-family is a unitary group and weakly continuous. Then the A_i-family must also be positive definite, which is verified as follows. Let $h_i \in \mathcal{H}(X), i = 1, \ldots, n$ be any set and $t_i \in T, n \geq 1$ and consider

$$0 \le \sum_{i,j=1}^{n} (Q_{t_i-t_j} v_i, v_j), \text{ since the } Q_t \text{ is positive definite,}$$

$$= \sum_{i,j=1}^{n} \left(A_{t_i-t_j} U_{t_i-t_j} U_{t_i}^* h_i, U_{t_j}^* h_j \right)$$

$$= \sum_{i,j=1}^{n} \left(U_{t_i}^* A_{t_i-t_j} h_i, U_{t_{ij}}^*, h_i \right), \text{ as } U_t^* \text{ and } A_s \text{ commute,}$$

as also U_t and U_s do,

$$= \sum_{i=1}^{n} \sum_{j=1}^{n} (A_{t_i-t_j}, h_i, h_j).$$

This shows that the $\{A_t, t \in T\}$ is positive definite, and since $Q_0|\mathcal{H}(X)$ is the identity, we get $A_0 U_0 = A_0$ as well.

To see the opposite direction is also valid, we need to apply a key theorem due to A. Grothendieck so that if $A : \mathcal{K} \to \mathcal{H}(X)$ has the above properties then $Q_t = A_t U_t$ is positive definite and it satisfies the hypothesis of the author's result (cf. Rao (1982), p. 330), so that the process is weakly harmonizable. $\square$

Remark 18. The first part of the result was given in an equivalent form by V. Mandreakar (1972) which refines some work by Gladyšev (1962–63). The measure $\mu_t(\cdot)$ defined by

$$\text{var}(X_t) = r(t, t) = \int_{\mathbb{R}} |a_t(\lambda)|^2 F(d\lambda) = \int_{\mathbb{R}} \mu_t(d\lambda), \qquad (11)$$

is called an "evolving spectrum", by Priestley (1965).

We recall that a set $S_\beta \subset \mathbb{R} \times \mathbb{R}$ is termed the *support* of a bimeasure β in $\mathbb{R} \times \mathbb{R}$, if $(x, y) \in \mathbb{R} \times \mathbb{R}$ belongs to S_β if and only if for each neighborhood $U_1 \times U_2$ of $(x, y), |\beta|(U_1 \times U_2) > 0$ where $|\beta|(\cdot, \cdot)$ is the variation of β, so that S_β is the smallest closed set outside of which β vanishes. With this concept we can state supports of harmonizable process bimeasures if their covariances are periodic, to indicate that the problem is useful and nontrivial.

Proposition 3.1.6 *Let $\{X_t, t \in \mathbb{R}\} \subset L_0^2(P)$ be weakly harmonizable with its covariance r being periodic in that $r(s + \alpha, t + \alpha) = r(s, t)$, for* *some $\alpha > 0$ and $(s, t) \in \mathbb{R}^2$. If $F(\cdot, \cdot)$ is its spectral bimeasure, then its support is contained in S_F, where*

$$S_F = \left\{(\lambda_1, \lambda_2) \in \mathbb{R}^2 : \lambda_1 - \lambda_2 = \frac{2\sigma k}{\alpha}, k \in Z\right\}. \tag{12}$$

In the opposite direction, if S above is the support of a bimeasure F of a weakly harmonizable process then it is periodically correlated.

Proof. It is known from the works of Gladysev and Hurd that the periodically correlated discrete parameter processes are always strongly harmonizable and we consider the continuous parameter case here.

From the structural analysis above (cf. Niemi's result on approximating weakly harmonizable processes by strongly harmonizable sequences on compact sets of $\mathbb{R}$), we get $\lim_n r_n(s, t) = r(s, t)$ uniformly for $(s, t) \in K \times K$, $K \subset \mathbb{R}$ compact. Since $r(\cdot, \cdot)$ is periodic, say with period α, we observe that for large n, $r_n(\cdot, \cdot)$ has the same property. If this is not true we must have for some $\varepsilon > 0$, and a point $(s, t) \in \mathbb{R} \times \mathbb{R}$, the inequality, for some $k \neq 0$,

$$\liminf_n |r_n(s + k\alpha, t + k\alpha) - r_n(s, t)| > \varepsilon. \tag{13}$$

Since $r(s + k\alpha, t + k\alpha) = r(s, t)$, and moreover we have, with $r_n \to r$ uniformly on the (compact) set $K = \{(s, t) : (s + k\alpha, t + k\alpha)\}$,

$$\begin{aligned}
0 < \varepsilon < &|r_n(s + k\alpha, t + k\alpha) - r_n(s, t)| \\
\leq &|r_n(s + k\alpha, t + k\alpha) - r(s + k\alpha, t + k\alpha)| + \\
&|r(s + k\alpha, t + k\alpha) - r(s, t)| + |r(s, t) - r_n(s, t)| \to 0
\end{aligned}$$

as n gets large which gives a contradiction. Hence X_t^n must also be periodically correlated.

Let F_n and F be the spectral measures of r_n and r. We claim that $F_n(A, B) \to F(A, B)$ for Borel sets A, B. To show this (there is no Helly-Bray), we consider X_t^n with covariance r_n.

$$X_t^n = \sum_{k=1}^n \phi_k(X_k, \phi_k), \quad n \geq 1,$$

$\{\phi_t, k \geq 1\}$ a CON system of the separable $\mathcal{H}(X)$ of the X_t-process. Then $X_t^n \in \mathcal{H}(X), X_t^n \to X$ in $L_0^2(P)$ as $n \to \infty$, uniformly in t on compacts. Let $\ell_k : Y \mapsto (Y, \phi_k), Y \in \mathcal{H}(X)$, be a linear mapping. If

ζ_m and Z are representing stochastic measures of X_t^n, X_t-processes then we get

$$\zeta_n(t) = \sum_{k=1}^{n} \phi_k \ell_k(Z(t)), \quad F_r(A, B) = (\zeta_n(A), \zeta_n(B)). \tag{14}$$

Since $\ell_k(Z(A)) = (Z(A), \phi_k)$ we get with Parsevel's equation that $Z(A) = \lim_n \zeta_n(A)$ in mean, so that

$$F(A, B) = (Z(A), Z(B)) = \lim_n (\zeta_n(A), \zeta_n(B)) = \lim_n F_n(A, B).$$

Now r_n is periodic and thus its support $S_{F_n} \subset S_F, n \geq 1$. Hence the above equation implies that F has its support in S_F.

In the converse direction $S_{F_n} \subset S_F$ for large enough n. The corresponding result of Hurd implies that X_t^n is periodically correlated and strongly harmonizable. Since $r_n \to r$ pointwise and (by Hurd's work) strongly harmonizable, r is periodic, so X_t is also periodically correlated, as a consequence. $\square$

It is perhaps appropriate at this point to consider a large class of second order processes, centered at their mean functions to start with and discuss their integral representations motivated by the Karhunan class, starting with vector harmonizability as indicated by Rozanov (1959), and the consequent (abstract) methods. This will also exhibit the basic structure and general potential of this class, as it was described in Chapter 1. (See Theorem 1.4.2 with full details there.)

Theorem 3.1.7 (Consistent Inversion) *Let $X : \mathbb{R} \to L_0^2(P)$ be weakly harmonizable with $Z : \mathcal{B}(\mathbb{R}) \to L_0^2(P)$ as its (stochastic) representing measure. Writing $Z(\lambda)$ for $Z((-\infty, \lambda))$, one has (the inversion)*

$$\underset{T \to \infty}{l.i.m.} \int_{-T}^{T} \frac{e^{-it\lambda_2} - e^{-it\lambda_1}}{-it} X(t) dt$$

$$= \frac{Z(\lambda_{2+}) + Z(\lambda_{2-})}{2} - \frac{Z(\lambda_{1+}) + Z(\lambda_{1-})}{2}, \tag{15}$$

l.i.m. being the $L^2(P)$-limit. Moreover the covariance bimeasure $F(\cdot)$ of $Z(\cdot)$ is given, for $A = (\lambda_1, \lambda_2), B = (\lambda_1', \lambda_2'),$ as:

$$F[A, B] = \lim_{0 \le T_1 < T_2 \to 0} \int_{-T_1}^{T_1} \int_{-T_2}^{T_2} \frac{e^{it\lambda_1} - e^{-i\lambda_2 t}}{-it} \cdot \frac{e^{is\lambda_1'} - e^{is\lambda_2'}}{is} r(s, t) ds dt,$$

(16)

provided A, B are continuity intervals of F in the usual sense and $r(\cdot, \cdot)$ is the covariance function of the X-process. In particular for each continuous $S : \mathbb{R} \to \mathbb{C}$, with $\frac{1}{T} \int_0^T S(t) dt \to a_0$ existing as $T \to \infty$, and $r(\cdot, \cdot)$ vanishes at the infinities, $\tilde{Y}(t) = S(t) + X(t)$, so that the signal $S(\cdot)$ is nonstochastic and the noise $X(\cdot)$ is strongly harmonizable, then the "estimator" $\hat{S}_T = \frac{1}{T} \int_0^T Y(t) dt \to a_0$ exists in $L^2(P)$, i.e., $E(|\hat{S}_T - a_0|^2) \to 0$ as $T \to \infty$, so that $Y(t)$ obeys the law of large numbers.

Proof. Using the dilation theorem, we reduce the result to the classical stationary case, so on an extension measure space $(\tilde{\Omega}, \tilde{\Sigma}, \tilde{P})$ of the given one, $L_0^2(\tilde{P}) \supset L_0^2(P)$, there is a stationary process $Y : \mathbb{R} \to L_0^2(\tilde{P})$ with $X(t) = QY(t), t \in \mathbb{R}$, Q being an orthogonal projection onto $L_0^2(P)$. Then we have

$$Y(t) = \int_{\mathbb{R}} e^{it\lambda} d\tilde{Z}(\lambda), \qquad Z(A) = Q\tilde{Z}(A), \quad A \in \mathcal{B},$$

with $Z : \mathcal{B} \to L_0^2(P)$ as an orthogonal measure of the X-process. Since Q and the integral commute, with $Z(\cdot) = Q\tilde{Z}(\cdot)$, it follows by the standard calculus that (15) results.

Regarding (16), we can express the left side of (15) as:

L.H.S.

$$= \lim_{T_1, T_2 \to \infty} E\left[\left(\int_{-T_1}^{T_1} \frac{e^{-is\lambda_2} - e^{-is\lambda_1}}{-is} X(s) ds \right) \right.$$
$$\left. \left(\int_{-T_1}^{T_2} \frac{e^{-it\lambda_2} - e^{-is\lambda_1}}{-it} \right) dt \right]$$

$$= F(A, B),$$

(17)

on using the standard vector analysis reasoning. Then we get on setting $\tilde{Y}(t) = S(t) + X(t), t \in \mathbb{R}, a_T = E(\hat{S}_T) = \frac{1}{T} \int_0^T S(t) dt, (T \to \infty)$

$$E\left(\left| \tilde{S}_t - a_0 \right|^2 \right) = \frac{1}{2T} \int_{-T}^{T} r_T(h) dh + 2|a_T - a_0|^2.$$

(18)

Since $r_T(h) \to r(h)$ as $\tilde{Y} \in$ class $(KF), |r(s,t)| \le M^2 < \infty$, it can be inferred that $(\tilde{Y} \in$ class $(KF)), r_T(h) \to r(h) = 0$ uniformly in h, and $E(\tilde{G}_T) \to a_0$ uniformly, as $T \to \infty$. This completes the sketch. $\square$

3.2 Linear Models for Weakly Harmonizable Classes

Let $\{X_t, t \in T\}$ and $\{Y_t, t \in T\}$ be centered second order processes, connected by a linear equation: $(T \subset \mathbb{R}$, usually an interval)

$$\Lambda X_t = Y_t, \quad t \in T \tag{19}$$

where in applications the $\{Y_t, t \in T\}$ is typically considered as a noise process and $\{X_t, t \in T\}$ as an observable process, both of second order and taken to be harmonizable with Λ as a difference/differential operator of the type

$$\Lambda X_t = \sum_{j=0}^{k'} \sum_{i=0}^{k} a_{ij} X^{(i)}(t - t_i) = Y_t, \qquad t \in T, \, a_{ij} \in \mathbb{C}, \tag{20}$$

where $X^{(i)}(\cdot)$ is the mean-square ith derivative existing in that for instance if $i = 1$, we have

$$E\left[\left|\frac{X(t+h) - X(t)}{h} - X^{(1)}(t)\right|^2\right] \to 0 \quad \text{as } h \to 0$$

and similarly for higher orders. Thus one has

$$Y_t = \Lambda X_t = \sum_{i=0}^{k} a_k \int_T X^{(i)}(t - \lambda) H(d\lambda), \quad t \in T, \tag{21}$$

with $H(\cdot)$ as a function of bounded variation on T $(T = \mathbb{R}$ is possible). Here taking $\Lambda = I + \Lambda_1$ so that is written as

$$Y_t = X_t + \Lambda_1 X_t = X_t + S_t, \quad \text{(say)} \tag{22}$$

in which $\{S_t, t \in T\}$ is considered as an unknown *stochastic signal* process and X_t the noise process with the Y_t as the output (or the observed) series. Then the problem is to extract (or estimate) the signal from noise, having only the output at our disposal.

Such general linear operations on second order processes were considered first by Karhunen (1947) in his thesis and then the models given by and were studied by Nagabhushanam (1951) for stationary processes X_t. Then the more general case for the weakly harmonizable processes (called V-*bounded* by him which we now know as weakly harmonizable by the analysis of Chapter 2), was established by Bochner (1956). The most interesting conclusion of Bochner's work is that if $\Lambda X_t = Y_t$ has a harmonizable solution, then one can also find another solution which is weakly stationary, generalizing Nagabhushanam's work:

Theorem 3.2.1 *In the model* $\Lambda X_t = Y_t, t \in T$, *let the output process* $\{Y_t, t \in T\}$ *be (weakly) stationary and the input* $\{X_t, t \in T\}$ *be weakly harmonizable where* Λ *for the above equation is given by*

$$\Lambda X_t = \sum_{j=0}^{l} \sum_{i=0}^{k} a_{ij} X^{(i)}(t - t_j) = Y_t, \quad a_{ij} \in \mathbb{C}, t \in T, \quad (23)$$

where the spectral characteristic of the filter λ, *namely* $\tilde{C}(\cdot) \colon \lambda \to \tilde{C}(\lambda)$ *is given by*

$$\tilde{C}(\lambda) = \sum_{j'=0}^{l} \sum_{j=0}^{k} a_{j'j} (i\lambda)^j e^{-j's\lambda} e^{jt\lambda}. \quad (24)$$

If $Q = \{\lambda \colon C(\lambda) = 0\}$, *and* F_y *is the spectral measure of the output process, satisfying*

$$\int_{\hat{T}-Q} |C(\lambda)|^{-2} dF_y(\lambda) < \infty, \quad (25)$$

and there is a stationary input $\{X_t^0, t \in T\}$ *so that* $\Lambda X_t^0 = Y_t^0, t \in T, \{Y_t^0, t \in T\}$ *is stationary whose spectrum coincides with that of* $\{Y_t, t \in T\}$ *on* $\hat{T} - Q$, *then the solution is given as:*

$$X_t^0 = \int_{\hat{T}-Q} e^{it\lambda} C(\lambda)^{-1} Z_y(d\lambda). \quad (26)$$

Remark 19. The process (26) satisfies actually the equation, as the proof above shows

$$\int_T \phi(t) Y_t \, dt = \int_T \left[\sum_{i=0}^{k} (-1)^j \int_{\hat{T}} \phi^{(j)}(t-u) H_j(du) \right] X_t \, dt, \phi \in C^k. \quad (27)$$

Proof. Under the condition (25), the process $\{X_t^0, t \in T\}$ given by (26) is well-defined, since $E(|X_t^0|^2)$ is the integral obeying (25). We then verify that the X_t^0 process also satisfies (23) as follows.

$$\Lambda X_t^0 = \int_{\hat{T}} C(\lambda)e^{it\lambda} Z_t^0(d\lambda), \text{ by (23) with } Z_t^0, \text{ as the stochastic}$$

spectral measure of X_t^0,

$$= \int_{\hat{T}-Q} C(\lambda)e^{it\lambda} \frac{1}{C(\lambda)} Z_y(\lambda)$$

$$= \int_{\hat{T}-Q} e^{it\lambda} Z_y(d\lambda) = Y_t^0, \text{ by definition.} \quad (28)$$

Thus the spectral measures of Y_t^0 and Y_t agree on $\hat{T} - Q$. We now need to verify (25), which is done as follows.

Since the distributional solution satisfies (27), for each ϕ with two continuous derivatives when its Fourier transform ψ has compact support and has two continuous derivatives, we get

$$\int_{\hat{T}} C(\lambda)e^{it\lambda}\ell(Z_x(d\lambda)) = \int_{\hat{T}} e^{it\lambda}\ell(Z_y(d\lambda)), \quad (29)$$

since $\ell(\cdot)$ and the integral commute. But $\ell \circ Z_x(\cdot), \ell \circ Z_y(\cdot)$ are signed measures, and the classical analysis shows that $C(\lambda)Z_x(d\lambda) = Z_y(d\lambda)$. This easily implies, ($C(\lambda) \neq 0$ on $\hat{T} - Q$):

$$\int_{\hat{T}-Q} \frac{\psi(\lambda)}{C(\lambda)} e^{it\lambda} Z_y(d\lambda) = \int_{\hat{T}-Q} \psi(\lambda)e^{it\lambda} Z_x(d\lambda).$$

But ψ is supported in $\hat{T} - Q$ and so the set $\hat{T} - Q$ can be replaced by $\hat{T}$, and one gets

$$\int_{\hat{T}} \left|\frac{\psi(\lambda)}{C(\lambda)}\right|^2 F_y(d\lambda) = E\left(\left|\int_{\hat{T}} \psi(\lambda)e^{it\lambda} Z_x(d\lambda)\right|^2\right), \quad (30)$$

since $Z(\cdot)$ has orthogonal increments and $E(|Z_y(A)|^2) = F_y(A)$, while $Z_x(A)$ and $Z_y(A)$ have zero means. With the inverse Hölder inequality applied to both sides of (30), one has the following computation if $\|\psi\|_\infty \leq 1$ is varied:

$$\int_{\hat{T}-Q} |C\lambda|^{-2} F(d\lambda) = \sup \left\{ \int_{\hat{T}-Q} |\psi(\lambda)|^2 |C(\lambda)|^{-2} F(d\lambda), \|\psi\|_\infty \leq 1 \right\}$$
$$\leq \sup \left\{ \|\psi(\cdot)e^{it(\cdot)}\|_\infty \|Z\|(\hat{T})^2, \|\psi\|_\infty \leq 1 \right\}$$
$$\leq \|Z_x\|(\hat{T})^2 < \infty. \tag{31}$$

Here $\|Z_x\|(\hat{T})$ is the semi-variation of Z_x. This establishes the theorem.
□

Remark 20. The key conclusion is that, if the output $\{Y_t, t \in T\}$ is stationary and if there is a weakly harmonizable solution process (in distribution) then omitting some frequencies suitably, we can get a stationary solution!

The preceding computation also gives the following characterization of a weakly harmonizable process which is worthy of recording here.

Theorem 3.2.2 *Let $\{X_t, t \in T\} \subset L_0^2(P)$ be a (centered) process. It is weakly harmonizable if and only if ($T = \mathbb{R}$ or $\mathbb{Z}$):*

(i) $E(|Y_t|^2) \leq M < \infty$, *and (ii)* $E\left(\left| \int_T \phi(t) X_t dt \right|^2 \right) \leq C\|\hat{\phi}\|_\infty^2$,
$$\tag{32}$$
for all summable $\phi : T \to \mathbb{R}$, $\hat{\phi}$ being the Fourier transform of ϕ. Thus $X_t = \int_{\hat{T}} e^{it\lambda} Z(d\lambda), t \in T$, iff (32) holds.

Sketch of Proof. Since the general argument is by now familiar we include an outline of the procedure leaving some details to the reader. In fact (31) gives (i), and (ii) follows with (32) and $T = \mathbb{R}$ as:

$$\int_{\mathbb{R}} \phi(t) X_t dt = \int_{\mathbb{R}} \left(\int_{\mathbb{R}} e^{it\lambda} \phi(t) dt \right) Z_y(d\lambda)$$
$$= \int_{\mathbb{R}} \hat{\phi}(\lambda) Z_y(d\lambda), \hat{\phi} \text{ is the Fourier transform of } \phi.$$

Hence

$$E\left(\left| \int_{\mathbb{R}} \phi(t) X_t dt \right|^2 \right) = E\left(\left| \int_{\mathbb{R}} \hat{\phi}(\lambda) Z_y(d\lambda) \right|^2 \right) \leq \|\hat{\phi}\|_\infty^2 \|Z_y\|^2(\mathbb{R}),$$
$$\leq C\|\hat{\phi}\|_\infty^2.$$

Thus (ii) holds. The converse is similar, using some earlier work. $\square$

For much of the analysis of the stationary or harmonizable processes, their spectral domain space plays a crucial role. Thus if $\{X_t, t \in T\} \subset L_0^2(P)$ is weakly harmonizable, with covariance $r(\cdot, \cdot)$ represented by a bimeasure $F(\cdot, \cdot)$ of finite Fréchet variation F, let

$$L^2(F) = \left\{ f : \hat{T} \to \mathbb{C} \;\middle|\; \int_{\hat{T}} \int_{\hat{T}} f(s)\overline{f(t)} F(ds, dt) = (f, f)_F < \infty \right\}.$$
(33)

If the process is stationary, then it is known that $\{L^2(F), (\cdot, \cdot)_F\}$ is a semi-inner product space and even complete for the (stationary) case. We now sketch a proof for the harmonizable case as follows.

Theorem 3.2.3 *Let $\{X_t, t \in T\}$ be a weakly harmonizable (scalar) process. Then its spectral domain $\{L^2(F), \|\cdot\|_F = \sqrt{(f, f)_F}\}$ is complete, so that it is a Hilbert space when equivalence classes are considered.*

Proof. We shall sketch the (standard) argument for completeness. First observe that each weakly harmonizable process is of Karhunen class, which was actually shown in Theorem 2.2.9 already. It is thus representable as:

$$r(s, t) = \int_{\hat{T}} f_s(\lambda)\overline{f_t(\lambda)}\mu(d\lambda), \quad s, t \in T$$

for Borel functions $f_t : \hat{T} \to \mathbb{C}$ and a (σ-finite Borel) measure μ on $\hat{T}$. Such a process is representable as

$$X_t = \int_{\hat{T}} f_t(\lambda)Z(d\lambda), \quad t \in T,$$
(34)

for an orthogonally valued measure on the Borel field of $\hat{T}$. But it was seen in the preceding chapter (cf. Theorem 2.7.3) that a (weakly) harmonizable process can be dilated to a stationary process on a suitable super Hilbert space containing the present one. Thus if $\{Y_t, t \in T\}$ is the stationary process in the 'super' Hilbert space $L_0^2(\tilde{P})$ and Q is the orthogonal projection, onto $L_0^2(P)$ so that $X_t = QY_t, t \in T$, with

$$Y_t = \int_{\hat{T}} e^{it\lambda} \tilde{Z}(d\lambda), t \in T,$$

for an orthogonally valued measure on $L^2(\tilde{P})$ onto $L_0^2(P)$, then,

$$X_t = QY_t = \int_{\hat{T}} \pi(e^{it(\cdot)})(\lambda)\tilde{Z}(d\lambda) = \int_{\hat{T}} f_t(\lambda)\tilde{Z}(d\lambda), \quad t \in T. \quad (35)$$

Letting $f_t(\cdot) = \pi(e^{it(\cdot)})$, a Borel function, $f_t \in L^2(\tilde{\mu})$, so that

$$X_t = \int_{\hat{T}} f_t(\lambda)\tilde{Z}(d\lambda), \quad t \in T,$$

and

$$r(s,t) = E(X_s\bar{X}_t) = \int_{\hat{T}} f_s(\lambda)\overline{f_t(\lambda)}\tilde{\mu}(d\lambda), t \in T.$$

Thus the (weak) harmonizable process is induced by a Karhunen process, $\{Y_t, t \in T\} \subset L^2(\tilde{\mu})$. Also, we have

$$\int_{\hat{T}} \int_{\hat{T}} e^{is\lambda - it\lambda'} F(d\lambda, d\lambda') = r(s,t) = (\pi e^{is(\cdot)}, \pi e^{it(\cdot)})_{\tilde{P}},$$

so that $e^{is(\cdot)} \to \pi(e^{is(\cdot)})$ gives an isometry between $L^2(F)$ and $L^2(\tilde{\mu})$, by extending the correspondence linearly onto the space determined by the trigonometric polynomials. This will yield the completeness of $L^2(F)$ as desired in the theorem. $\square$

As a consequence of the above result we have the following proper class containments of interest in the theory of $L^2(P)$-processes:

Stationary class $\subset$ Strongly harmonizable class

$\subset$ Weakly harmonizable class $\subset$ Karhunen class.

In general, the defining family $\{f_t, t \in T\}$ of the general class of the last type is not explicit in contrast to the $\{e^{it(\cdot)}, t \in T\}$ family.

3.3 Application to Signal Extraction from Noise, and Sampling

Here we consider a signal extraction from observations contaminated by (harmonizable) noise processes which are possibly (harmonizably) correlated with the signal. Thus let

$$X_t = Y_t + Z_t, \quad t \in T, \quad (36)$$

where $\{Y_t, t \in T\}$ is the (stochastic) signal and $\{Z_t, t \in T\}$ is a harmonizable noise when both processes are centered uncorrelated or harmonizably correlated (in the general model). It is desired that the signal Y_t

at $t = a$ be estimated based on a realization of the process when the co-variance structure of the X_t-process is known from previous knowledge. Thus let r_y, r_z be the covariance functions of the X_t, Y_t processes (i.e. $r_y(s, t) = E(Y_s \bar{Y}_t)$ and $r_z(s, t) = E(Z_s \bar{Z}_t)$, and $r_{yz} \colon (s, t) \to E(Y_s \bar{Z}_t)$ with F_y, F_z and F_{yz} as their spectral measures. Also let

$$k(\lambda) = F_y(\lambda, \hat{T}) + F_{yz}(\lambda, \hat{T}), [F_{yz}(A, B) = \bar{F}_{zy}(B, A)],$$

$$h(\lambda) = F_y(\lambda, \hat{T}) + F_z(\lambda, \hat{T}) + F_{yz}(\lambda, \hat{T}) + F_{zy}(\lambda, \hat{T}), \qquad (37)$$

where $F_y(\cdot, \hat{T}), F_z(\cdot, \hat{T}), F_{yz}(\cdot, \hat{T})$ are "marginal" spectral measures of F_y, F_z, F_{yz}. The problem of obtaining the least squares estimator $\hat{Y}_z$ of Y_z based on an observation of $\{X_t, t \in T\}$ can be presented as follows. Since $F_{yz}(A, B) = \bar{F}_{zy}(B, A)$ the solution is given (assuming the process to be strongly harmonizable, (leaving the weakly harmonizable extension to the reader) by the following:

Theorem 3.3.1 *Let the signal plus noise (correlated) be given by the additive model* (36). *Then the least squares optimal signal estimator* $\hat{Y}_a$ *of* Y_a *of* (36), *based on one realization* $\{X_t, t \in T\}$, *is given by*

$$\hat{Y}_a = \int_{\hat{T}} g_a(\lambda) Z_x(d\lambda), \qquad (38)$$

where $g_a(\cdot)$ *is the unique solution in* $L^2(F_z)$ *of the integral equation:*

$$\int_{\hat{T}} g_a(\lambda) h(\lambda) d\lambda = \int_{\hat{T}} e^{ia\lambda} K(d\lambda), \qquad (39)$$

the functions $h(\cdot), K(\cdot)$ *and* $Z_x(\cdot)$ *are defined by* (37) *with* $Z_x(\cdot)$ *as the stochastic measure of the* X_t-*process. If the derivatives* h', K' *exist, then*

$$g_a(\lambda) = K'(\lambda) e^{ia\lambda / h'(\lambda)}, \text{ for a.a.}(\lambda). \qquad (40)$$

Proof. Let $H_z = \bar{sp}\{X_t, t \in T\} \subset L_0^2(P)$ be the linear span. We need to find $Y_a \in H_Z$ such that

$$E(|Y_a - \hat{Y}_a|^2) = \|Y_a - \hat{Y}_a\|_p^2 = \inf\{\|Y_a - \xi\|_p^2 : \xi \in H_z\}. \qquad (41)$$

Since H_z is closed and convex, there exists a unique $\hat{Y}_a$ satisfying the above equation, and the Hilbert space geometry implies that $\hat{Y}_a = QY_a$, Q being the orthogonal projection onto H_z, and $Y_a - \hat{Y}_a \perp H_z$ so that

$(Y_a - \hat{Y}_a, X_t) = 0, t \in T$. Since each element of H_z is the l.i.m. of elements such as $\sum_{i=1}^{n} a_i^n X_{t_i}$, we get

$$\hat{Y}_a = \int_T g_a(\lambda) Z_x(d\lambda) \qquad (42)$$

for some $g_a \in L^2(F_z)$, and then $(Y_a, X_t) = (\hat{Y}_a, X_t), t \in T$, so that with (42) we have:

$$\int_{\hat{T}} \int_{\hat{T}} e^{is\lambda - it\lambda'} F_{yz}(d\lambda, d\lambda') = \int_{\hat{T}} \int_{\hat{T}} g_a(\lambda) e^{-it\lambda'} F_z(d\lambda, d\lambda'). \qquad (43)$$

But (36) implies $F_{yz} = F_y + F_{yz}$ and $F_x = F_y + F_z + F_{yz} + \bar{F}_{zy}$. With these, one can simplify the above after setting $t = 0 (\in I)$ to get

$$\int_{\hat{T}} \int_{\hat{T}} g_a(\lambda)[F_y(d\lambda, d\lambda') + F_z(d\lambda, d\lambda') + 2\mathrm{Re}\, F_{yz}(d\lambda, d\lambda')]$$

$$= \int_{\hat{T}} \int_{\hat{T}} e^{is\lambda}(F_y + F_{yz})(d\lambda, d\lambda'). \qquad (44)$$

Interchanging the order of integration, which is valid, one gets

$$\int_{\hat{T}} g_a(\lambda) h(d\lambda) = \int_{\hat{T}} e^{is\lambda} h(d\lambda). \qquad (45)$$

This establishes the first (the main) part.

The general case follows by truncation for $0 < \varepsilon < |g_a| < N < \infty$, and verifying that on $[0, \varepsilon]$ and $[N, \infty]$, the function g_a is such that the integrals are bounded, and the result then is seen to follow easily. The details can be left to the reader. $\square$

Remark 21. In the stationary case, (with Y_t, Z_t stationarily correlated) the work reduces to

$$h(\lambda) = (F_y + F_z + F_{yz} + \bar{F}zy)(\lambda, \lambda) \qquad (46)$$

$$K(\lambda) = F_y(\lambda, \lambda) + F_{yz}(\lambda, \lambda), \qquad (47)$$

and setting $\hat{F}_z(\lambda) = F_z(\lambda, \lambda)$, etc. the result reduces to

$$g_z(\lambda) = \frac{[\hat{F}_y(\lambda) + \hat{F}_{yz}(\lambda)]e^{is\lambda}}{\hat{F}_y(\lambda) + \hat{F}_z(\lambda) + 2\mathrm{Re}\, \hat{F}_{yz}(\lambda)}. \qquad (48)$$

In this form Grenander ((1950), p. 275) has first obtained this result. The method of calculation is a modification of the original one that is given in Yaglom (1962).

An important practical as well as a theoretical question is the physical realizability of the filter equation for Λ. This means if $\Lambda X_t = Y_t, t \in T$ and $\{Y_t, t \in T\}$ is harmonizable, must the input process $\{X_t, t \in T\}$ be of the harmonizable class? Here the process is termed physically realizable if the solution process depends on the past and present only. Thus if

$$Y_n = \Lambda X_n = \sum_{i=1}^{k} a_i X_{n-i}, a_i \in \mathbb{C} \tag{49}$$

then we need to find conditions on a_i so that $\{Y_n, n \in \mathbb{Z}\}$ is harmonizable, and the $X_n, n \in \mathbb{Z}$ is obtainable from

$$\int_{-\pi}^{\pi} e^{in\lambda} Z_y(d\lambda) = \int_{-\pi}^{\pi} \tilde{C}(\lambda) e^{in\lambda} Z_y(d\lambda), \tag{50}$$

where $\tilde{C}(\lambda) = \sum_{j=1}^{k} a_i e^{-ij\lambda}$, the characteristic equation of the filter given by (49). We present conditions on the solution of (49) so that the result depends just on the past and present only and hence that the filter is physically realizable. Here the zeros of the characteristic equation (of the filter) $\tilde{C}(\lambda) = \sum_{j=1}^{k} a_j e^{-ij\lambda}$, play an important role in the analysis of the problem.

Proposition 3.3.2 *Let $\{Y_n, n \in \mathbb{Z}\}$ be a strongly harmonizable output of the filter equation (49), namely of*

$$Y_n = \Lambda X_n = \sum_{i=1}^{k} a_i X_{n-i}, \quad a_i \in \mathbb{C}. \tag{51}$$

If $Q = \{\lambda : \sum_{j=1}^{k} a_j e^{-ij\lambda} = 0\} \subset \hat{T} = (-\pi, \pi)$, is the zero set of the (filter) characteristic equation (49) of the process, then a strongly harmonizable solution of it exists iff (i) $|F_y|(Q \times Q) = 0$ and (ii) $\tilde{C}(\cdot)$ satisfies on Q^c the "inverse integrability" in the sense that

$$\int_{Q^c} \int_{Q^c} |\tilde{C}(\lambda)\bar{\tilde{C}}(\lambda')|^{-1} |F_y|(d\lambda, d\lambda') < \infty \tag{52}$$

where $|F_y|(\cdot, \cdot)$ is the variation measure of F_y of the output which exists. Moreover, if the roots of the characteristic equation

$$p_k(z) = \sum_{j=0}^{k} a_j z^j = 0 \tag{53}$$

lie outside the unit circle of $\mathbb{C}$, *then the filter is physically realizable so that (the solution depends only on the past and present):*

$$X_n = \sum_{j=0}^{\infty} b_j Y_{n-j}, \tag{54}$$

the b_j *are coefficients of* $(p_k(z))^{-1}$, *expanded in power series around each root of* $p_k(z) = 0$.

Proof. Under the present hypothesis, the solution is given by

$$X_n = \int_{Q^c} e^{in\lambda} [C(\lambda)]^{-1} Z_y(d\lambda), n \in \mathbb{Z}, \tag{55}$$

where we used the condition that $|F_y|(Q \times Q) = 0$. It then follows that X_n satisfies (51).

For the converse, after squaring and taking expectation with (50), one gets the following for any Borel sets A, B of $\hat{T}$:

$$\int_A \int_B |F_y|(d\lambda, d\lambda') = \int_A \int_B \tilde{C}(\lambda)\bar{\tilde{C}}(\lambda')|F_z|(d\lambda, d\lambda'). \tag{56}$$

Taking $A = B = Q$, since $\tilde{C}(\cdot)$ vanishes on Q, (i) follows. Next with $A = B = Q^c$ in (56), and $\tilde{C}(\lambda) \neq 0$ on Q^c one has

$$\int_{Q^c} \int_{Q^c} |\tilde{C}(\lambda)\bar{\tilde{C}}(\lambda')|^{-1} F_y(d\lambda, d\lambda') = \int_{Q^c} \int_{Q^c} |F_x|(d\lambda, d\lambda') < \infty, \tag{57}$$

so that (52) holds.

When the conditions (i) and (ii) of the hypothesis hold, then X_n is given by (55). Since $\tilde{C}(\lambda) = p_k(e^{-i\lambda})$ has its zeroes outside the unit circle by assumption, $(p(t))^{-1} = \sum_{j=0}^{\infty} b_j t^j$ holds, the series converging uniformly. Hence (55) becomes:

$$X_n = \int_{Q^c} e^{in\lambda} \sum_{j=0}^{\infty} b_j e^{ij\lambda} Z_y(d\lambda) = \sum_{j=0}^{\infty} b_j Y_{n-j},$$

so that the X_n depend only on the past and present of $\{Y_n, n \in \mathbb{Z}\}$. $\square$

Remark 22. 1. This result in the stationary case was obtained by K. Nagabhushanam (1951) in his thesis under H. Cramér, and was extended by J. P. Kelsh (1978) to the strongly harmonizable case. As he showed, further conditions should be imposed for the uniqueness of solutions. If $Q = \phi$, there is uniqueness, already obtained by Nagabhushanam. He showed that if $Q \neq \phi$, there is only one solution whose exceptional set Q should satisfy $|F_z|(Q \times Q) = 0$. However, there is no recipe available to find the unique solution outside of the "bad" $Q \neq \phi$, and no recipe as yet is found to compute the desired solution.

2. The preceding result and remarks are on strongly harmonizable classes. What about the weakly harmonizable class? The results are also not complete, but we can present the following result, indicating that some further ideas are needed. It is a direct extension of the above result and shows the possible directions that one may proceed.

Proposition 3.3.3 *Let $\{Y_n, n \in \mathbb{Z}\}$ be weakly harmonizable and suppose it satisfies the filter equation* (51) *above. Then there exists a weakly harmonizable solution of that equation iff the following two conditions hold for its stochastic spectral measure Z_y:*

(i) $\|Z_y\|(Q) = 0$, $\|Z_y\|(\cdot)$ being the semi-variation of Z_y, of the output series $\{Y_n, n \in \mathbb{Z}\}$, and

(ii) $\int_{Q'} (\tilde{C}(\lambda))^{-1} Z_y(dy)$ exists as a D-S integral. [As before Q is the set of all zeros of the equation $\sum_{i=1}^{k} a_j e^{ij\lambda} = 0$, as in Prop. 3.3.2.] The filter is moreover physically realizable when the roots of the characteristic equation $p_k(z) = 0$ lie outside the unit circle.

As noted above, this is a direct refinement of the preceding result and will be left to the reader as an exercise. To use the MT-integration freely in our work, we present the following with strict integrals.

Theorem 3.3.4 *Let (Ω_i, Σ_i) be a measurable space, $\beta \colon \Sigma_1 \times \Sigma_2 \to \mathbb{C}$ be a bimeasure, $f_i \colon \Omega_i \to \mathbb{C}$ be Σ_i-measurable, $i = 1, 2$, with (f_1, f_2) as strictly β-integrable in that f_1, is $\beta(\cdot, F)$ and f_2 is $\beta(E, \cdot)$-integrable for $E \in \Sigma_1, F \in \Sigma_2$ where $_{f_1}\beta(E, B) = \int_E f_1(\omega_1)\beta(d\omega_1, B)$ and likewise $\beta_{f_2}(A, F) = \int_A f_2(\omega_2)\beta(A, d\omega_2)$ exist. Suppose now*

$$\int_E \int_F^* (f_1, f_2) d\beta = \int_{s_1} \int_{s_2} (\chi_E f_1, \chi_F f_2) d\beta$$

$$= \int_E f_1(\omega_1) \beta_{f_2}(d\omega_1, F),$$

and similarly with F, and both agree with value denoted as $\int_E \int_F^* (f_1, f_2)$ $d\beta$. Then for any bounded measurable $h \colon \Omega_1 \to \mathbb{C}$ and $k \colon \Omega_2 \to \mathbb{C}$, we *have the common value as:*

$$\int_A \int_B^* (fh, gk)d\beta = \int_A \int_B^* (h, k)d\mu, A \in \Sigma_1, B \in \Sigma_2. \tag{58}$$

This manipulation with the MT-integration is needed for our work here. The details from Morse-Transue are used and the standard but somewhat tedious discussions are referred to the paper by D. K. Chang and M. M. Rao (1983). We shall use this now in presenting the Cramér class along with the Karhunen formulation in the following account of the results of both the authors, leading to further analysis.

Proposition 3.3.5 *Let* $\{X_t, t \in \mathbb{R}\} \subset L_0^2(P)$ *be strongly harmonizable, whose spectral function* F_x *satisfies*

$$\int \int_{\mathbb{R}^2} e^{s\lambda + t\lambda'} |dF_x|(\lambda, \lambda') < \infty, \text{ for } s, t \in \mathbb{R}, \tag{59}$$

where $|dF_x|(\lambda, \lambda')$ *denotes* $d|F_x|(\lambda, \lambda')$, *the variation measure. Then for any bounded (distinct) infinite set* $\{t_i, i \geq 1\} \subset \mathbb{R}$, *of observation points, the sample* $\{X_{t_i}, i \geq 1\}$ *determines the process itself on the line* $\mathbb{R}$.

Proof. Since the covariance function $r(\cdot, \cdot)$ of the process has its spectral function F_x obeying (59), $r(\cdot, \cdot)$ is continuously differentiable on $\mathbb{R}^2$. If $\{t_i, i \geq 1\} \subset \mathbb{R}$ is a bounded infinite set it has a convergent subset having the limit t_0, so that the process is an analytic random function around t_0 (cf., e.g. Loève (1985), p. 471) and X_t can then be expressed as a mean-square convergent series:

$$X_t = \sum_{n=0}^{\infty} \frac{(t - t_0)^n}{n!} X_{t_0}^{(n)}, \tag{60}$$

$X_{t_0}^{(n)}$ being the nth derivative of X_t at $t = t_0$. Varying $t_0 \in \mathbb{R}$, and repeating the familiar argument one concludes that, using the analyticity of the process that results, and using the fact that $\mathbb{R}$ is connected, the process X_t given by (60) is analytic. Then by the classical principle of analytic continuation (cf. Hille and Phillips (1957), p. 98 and Thm. 3.14.1 there), we can conclude that $\{X_t, t \in \mathbb{R}\}$ is the only process that agrees with $\{X_{t_i}, i \geq 1\}$. Since the $\{t_i, i \geq 1\}$ is arbitrary in $\mathbb{R}$, we have the stated conclusion. $\square$

Let $\mathcal{B}(A)$ denote the Borel σ-algebra of $A \subset \mathbb{R}$. We then have:

Proposition 3.3.6 *Let* $\{X_t, t \in \mathbb{R}\} \subset L_0^2(P)$ *be weakly harmonizable with spectral function (or distribution) F so that its covariance is*

$$r(s,t) = \int_{\mathbb{R}} \int_{\mathbb{R}} e^{is\lambda - it\lambda'} F(d\lambda, d\lambda'), \text{ for } s, t \in \mathbb{R}, \tag{61}$$

and for each $\varepsilon > 0$, there is a bounded Borel set $A^\varepsilon \subset \mathbb{R}$, such that

$$\sup_{B \subset A^\varepsilon} F(B, B) < \varepsilon \text{ (bounded-Borel)}. \tag{62}$$

If σ_0 is defined as $\sigma_0 = \sup\{|x - y| : x, y \in A^\varepsilon\}$, the diameter of A^ε, then for $h > \sigma_0$,

$$E(|X_t - X_\varepsilon^t|^2) \leq \frac{C(t)}{n_t(t_1 - \sigma_0)} + 4\varepsilon, \tag{63}$$

and X_ε^t is given as:

$$X_\varepsilon^t = \sum_{k=-n}^{n} X\left(\frac{k\pi}{h}\right) \frac{\sin(th - \varepsilon\pi)}{(th - k\pi)}, \tag{64}$$

for some $0 < C(t) < \infty$, where $C(t)$ is bounded in bounded t-sets, and if the support S of F is bounded, $S \subset A^\varepsilon \times A^\varepsilon$, then (64) is automatically satisfied and $\varepsilon = 0$ can be taken in (63).

This result is known in the strongly harmonizable case and the proof extends to the present case with simple modifications which will be left to the reader. Here we include an adjunct to the general operator representation, given as Theorem 3.1.5, which is of interest in several special applications. They are sometimes called a *slowly changing class* which class is more general than stationary but is included in the (weak) harmonizable class. These were studied by R. Joyeux (1987) and extended by R. J. Swift (1997) much further.

If $\{X(t), t \in \mathbb{R}\}$ is weakly stationary and $A(t)$ is slowly varying nonstochastic function, then $Y(t) = A(t)X(t), t \in \mathbb{R}$, is termed a modulated process which is of interest in applications, when the $X(t)$-process is (weakly) stationary and more generally (weakly or strongly) harmonizable.

Definition 3.3.7 *Let* $X : \mathbb{R} \to L_0^2(P)$ *be a mapping and* $A(t, \cdot)$ *be a family of measurable scalar functions, that are Fourier transforms:*

$$A(t, \lambda) = \int_{\mathbb{R}} e^{itx} H(\lambda, dx), \lambda \in \mathbb{R} \tag{65}$$

and $r(s, t) = E(X_s \bar{X}_t)$ *admits a representation:*

$$r(s, t) = \int_{\mathbb{R}} \int_{\mathbb{R}} A(s, \lambda) \bar{A}(t, \lambda') e^{(i\lambda s - i\lambda' t)} dF(\lambda, \lambda'), \tag{66}$$

with $F(\cdot, \cdot)$ *as a function of finite Fréchet variation. Such a mapping* X *is termed an* oscillatory weakly harmonizable process.

This class is an abstraction of a class, termed "oscillatory" and some statistical applications are indicated by M. B. Priestley (1981), and followed up by R. Joyeux (1987). They are further extended to the above form by Swift (1997). We include an account to indicate some useful applications of the work.

This general class is of interest in applications where stationary and even harmonizable classes should be generalized, as seen in the discussion of Priestly's analysis, and the consequent extensions. The following general characterization, due to Swift (1997), is of importance in this study. It sharpens Theorem 3.1.5 in some sense:

Theorem 3.3.8 *The process* $\{X_t, t \in \mathbb{R}\} \subset L_0^2(\mathbb{R})$ *is oscillatory weakly harmonizable iff it is representable as*

$$X(t) = A(t)T(t)Y(0), t \in \mathbb{R}, \tag{67}$$

where $Y(0) \in \mathcal{H}(X) = \bar{sp}\{X_t, t \in \mathbb{R}\}$ *and* $A(t)$ *is a closed densely defined linear operator in* $\mathcal{H}(X), t \in \mathbb{R}$. *Here* $\{T(t), t \in \mathbb{R}\} \subset \mathcal{H}(X)$ *is a weakly continuous family of positive definite contractive maps commuting with* $\{A(t), t \in \mathbb{R}\}$.

Proof. In the forward direction, the oscillatory weak harmonizability gives

$$X(t) = \int_{\mathbb{R}} A(t, \lambda) e^{it\lambda} dZ(\lambda),$$

with $E(Z(B_1)\overline{Z(B_2)}) = F(B_1, B_2)$, giving $F(\cdot, \cdot)$ to have Fréchet variation bounded.

Also, the family $Y : t \mapsto \int_{\mathbb{R}} e^{it\lambda} dZ(\lambda)$ defines a weakly harmonizable process. But by Theorem 2.3.3 above, there is a weakly continuous contractive family $\{T(t), t \in \mathbb{R}\}$ of positive type on $\mathcal{H}(X)$, the Hilbert space determined by the $X(t)$-process, such that $Y(t) = T(t)Y_0$. Now by the spectral representation of the $T(t)$'s

$$T(t) = \int_{\mathbb{R}} e^{it\lambda} d\tilde{E}(\lambda), t \in \mathbb{R}, \tag{68}$$

and let $Z(A) = \tilde{E}(A)Y_0, A \in \mathcal{B}(\mathbb{R})$, so that $Z(\cdot)$ is well-defined, and let $a(t) = \int_{\mathbb{R}} A(t, \lambda)\tilde{E}(d\lambda), t \in \mathbb{R}$. Then $\{a(t), t \in \mathbb{R}\}$ is a *densely defined closed set of operators* on $\mathcal{H}(X)$ whose domain includes $\{Y(t), t \in \mathbb{R}\}$. But $T(t) \smile \tilde{E}(D), t \in D$, (commute) we get that $a(t)$ and $\tilde{E}(D)$ families also commute, so that $a(t) \smile T(s), s \in \mathbb{R}, t \in \mathbb{R}$. Consequently,

$$a(t)T(t)Y_0 = a(t) \left(\int_{\mathbb{R}} e^{iut} \tilde{E}(du)Y_0 \right)$$
$$= \int_{\mathbb{R}} \int_{\mathbb{R}} A(t, \lambda)\tilde{E}(d\lambda) \left(\int_{\mathbb{R}} e^{iwt} E(d\omega)Y_0 \right)$$
$$= \int_{\mathbb{R}} A(t, \lambda)e^{i\lambda t}\tilde{E}(d\lambda)Y_0 = \int_{\mathbb{R}} A(t, \lambda)e^{i\lambda t}Z(d\lambda) = X(t).$$

This gives (67) since $\tilde{E}$ and the integral commute. Thus if $X(t)$ is oscillatory weakly harmonizable, then $X(t) = a(t)T(t)Y(0)$ belongs to $\mathcal{H}(X) = \bar{sp}\{\tilde{X}(t), t \in \mathbb{R}\}, (Y_0 = Y(0))$, being a closed densely defined operator on $\mathcal{H}(X)$, and $T(s) \smile a(t), s, t \in \mathbb{R}$, the $T(s)$ family being a positive definite contractive set on $\mathcal{H}(X)$, admits an integral representation (von Neumann and F. Riesz) so that we have

$$X(t) = a(t)T(t)Y(0) = \int_{\mathbb{R}} g(t, \lambda)E(d\lambda) \int_{\mathbb{R}} e^{iwt} E(d\omega)Y(0)$$
$$= \int_{\mathbb{R}} e^{i\lambda t} g(t, \lambda)\tilde{E}(d\lambda)Y(0)$$
$$= \int_{\mathbb{R}} e^{i\lambda t} g(t, \lambda)dZ(\lambda). \tag{69}$$

This is just the representation of oscillatory weakly harmonizable process. $\square$

In this connection, it may be of interest to present simple conditions to sample the random field $\{X_t, t \in \mathbb{R}\}$ using the above ideas.

Theorem 3.3.9 *Let $\{X_t, t \in \mathbb{R}\}$ be a centered Cramér type process:*

$$X_t = \int_{\mathbb{R}} g(t, \lambda) dZ(\lambda), \qquad t \in \mathbb{R}, \tag{70}$$

$E(Z(A)) = 0, E(Z(A)Z(B)) = \int_A \int_B d^2\rho(x, y), A, B \in \mathcal{B}(\mathbb{R})$, where $\rho(\cdot, \cdot)$ *is a covariance function of bounded variation on bounded sets of* $\mathbb{R}^2$. *If* $L^2(X) \subset L^2(P)$ *and* $L^2(\rho)$ *the corresponding Hilbert space determined by* ρ, *let* $g^{(n)}(\cdot, \lambda)$ *exist and* $g^{(n)}(t, \cdot) \in L^2(\rho), n \geq 1$. *If for* $\{t_i, i \geq 1\} \subset \mathbb{R}$ *of nonperiodic points,* $g^{(n)}(t, \cdot)[\in L^2(\rho)]$ *are considered and if* $\mathcal{M} = \bar{sp}\{X(t_i), i \geq 1 \text{ distinct}\}$, *then* $L^2(\rho) = \mathcal{M}$.

Proof. The hypothesis implies that the covariance $r(\cdot, \cdot)$ is given by

$$r(s, t) = \int_{\mathbb{R}} \int_{\mathbb{R}} g(s, x)\overline{g(t, y)}d^2\rho(x, y)$$

and $g^{(n)}(t, \cdot) \in L^2(\rho), n = 0, 1, 2, \ldots$. This implies $\frac{\partial^{m+n} r}{\partial s^n t^m}$ exists, and then the covariance $r(\cdot, \cdot)$ is infinitely differentiable and so also is $X(\cdot)$.

Let $\tilde{L}^2(\rho) \subset L^2(\rho)$ be the subspace corresponding to $L^2(X)$ by the resulting isometry. Since $\{g(t, \cdot), t \in \mathbb{R}\}$ determines $L^2(\rho)$, it follows from the Class C structure, that $L^2(\rho) = \tilde{L}^2(\rho)$. Since $g(\cdot, \cdot)$ is infinitely differentiable, we easily get

$$\frac{\partial^n r(s, t)}{\partial s^n}\bigg|_{s=t_0} = \frac{\partial}{\partial s}\left(\int_{\mathbb{R}} \int_{\mathbb{R}} g^{(n-1)}(s, x)\bar{g}(t, y) \, d^2\rho(x, y)\right)\bigg|_{s=t_0}$$

and $g^{(n)}(t_0, \cdot) \in \mathcal{N}$, the closed subspace of $L^2(\rho)$ determined by $\{g(t_i, \cdot)\}$. From this, it is easily seen that $g^{(n)}(t_i, \cdot) \in \mathcal{M}$. But we also have

$$g(t, x) = \sum_{n=0}^{\infty} \frac{(t - t_0)^n}{n!} g^{(n)}(t_0, x). \tag{71}$$

This gives from the fact that $X(t) \leftrightarrow g(t, \cdot)$, as is known, we get

$$X_t = \sum_{n=0}^{\infty} \frac{(t - t_0)^n}{n!} X^{(n)}(t_0), \tag{72}$$

in the sense of mean-squared convergence. The isometry here leading $X_t \leftrightarrow g(t, \cdot)$ implies that $g(t, \cdot) \in \mathcal{M}$ and $L^2(\rho) = \mathcal{M}$, as desired. $\square$

As a consequence, we get the statement for the harmonizable case:

Corollary 3.3.10 *Let $\{X_t, -\infty < t < \infty\}$ be a weakly harmonizable process with ρ as its spectral measure. Suppose its moment generating function exists, then for any sample $\{X(t_i), i \geq 1\}$ as in the theorem above, the set $\{X(t_i), i \geq 1\}$ determines $L^2(\rho)$.*

The hypothesis implies that the process must be strongly harmonizable, and $\rho(\cdot)$ has all moments. The result is then a simple consequence of the theorem.

We next consider some other conditions that generalize the strongly harmonizable (hence weakly stationary) classes, of some interest in applications.

3.4 Class (KF) and Nonstationary Processes Applications

It was pointed out by H. Niemi that there exist weakly harmonizable processes that are not of class (KF), although all strongly harmonizable ones are in it. This is of interest here since the class (KF) is based on summability methods, not completely dependent on Fourier analysis. The positive result is as follows:

Theorem 3.4.1 *If $X \colon \mathbb{R} \to L_0^2(P)$ is strongly harmonizable, then it is in class (KF) so that it has an associated spectral function.*

Proof. Since X is (strongly) harmonizable there exist a stochastic measure $Z(\cdot)$ and a bounded bimeasure F, such that

$$X(t) = \int_{\mathbb{R}} e^{it\lambda}\, dZ(\lambda); \quad F(A, B) = \big(Z(A), Z(B)\big), \qquad (73)$$

with $Z \colon \mathcal{B}(\mathbb{R}) \to L^2(P)$, so $F \colon \mathcal{B}(\mathbb{R}) \times \mathcal{B}(\mathbb{R}) \to \mathbb{C}$, and $K \colon (s, t) \to E(X(s)\bar{X}(t))$.

Then we get for $T > 0$

$$r_T(h) = \frac{T - h}{T} \cdot \frac{1}{T - h} \int_0^{T-h} K(s, s + h)\, ds \qquad (74)$$

and we assert that $\lim_{T \to 0} r_T(h)$ exists. For this consider (74) as

$$\frac{1}{T} \int_0^T K(s, s + h)\, ds = E\left(\frac{1}{T} \int_0^T ds \int_{\mathbb{R}} e^{is\lambda} Z(d\lambda) \int_{\mathbb{R}} e^{-i(s+h)\lambda'} Z(d\lambda') \right),$$

$$(75)$$

and we note that the right side has a limit as $T \to \infty$. For, let $\mathcal{X} = \mathcal{Y} = L_0^2(P)$ and $\mathcal{Z} = L^1(P)$. Now $Z: \mathcal{B} \to \mathcal{X}$, $\tilde{Z} = Z: \mathcal{B} \to \mathcal{Y}$ are stochastic measures, we can define a product measure $Z \otimes \tilde{Z}$ on $\mathcal{B} \times \mathcal{B} \to \mathcal{X}$ as a Dunford-Schwartz integral which satisfies

$$
\int_{\mathbb{R} \times \mathbb{R}} f(s,t)(Z \otimes \tilde{Z})(ds, dt) = \int_{\mathbb{R}} Z(ds) \int_{\mathbb{R}} f(s,t)\tilde{Z}(dt)
$$

$$
= \int_{\mathbb{R}} \tilde{Z}(dt) \int_{\mathbb{R}} f(s,t) Z(ds) \qquad (76)
$$

for $f \in C_0(\mathbb{R} \times \mathbb{R})$, (cf. N. Dinculeanu (1974) and M. Duchoň and I. Kluvánek (1967), for such product measures with finite semi-variation). Here

$$
\|Z \otimes \tilde{Z}\|(\mathbb{R} \times \mathbb{R}) \le \left(\|Z\|(\mathbb{R})\right)^2 < \infty
$$

implying that $Z \otimes \tilde{Z}$ is again a stochastic measure. So (76) gives

$$
\int_{\mathbb{R}} e^{is\lambda} Z(d\lambda) \int_{\mathbb{R}} e^{-i(s+h)\lambda'} Z(d\lambda') = \int_{\mathbb{R} \times \mathbb{R}} e^{is(\lambda-\lambda')-ih\lambda'} Z \otimes Z(d\lambda, d\lambda')
$$
$$(77)$$

and the right side belongs to $L^1(P)$. Using the same kind of argument one gets with $\mu: \mathcal{B}((0,T)) \to \mathbb{R}^+$ as Lebesgue measure,

$$
\int_0^T \mu(dt) \int_{\mathbb{R} \times \mathbb{R}} f(t,\underline{\lambda}) Z \otimes Z(d\underline{\lambda}) = \int_{\mathbb{R} \times \mathbb{R}} Z \otimes Z(d\underline{\lambda}) \int_0^T f(t,\lambda)\mu(dt).
$$
$$(78)$$

Writing $\mu(dt)$ as dt, (76)–(78) yield:

$$
E\left(\frac{1}{T} \int_0^T ds \int_{\mathbb{R} \times \mathbb{R}} e^{is(\lambda-\lambda')-ik\lambda'} Z \otimes Z(d\lambda, d\lambda') \right)
$$

$$
= E\left(\int_{\mathbb{R} \times \mathbb{R}} e^{-ih\lambda'} \left[\frac{e^{iT(\lambda-\lambda')}-1}{T(\lambda-\lambda')} \chi_{[\lambda \ne \lambda']} + \delta_{\lambda\lambda'} \right] Z \otimes Z(d\lambda, d\lambda') \right).
$$
$$(79)$$

But the quantity inside $E(\cdot)$ is bounded for all T, and the dominated convergence for these vector integrals holds, so that we can let $T \to \infty$ and simplify (79) to get

$$
\lim_{T \to \infty} \frac{1}{T} \int_0^T K(s, s+h)\, ds = E\left(\int_{\mathbb{R}} e^{-ih\lambda'} \delta_{\lambda\lambda'} Z \otimes Z(d\lambda, d\lambda') \right)
$$

$$
= \int_{[\lambda=\lambda']} e^{-ih\lambda} F(d\lambda, d\lambda'),
$$

where F is the bimeasure of Z. Hence $\lim_{T\to\infty} r_T(h) = r(h)$ exists and $r(h) = \int_{\mathbb{R}} e^{-iht} G(d\lambda)$, $G\colon A \mapsto \int_{\pi^{-1}(A)} \delta_{\lambda\lambda'} F(d\lambda, d\lambda')$, $A \in \mathcal{B}$, is a positive definite measure of X in class (KF), as asserted. $\square$

The preceding result implies the following (major) result (an inversion theorem) obtained by Rozanov ((1959), Thm. 3.2) differently. The first part is a type of [Fourier] inversion formula.

Theorem 3.4.2 *Let $X\colon \mathbb{R} \to L_0^2(P)$ be a weakly harmonizable process with $Z\colon \mathcal{B} \to L_0^2(P)$ as its representing stochastic measure. Then for any $-\infty < \lambda_1 < \lambda_2 < \infty$, with $Z(\lambda)$ for $Z((-\infty, \lambda))$, we have*

$$\underset{T\to\infty}{l.i.m.} \int_{-T}^{T} \frac{e^{-it\lambda_2} - e^{-it\lambda_1}}{-it} X(t) dt$$

$$= \frac{Z(\lambda_2+) + Z(\lambda_2-)}{2} - \frac{Z(\lambda_1+) + Z(\lambda_1-)}{2} \quad (80)$$

where l.i.m. is the $L^2(P)$-limit. Also, the covariance bimeasure F of Z is given for $A = (\lambda_1, \lambda_2)$, $B = (\lambda_1', \lambda_2')$ as:

$$\lim_{0 \leq T_1, T_2 \to \infty} \int_{-T_1}^{T_1} \int_{-T_2}^{T_2} \frac{e^{-i\lambda_2 s} - e^{-i\lambda_1 s}}{-is} \cdot \frac{e^{-i\lambda_2' t} - e^{-i\lambda_1' t}}{-it} r(s,t) ds\, dt$$

$$= F(A, B) \quad (81)$$

where A, B are continuity intervals of F, and $r(\cdot, \cdot)$ is the covariance function of the X-process. In particular, if $r(\cdot, \cdot)$ vanishes at $(\pm\infty, \pm\infty)$, and $S\colon \mathbb{R} \to \mathbb{C}$ is continuous such that $\frac{1}{T} \int_0^T S(t) dt \to a_0$, then the signal '$a_0$', can be estimated as $\hat{S}_T = \frac{1}{T} \int_0^T Y(t)\, dt \to a_0$ as $T \to \infty$ in that $E(|\hat{S}_T - a_0|^2) \to 0$ as $T \to \infty$, with $Y(t) = X(t) + S(t)$.

Sketch of Proof. We reduce this result to the stationary case by using the dilation theorem. Thus there is an enlarged probability space $(\tilde{\Omega}, \tilde{\Sigma}, \tilde{P})$ of (Ω, Σ, P) with $L_0^2(\tilde{P}) \supset L_0^2(P)$, and a stationary $Y\colon \mathbb{R} \to L_0^2(\tilde{P})$, such that $X(t) = QY(t), t \in \mathbb{R}$ where $Q\colon L_0^2(\tilde{P}) \to L_0^2(P)$, the orthogonal projection, and there is an orthogonally-valued measure $\tilde{Z}\colon \mathcal{B} \to L_0^2(\tilde{P})$ such that

$$Y(t) = \int_{\mathbb{R}} e^{it\lambda} \tilde{Z}(d\lambda), \quad t \in \mathbb{R} \quad (82)$$

and $Z(A) = Q\tilde{Z}(A), A \in \mathcal{B}$, with $Z\colon \mathcal{B} \to L_0^2(P)$ representing the X-process. Since Q is bounded, it commutes with l.i.m. as well as the integral. Then the standard familiar argument shows that, applying Q to both sides of (80), the result follows as stated.

Regarding (81), consider the left side with (80), and we get

$$
\begin{aligned}
\text{L.H.S.} &= \lim_{T_1,T_2 \to \infty} E\left[\left(\int_{-T_1}^{T_2} \frac{e^{-is\lambda_2} - e^{-is\lambda_1}}{-is} X(s)\, ds\right)\right.\\
&\qquad\qquad \left.\cdot \left(\int_{-T_1}^{T_2} \frac{e^{-it\lambda_2} - e^{-it\lambda_1}}{-it} X(t)\, dt\right)\right]\\
&= E\left[\left(\left(\frac{Z(\lambda_2+) + Z(\lambda_2-)}{2}\right) - \left(\frac{Z(\lambda_1+) + Z(\lambda_1-)}{2}\right)\right)\right.\\
&\qquad\quad \left.\cdot \left(\left(\frac{Z(\lambda_2+) + Z(\lambda_2-)}{2}\right) - \left(\frac{Z(\lambda_1+) + Z(\lambda_1-)}{2}\right)\right)\right]\\
&= F(A, B)
\end{aligned}
\tag{83}
$$

by the continuity hypothesis on F, after expanding and simplifying.

Finally, let $a_T = E(\hat{S}_T) = \frac{1}{T}\int_0^T S(t)\, dt$, with $\tilde{Y}(t) = X(t) + S(t), t \in \mathbb{R}$. Since now $\tilde{Y} \in \text{class}\,(KF)$ and $a_T \to a_0$ as $T \to \infty$,

$$
\begin{aligned}
E(|\hat{S}_T - a_0|^2) &= \frac{2}{T}\int_0^T\int_0^T r(s,t)\, ds\, dt + 2|a_T - a_0|^2\\
&= \frac{1}{2T}\int_{-T}^T r_T(h)\, dh + 2|a_T - a_0|^2,
\end{aligned}
$$

where $r_T(h) = \frac{1}{T}\int_0^{T-|h|} K(s, s+|h|)\, ds \to r(h)$ as $T \to \infty$ since $\tilde{Y} \in$ class (KF). Since X_t is V-bounded, one can invoke the known Cesaro-summability, to conclude that $r(h) = 0, h \in \mathbb{R}$, (and tends uniformly on compact subsets of $\mathbb{R}$). It follows finally that $E(|\hat{S}_T - a_0|^2) \to 0$ if $T \to \infty$, as asserted. $\square$

It is also of interest here to sketch the vector (or multivariate) versions of the preceding analysis. Thus let $L_0^2(P, \mathbb{C}^k)$ be the space of k-vector complex measurable functions (r.v.'s for us) on (Ω, Σ, P), centered and square integrable, i.e., $E(f) = 0$, and $\|f\|^2 = (f, f)$, for $f = (f_1, f_2, \ldots, f_k) \in L_0^2(P, \mathbb{C}^k)$, where

$$
(f, g) = \int_\Omega ((f(\omega), g(\omega))\, dP(\omega) = \sum_{i=1}^k \int_\Omega f_i(\omega)\bar{g}_i(\omega)\, dP(\omega), \tag{84}
$$

and let $\mathfrak{X} = L_0^2(P, \mathbb{C}^k)$, which becomes a Hilbert space with (84) as its inner product. Then $X: G \to \mathfrak{X} \left(= L^2(P, \mathbb{C}) \right)$ is *weakly* or *strongly* harmonizable (vector) random field if for each $a = (a_1, \ldots, a_k), a_i \in \mathbb{C}$, the scalar process $Y_a = a \cdot X = \sum_{i=1}^k a_i X_i: G \to L_0^2(P)$ is weakly or strongly harmonizable (scalar) random field, $a \in \mathbb{C}^k$. In an analogous way, the vector Karhunen or class(C) as well as stationary random fields are defined from the scalar concepts. Letting $R(g, h)$ be the covariance matrix of the vector process $X: G \to L^2(P, \mathbb{C}^n)$ we can easily obtain the multivariate version of the random field as follows.

Theorem 3.4.3 *If G is an LCA group and $X: G \to \mathfrak{X} = L_0^2(P, \mathbb{C}^k)$ is a weakly continuous centered random field, then it is weakly harmonizable iff there exists a vector stochastic measure on $\hat{G}$ with $\mathfrak{X}$, denoted $\tilde{Z} = (Z_1, \ldots, Z_k): \mathcal{B}(\hat{G}) \to L_0^2(P, \mathbb{C}^k)$, such that*

$$X(g) = \int_{\hat{G}} \langle g, s \rangle \tilde{Z}(ds), \quad g \in G, \qquad (85)$$

$\hat{G}$ being the dual group of G. Further X is strongly harmonizable if also the matrix $F = (F_{ij}), i, j = 1, \ldots, k$ with $F(A, B) = ((Z_j(A), Z_k(B)), i, j = 1, \ldots, k$ is a matrix function of bounded variation on $\hat{G}$, the dual of G.

The covariance matrix $R = (Z, Z) = (Z_i(\cdot), Z_j(\cdot))$ is equivalently given (or representable) as:

$$R(g, h) = \int_{\hat{G}} \int_{\hat{G}} \langle g, s \rangle \overline{\langle h, t \rangle} F(ds, dt), \quad g, h \in G, \qquad (86)$$

where the integral in (86) is the MT-integral, or the LS-integral according as the random field X is weakly or strongly harmonizable. Here F is a positive definite matrix function of bounded variations on $\hat{G}$ according as the MT or Lebesgue-Stieltjes concepts apply.

Conversely, if $R(\cdot, \cdot)$ is a positive definite matrix that is representable as (86), then it is the covariance function of an n-variate harmonizable random field.

Outline of proof. It is enough to sketch the ideas of proof and the full details can be easily filled in from it by the interested reader.

Let $Y_a = a \cdot X$, for $a \in \mathbb{C}^k$, as a scalar product. Now Y_a is weakly harmonizable when X is, and thus is representable as

$$Y_a(g) = \int_{\hat{G}} \langle g, s \rangle Z_a(ds), \quad g \in G \tag{87}$$

relative to a stochastic measure Z_a on the dual $\hat{G} \to \mathcal{H}$, and that $Z_{(\cdot)}(A) \colon \mathbb{C}^k \to \mathcal{H}$ is linear and continuous. By reflexivity of $\mathcal{H}$, $Z_a(A) = a.\tilde{Z}(A)$ with the 'dot product' notation and that $\tilde{Z} \colon \mathcal{B}(\hat{G}) \to \mathcal{H}$ defines a stochastic measure. Then

$$Y_a(g) = a \cdot X(g) = \int_{\hat{G}} \langle g, s \rangle a \cdot \tilde{Z}(ds), = a \cdot \int_{\hat{G}} \langle g, s \rangle Z(ds), \tag{88}$$

the last integral being an element of $\mathcal{X}$. Since the a is an arbitrary vector, this implies (85).

If X is strongly harmonizable, then Y_a has the same property, and the covariance bimeasure $F_a = a \cdot F$ where

$$F(A, B) = (Z_i(A), Z_j(B)), \text{ where } i, j = 1, \ldots, k$$

and the components of $F(= F_{ij})$ are of bounded variation. The result follows from the one-dimensional case. Finally the same holds for the weakly harmonizable case, using the MT-integrals, and all the statements hold with this change. $\square$

The following result of a Karhunen process, is an easy extension of the above work which is considered already for a useful special case in Gikhman and Skorokhod (1974). We include it here for applications as well as the completeness of the analysis.

Proposition 3.4.4 *Let $X \colon S \to L_c^2(P)$ be a Karhunen process on a locally compact space S relative to a regular or Radon measure F on S with functions $\{g_t, t \in S\} \subset L^2(F)$, based on a Radon measure space $(S, \mathcal{B}, F)$. Then there exists a locally bounded regular or Radon stochastic measure $Z \colon \mathcal{B} \to L_c^2(P)$ where $\mathcal{B}_c \subset \mathcal{B}$ is the δ-ring of bounded (Borel) sets such that (Z is orthogonally valued)*

$$(i) \quad E(Z(A)\bar{Z}(B)) = F(A \cap B), \quad A, B \in \mathcal{B}_c \tag{89}$$

and with it one has the D-S integral representations:

$$(ii) \quad X(t) = \int_S g(\lambda) Z(d\lambda), \quad t \in S. \tag{90}$$

Conversely, if $X \colon S \to L_c^2(P)$ is given by (90) for a $Z(\cdot)$ of the above type, with g and F satisfying (89) and (90), then $X(t), t \in S$ is of

Karhunen class. Also the linear hulls $\mathcal{H}_X = \mathcal{H}_Z$ *iff the* $g_s \in L^2(F)$ *form a dense set.*

Proof Sketch.   This is an extension of the earlier one with the D-S integration given for bounded operators, but is extended (and shown to be valid) for vector measures $Z(\cdot)$ of locally finite type (by Thomas (1970)). Thus for $T \in B(L_0^2(P))$

$$TX(t) = \int_S g(\lambda)(T \circ Z)(d\lambda)$$

$\tilde{Z} = T \circ Z$ is seen to be a locally finite stochastic measure. It is then inferred with our earlier discussion that TX is weakly of class (C). The opposite direction is, however, more restricted.

Conversely, let $\{X(t), t \in S\}$ be weakly of class (C) and the 'accompanying' g-functions are bounded Borel and $M(S)$ is the uniformly closed algebra based on $\{g_s, s \in S\}$, then each $g_s \in L^2(F_X)$, and if $Tg_s = X(s) = \int_S g_s(\lambda)Z(d\lambda)$, we extend T linearly to the closed algebra determined by $\{g_t, t \in S\}$, then $M(S) \subset L^2(F_x)$, and $Tg_t = X(t) = \int_S g(\lambda)Z(d\lambda)$; we extend T linearly to $M(S)$, so $T \in B(M(S), \mathcal{H})$, with sup norm for $M(S)$.

$$\|Tf\| \leq \|f\|_{2,\mu}, \quad f \in M(S),$$

for a finite measure μ on S. Then one can follow the dilation theorem's argument and can complete the proof. $\square$

The preceding analysis can be summarized in the following simpler form for reference and extensions.

Theorem 3.4.5 *Let S be locally compact and $X: S \to L_0^2(P)$ be a Karhunen field relative to a Radon measure $F: \mathcal{B}(S) \to \mathbb{R}^*$ and a class $\{g_s(\cdot), s \in S\} \subset L^2(S, \mathcal{B}(S), F)$, the space of scalar square integrable functions (for F) on $(S, \mathcal{B}(S))$. Then there is a regular (or Radon) stochastic measure $Z: \mathcal{B}_0 \to L_0^2(P)$, with $\mathcal{B}_0 \subset \mathcal{B}$, the δ-ring of bounded sets such that*

(i) $E(Z(A)Z(B)) = F(A \cap B)$, $A, B \in \mathcal{B}_0$, *and*
(ii)

$$X(t) = \int_S g_t(\lambda)Z(d\lambda), \quad t \in S, \tag{91}$$

where the integral in (91) is the standard (or DS) vector integral, and $\{g_t, t \in S\} \subset L^2(F)$.

In the other direction, if $Q\colon \mathcal{H} \to \mathcal{H}$ is a bounded projection operator, then $\tilde{X}(t) = QX(t), t \in S$, is weakly of class (C), when the $X(t)$-process is of Karhunen class.

Since the $g_t(\cdot)$ function is not necessarily $g_t(\lambda) = e^{it\lambda}$, for $\lambda \in \mathbb{R}$, the classes of Hilbert spaces containing the Cramér processes and the Karhunen classes associated with the processes do not necessarily have the familiar inclusions as in the stationary and harmonizable families, and the Sz.-Nagy-Naĭmark type simple inclusions are not always valid. The Fourier characters $\{e^{it(\cdot)}\}$ and the general $\{g(t, \cdot)\}$-classes vary differently and in the generalization, the "charm" of harmonic classes is lost! This distinguishes the harmonizable families (weak or strong) from the Cramér class. These differences make the study of second order processes interesting and this will be illustrated below.

3.5 Further Classifications and Representations of Second Order Processes

In this section, we present some useful integral representations of not necessarily stationary or even harmonizable processes that extend our analysis of classes already treated above. Recall that a centered second order Khintchine (or K-) stationary process $\{X_t, t \in \mathbb{R}\} \subset L_0^2(P)$, admits the representation for its covariance $r\colon (s,t) \mapsto E(X_s \bar{X}_t)$ as $\int_{\mathbb{R}} e^{i(s-t)\lambda} \, dZ(\lambda)$, for an orthogonally valued measure $Z\colon \mathcal{B}(\mathbb{R}) \to L_0^2(P)$, so that $(Z(A), Z(B)) = \alpha(A \cap B), A, B \in \mathcal{B}(\mathbb{R})$, and $r(s,t) = \tilde{r}(s-t)$.

Suppose, more generally, $V_s\colon X_t \to X_{s+t}, s, t \geq 0$, and for some linear combination $V_s(\sum_{j=1}^n a_j X_{t_j}) = \sum_{j=1}^n a_j X_{s+t_j}$ holds. If this is to be valid for each linear combination $(a_i \in \mathbb{C})$ we must have (i) $\sum_{j=1}^n a_j X_{t_j} = 0 \implies \sum_{j=0}^n a_j X_{t_j+s} = 0, n \geq 2$, and for the operators $\{V_s, s \geq 0\}$ to be bounded on $L_0^2(P)$, one must have

$$\|V_s(\textstyle\sum_{j=1}^n a_j X_{t_j})\|_2 \leq C_s \|\textstyle\sum_{j=0}^n a_j X_{t_j}\|_2, \quad n \geq 2, \quad C_s > 0,$$

so that we also will find $V_{s_1} V_{s_2} = V_{s_1+s_2}, V_0 = I$, and that $\{V_s, s \geq 0\}$ forms a semi-group with $V_0 = I$, the identity. But for our analysis here, integral representations of these operators is desired. These are called "subnormal" operators on an $L_0^2(P)$. That such an extension is possible on enlarging the basic probability space so that $L_0^2(P)$ can be assumed to carry these processes was established by Bram (1955). Using this, we have

$$X_t = V_t X_0 = \iint_\Delta e^{it\lambda} E(d\lambda) X_0 = \iint_\Delta e^{it\lambda} dZ_{X_0}(\lambda), \qquad (92)$$

for a Borel set $\Delta \subset \mathbb{R}^2$ and $Z_{X_0} : \mathcal{B}(\Delta) \to L_0^2(P)$ is orthogonally valued (vector) measure. Then the covariance $r(\cdot, \cdot)$ is given by

$$r(s, t) = \int\!\!\int_\Delta e^{i(s\lambda + t\bar{\lambda}')} d\beta(\lambda, \lambda'), \qquad (93)$$

with $\beta(A) = E(|Z(_{X_0}(A))|^2)$, (93) being analogous to Khintchine's form. However, the general operator theory of semi-groups allows us to express $V_t = R_t U_t = U_t R_t$ where the $\{U_t, t \in \mathbb{R}\}$ is a unitary group with $U_{-t} = U_t^*, t > 0$ and $\{R_t, t \geq 0\}$ is a positive self-adjoint semi-group. The R_t, U_t families commuting with all Borel functions of these operators. Thus

$$V_{t+s} = V_t V_s = R_t U_t R_s U_s = R_{s+t} U_{s+t}, \qquad (94)$$

both the R_t and U_t families are strongly continuous semi-groups (the U_t being a group) of operators in $L_0(P)$ and we then have

$$\begin{aligned}
r(s, t) = (X_s, X_t) &= (R_s U_s X_0, R_t U_t X_0) \\
&= (U_s X_0, R_s R_t U_t X_0) \\
&= (U_t^* U_s X_0, R_{s+t} X_0) \\
&= \overline{(R_{s+t} X_0, U_{s-t} X_0)} \\
&= E(\bar{X}_{s-t} X_{s+t}) = \tilde{r}(s - t, s + t) \qquad (95)
\end{aligned}$$

and r is positive definite on $\bar{R}^+$; so $\tilde{r}$ is now defined on the cone $\{(s, t) \in \mathbb{R} \times \mathbb{R} : |t| \leq s\}$. Also $r(\cdot, \cdot)$ and $\tilde{r}(\cdot, \cdot)$, are related by

$$\tilde{r}(s, t) = r\left(\frac{s + t}{2}, \frac{s - t}{2}\right), \qquad s > t > 0. \qquad (96)$$

A characterization of $\tilde{r}$ needs a nontrivial extension of the classical Bochner's result. This was obtained by A. Devinatz (1954). Thus if $C = \{(s, t) \in \mathbb{R}^+ \times \mathbb{R} : |t| < s/2\}$, then we have the following:

$$\sum_{i,j=1}^{n} a_i \bar{a}_j \tilde{r}(s_i + s_j, t_i - t_j) = \sum_{i,j=1}^{n} a_i \bar{a}_j (R_{s_i + t_j} X_0, U_{s_i - t_j} X_0)$$

$$= \sum_{i,j=1}^{n} a_i \bar{a}_j (R_{s_i} U_{s_i} X_0, R_{t_j} U_{t_j} X_0)$$

since the $V_t = R_t U_t$ commute,

$$= \sum_{i=1}^{n} (a_i R_{s_i} U_{s_i} X_0, a_j R_{s_j} U_{s_i} X_0) \geq 0.$$

$$(97)$$

The nontrivial converse implication is given by Devinatz' result.
    With these properties and that of his key theorem, we have

$$X_t = V_t X_0 = R_t U_t X_0 = \int_{\mathbb{R}^+} \lambda^t dE_{1\lambda} \int_{\mathbb{R}} e^{it\lambda'} dE_{\lambda'} X_0, \ (R_t = R_1^t)$$

$$= \int_{[\mathrm{Re}\, \lambda > 0]} e^{t(\log \lambda + i\lambda')} d\tilde{E}_{\lambda\lambda'} X_0. \ (\tilde{E}_{\lambda\lambda'} = E_{1\lambda} E_{\lambda'}). \quad (98)$$

The commutativity of $E_{1\lambda}$ and $E_{\lambda'}$ is used. We may summarize the above work in the following (general) result:

Theorem 3.5.1 *Let* $\{X_t, t \geq 0\} \subset L_0^2(P)$ *be a weakly continuous process admitting a right translation in that* $\tau_s \colon X_t \to X_{s+t}$ *is a bounded linear subnormal mapping on* $\overline{sp}\{X_t, t \geq 0\} \subset L_0^2(P)$, *and it has an extension to be normal on* $L_0^2(P)$, *possibly enlarging the measure space by adjunction if necessary, so that one has*

$$r(s,t) = (X_s, X_t) = \int_S g(s,\lambda) \overline{g(t,\lambda)} d\alpha(\lambda), \quad s, t \in T. \quad (99)$$

Then there exists a family of commuting set of bounded operators $\{B_t, t \in T\}$ *on* $L_0^2(P)$ *such that for a fixed* $t_0 \in T$,

$$X_t = B_t X_{t_0} = \int_S g(t,\lambda) dZ_{X_{t_0}}(\lambda), \quad t \in T, \quad (100)$$

holds for a unique orthogonal random measure Z_{X_0} *on* S *into* $L_0^2(P)$ *and the integral* (100) *is in the Dunford-Schwartz sense.*

These considerations, utilizing somewhat general but popular and interesting classes of operators from the Hilbert space theory, contribute several new avenues of analyzing problems originating from stationarity. Here is another example. Let $\{B_t, t > 0\}$ be a class of linear weakly continuous semi-group of contractions on $L_0^2(P)$ and let $X_t = B_t X_0$ for an $X_0 \in L_0^2(P)$ so that $r(s,t) = E(X_s \bar{X}_t)$, $X_t \in L_0^2(P)$. The process $\{X_t, t \geq 0\} \subset L_0^2(P)$ may not admit a shift. It is called *conservative* if $r(s+t, t+h) = r(s,t)$ for $s, t > 0, h > 0$ and is *dissipative* if for each finite set $t_1, \ldots, t_n$ and $h > 0$, the process $Y_n(h) = \sum_{i=1}^n a_i X_{t_i+h}$ has a decreasing variance function in h, i.e.,

$$\sigma_n^2(s) = E(|Y_n(s)|^2) \geq \sigma_n^2(s+h), \quad n > 0, s > 0, h > 0.$$

The existence of such processes can be obtained from the Kolmogorov (Bochner) projective limit theorem. The following property is interesting. (It is given for information and completeness.)

Proposition 3.5.2 *Let* $\{X_t = B_t X_0, t \geq 0\}$ *be a dissipative* $L_0^2(P)$-*valued weakly continuous process. Then it has a stationary dilation in a super* $L_0^2(P) \supset \mathcal{L} = \overline{sp}\{X_t, t > 0\}$, *and the* X_t *has the integral representation*

$$X_t = \int_{\mathbb{R}^+} e^{it\lambda} dZ(\lambda), \quad t \geq 0, \tag{101}$$

where $Z(\cdot)$ *is a not necessarily orthogonally valued measure with the vector integral* (101) *in the Dunford-Schwartz sense. Thus a dissipative process is weakly, not necessarily strongly, harmonizable, but may still be taken (enlarging* (Ω, T, P)) *as* $\mathcal{L} \subset L_0^2(P)$.

The following key result in another direction is due to Pitcher (1963), slightly extended, and gives the structure of the set of admissible means of a class of second order processes including Gaussians.

Theorem 3.5.3 *Let* $X = \{X_t, t \in T = [a,b]\} \subset L_0^2(P)$ *be a centered second order process with a continuous covariance function* r. *If* $\{\phi_n, n \geq 1\}$ *and* $\{\lambda_n, n \geq 1\}$ *are the eigenfunctions and eigenvalues of the kernel* r, *let* $\{X_n, n \geq 1\}$ *and* $\{f_n, n \geq 1\}$ *be defined by:*

$$X_n = \lambda_n^{-\frac{1}{2}} \int_T X_t \phi_n(t) \, dt$$

$$f_n = \lambda_n^{-\frac{1}{2}} \int_T f(t) \phi_n(t) \, dt, \quad f \in M_P = M_p(X) \tag{102}$$

so that $\{X_n, n \geq 1\} \subset L^2(P)$ *are orthonormal and* $f = \{f_n, n \geq 1\} \subset \ell^2$. *Let* $\{P_n, n \geq 1\}$ *be the* n-*dimensional distributions of*

$\{X_n, n \geq 1\}$ *with densities* $\{p_n, n \geq 1\}$ *relative to the Lebesgue measure in* $\mathbb{R}^n, n \geq 1$. *Suppose that (i)* $p_n > 0$ *a.e. (Leb),* $n \geq 1$, *(ii)* $\lim_{|t_j| \to \infty} p_n(t_1, \ldots, t_n) = 0$, $1 \leq j \leq n$ *for almost all* $t_i, i \neq j$ *(iii)* $\frac{\partial p_i}{\partial t_i}$, $1 \leq i \leq n$ *exists for* $n \geq 1$, *and (iv) there exists* $K_n < \infty$ *such that for* $n \geq 1$, *we have:*

$$\sum_{j=1}^{n} \int_{\mathbb{R}^n} \left(\frac{\partial \log p_n}{\partial t} \right)^2 dP_n \leq K_n < \infty. \tag{103}$$

Then the set $M_1(X) = \{ f \in M(X) : f = \{f_n, n \geq 1\} \in \ell^2 \}$ *is a positive cone. Further, if* $\sum_{n \geq 1} \lambda_n^{-\frac{1}{2}} < \infty$, *then* $M_1(X) = M(X)$. *If also the* P_n *are symmetric about the origin of* $\mathbb{R}^n, n \geq 1$, *then* $M_1(X)$ *becomes a linear set.*

Remark. The result (conclusion here) fails if (103) is replaced by uniform boundedness of the terms. Here we just consider nonstochastic signals of the process, to give a feeling for the subject.

Proof. First, observe that $X(t)$ has the $L^2(P)$-convergent (Karhunen) representation

$$X(t) = \sum_{n \geq 1} X_n \frac{\phi_n(t)}{\sqrt{\lambda_n}}, \quad t \in T \tag{104}$$

and by Mercer's theorem, $f = (f_1, f_2, \ldots) \in \ell^2$ since

$$\sum_{n=1}^{\infty} |f_n|^2 = \int_T \int_T f(t) \overline{f(s)} r(s, t) \, ds \, dt < \infty.$$

Let $f = (f_1, f_2, \ldots) \in \ell^1 \subset \ell^2$ and for $a \geq 0$, set

$$aY_n^2(\omega) = p_n \big(X_1(\omega) - af_1, \ldots, X_n(\omega) - af_n \big) / p_n \big(X_1(\omega), \ldots, X_n(\omega) \big).$$

Then $\{aY_n, \mathcal{F}_n, n \geq 1\}$, $\mathcal{F}_n = \sigma(X_1, \ldots, X_n)$ is a positive super martingale, $\|aY_n\|_2^2 = 1$. Thus by a standard martingale convergence theorem $aY_n \to aY_\infty$ a.e. $[P]$, as well as in $L^1(P)$. Let $V_n(\alpha) : L^2(P) \to L^2(P)$ be defined by the equation,

$$(V_n(\alpha)g)(X_1, \ldots, X_n) = aY_n \cdot g(X_1 - \alpha f_1, \ldots, X_n - \alpha f_n) \tag{105}$$

where g is a bounded 'tame' function — one that depends on a finite number of coordinates. It is seen that $V_n(\alpha)h \to V(\alpha)h$ a.e., as well as in $L^1(P)$, for $h \in \bigcup_n L^\infty(\mathcal{F}_n) = \mathcal{M}$ (say), and that $\{V_n(\alpha), \alpha \geq 0\}$ is a strongly continuous semi-group of isometries on $L^2(\mathcal{F}_n)$, $n \geq 1$, and then the $V(\alpha)$ is defined on $L^2(\mathcal{F}_n)$ as a strongly continuous semi-group of isometries with ${}_\phi Y_\infty$ in place of ${}_\alpha Y_n$. [We omit the standard (not obvious) computations here, and refer to the details given in Pitcher (1963).] The $\{V(\alpha), \alpha \geq 0\}$ is a strongly measurable semi-group of isometries on $L^2(\mathcal{F}_\infty)$, where $\mathcal{F}_\infty = \sigma(\bigcup_{n \geq 0} \mathcal{F}_n)$. It can be verified (nontrivially) that for 'tame' h:

$$\|h\|_2 = \|V(1)h\|_2 = \|V(\alpha)V(1-\alpha)h\|_2 \leq \|V(\alpha)h\|_2 \leq \|h\|_2, \ h \in \mathcal{M}. \tag{106}$$

The crucial step is to verify the semi-group property of the set $\{V(\alpha), \alpha \geq 0\}$. There does not seem to be a simple direct method to see this, and it depends on an approximation theorem of Trotter's (1958) applied to the resolvent set $\{R_n(\lambda), \lambda > 0\}$, $R_n(\lambda) = \int_0^\infty e^{-\lambda\alpha} V(\alpha)d\alpha, \lambda > 0$, which is strongly continuous in $L^1(P)$. This class satisfies

(a) $R(\lambda) - R(\lambda') = (\lambda' - \lambda)R(\lambda)R(\lambda), \lambda, \lambda' > 0$,
(b) $\|\lambda^n R(\lambda)^n\| \leq K < \infty, n \geq 1, \lambda > 0$
(c) $\lim_{\lambda \to \infty} \lambda R_\lambda(\lambda) = I$ in $L^2(\mathcal{F})$.

Then since $R_n(\lambda) \to R(\lambda)$ strongly, one verifies with some detailed analysis that $\|\lambda R(\lambda)g - g\|_2 = \lim_n \|\lambda R_n(\lambda)g - g\| \leq c/\lambda \to 0$ as $\lambda \to \infty$ for g a bounded function defined on Ω, where $g(\cdot)$ depends on only a finite number of coordinate functions (varying with g). Such g's form a dense set in $L^2(\mathcal{F})$. With this approximation, the result follows. $\square$

Remark. The omitted (nontrivial) details are given in the paper (Rao (1975), Inference in Stochastic Processes-VI) and may be consulted for discussions and related computations, as it also answers Skorokhod's question on Pitcher's work as well.

The preceding result can be stated in a somewhat different form, and it is also due to Pücher (1962) which is given here for reference, and this form is useful for the structure theory.

Proposition 3.5.4 *Let $\{X_t, t \in T\} \subset L_0^2(P)$ be a centered process with covariance r, and $T = [a, b] \subset \mathbb{R}$. Then the integral operator $R: L^2(T, dt) \to L^2(T, dt)$, given by $(Rg)(s) = \int_T r(s, t)g(t)\,dt$ is positive definite and compact (so takes bounded sets into compact sets), and the set of means M_P of the process, satisfies $M_P \subset R(L^2(T\,dt))$ so that $f \in M_P$ is of the form $f = R^{\frac{1}{2}}h$ for some element $h \in L^2(T, dt)$.*

The processes considered thus far are scalar-valued, and their vector versions are *not* trivial extensions, as their analysis due to H. Wold (a thesis under H. Cramer's direction) in the stationary case and its deeper analysis by N. Wiener (and P. R. Masani) later showed. Here we like to indicate possible extensions from our general viewpoint.

Definition 3.5.5 *If* $X_t = (X_t^{(1)}, \ldots, X_t^{(n)}), t \in \mathbb{R}$ *or* $\mathbb{Z}$, *is an n-dimensional centered process, it is weakly (or strongly) vector harmonizable whenever for each (complex) vector* $a = (a_1, \ldots, a_n)$, *the scalar process* $\tilde{X}_t = a \cdot X_t (= \sum_{i=1}^{n} a_i X_t^i)$ *is respectively weakly (or strongly) harmonizable.*

This entails the covariance matrix $R(s,t)$ of the X_t-vector process with mean zero to have the integral representation

$$r(s,t) = \int \int_{T \times T} e^{is\lambda - it\lambda'} F(d\lambda, d\lambda') \tag{107}$$

for a (unique) positive (semi-)definite matrix $F(\cdot, \cdot) = \big(F_{ij}(\cdot, \cdot)\big)$, also an $n \times n$ [positive (semi-)definite] matrix of bimeasures of Fréchet (or Vitali) variations on the (product) σ-ring $\mathcal{B}(T) \times \mathcal{B}(T)$.

The following extensions of the scalar case are of interest, for the vector-valued process, and is given by Mehlman (1992).

Theorem 3.5.6 *(Wold Decomposition) For a vector process* $X_t \in [L_0^2(P)]^n$ *there exists a unique decomposition as:*

$$X_t = R_t + S_t, \quad E(R_t S_t^*) = 0, \quad t \in \mathbb{R}, \tag{108}$$

where the R_t-*process is purely nondeterministic and the* S_t-*process is deterministic. Moreover, if the* X_t-*process is weakly harmonizable, then the* R_t *and* S_t-*processes have the same property, where deterministic means* $\mathcal{H}_X(-\infty) = \mathcal{H}_X(+\infty)$ *if* $\mathcal{H}_X(t) = \overline{sp}\{X_s^i, 1 \leq i \leq n, s \leq t\} \subset L_0^2(P), -\infty \leq t \leq \infty.$

Proof. The notations and matrix multiplication complicate the layout more than the theory itself. Thus let

$$I_t = \pi X_t \quad \text{and} \quad R_t = X_t - S_t \tag{109}$$

where $\pi \colon [\mathcal{H}_X^-(\infty)]^n \to [\mathcal{H}_X^-(-\infty)]^n$. Now if X_t has a representation as the integral $(T = \mathbb{R}$ or $(-\pi, \pi)$ for continuous or discrete indexed processes), then

$$X_t = \int_T e^{it\lambda} dZ(d\lambda). \tag{110}$$

Thus $S_t = \int_T e^{it\lambda}(\pi \circ Z)(d\lambda)$, is weakly harmonizable. But this and (110) together imply in (109) that the R_t-process also is weakly harmonizable. Then the X_t-process has the weakly harmonizable decomposition, generalizing the stationary case. □

Extending the stationary case of the basic analysis by Wiener and Masani, to a class of harmonizable process Mehlman (1992) has presented some results which may be continued in taking the Wiener's fundamental ideas to the harmonizable fields and beyond. For now, we leave this (vector) process analysis to future researchers and add some complements based on the above sections.

3.6 Complements and Exercises

1. Recall that by a stochastic process $\{X_t, t \in T\}$ one means that on a probability space (Ω, Σ, P), for each $\omega \in \Omega$ we observe $X_t(\omega)$ at $t \in T$, and in the case of real process one can take $\Omega = \mathbb{R}^T$ and $X_t(\omega) = \omega(t)$ and $f: \mathbb{R} \to \mathbb{R}, Y_t = X_t + f(t)$ a translate of X_t by $f(t), T_f X = X + f = Y$ where $f(\cdot)$ is termed a (nonstochastic) mean and $X(t)$ is the (stochastic) noise. If $P_f(= P \circ T_f^{-1})$, then f is called an admissible mean of X, if P_f is P-continuous. The class M_p of *admissible means* of X is needed for an analysis of the process if the covariance behavior is known or can be assumed given.
   Let $\{X_t, t \in [a,b] \subset \mathbb{R}\}$ be a second order process with mean zero and continuous covariance $r(\cdot, \cdot)$. If $R: L^2([a,b], dt) \to L^2([a,b], dt)$ is given by $(Rg)(s) = \int_a^b r(s,t)g(t)\, dt$ then the operator R is positive definite and compact, show the set $M_p \subset R^{\frac{1}{2}}(L^2(a,b))$, so that $f \in M_p \implies f = R^{\frac{1}{2}}h$ for some $h \in L^2([a,b], dt)$. [This interesting characterization of M_p is due to Pitcher (1963).]

2. The significance of the linearity of the set of admissible means M_p of the preceding problem is noted by the following (negative) property: Thus if (Ω, Σ, P) is a Gaussian (function space represented) measure space $[P(\Omega) = 1]$, and M_p as in the above problem, let $f_0 \in \Omega - M_p$ then P and P_{f_0} are singular, i.e., $\alpha f_0 \notin M_p, \alpha \neq 0$, and $P_{af_0} \perp P_{bf_0}$ for $a \neq b$. Consider the mixture Q on Σ:

$$Q = \sum_{k=-\infty}^{\infty} \alpha_k P_{k f_0}, \alpha_k > 0, \sum_{k=-\infty}^{\infty} \alpha_k = 1.$$

So $f_0 \in M_Q$ show $t f_o \in M_Q$ only for integral t and M_Q is not convex. [This is a modification of an example of Skorokhod's (1970).]

3. This problem explains the "admissible means" property further. Let $X = \{X(t), t \in [a, b] = T \subset \mathbb{R}\} \subset L_0^2(P)$ with continuous covariance r, and $\{\lambda_n, n \geq 1\}, \{f_n, n \geq 1\}$ as its eigenvalues and the corresponding eigenfunctions. Let $X_n = \lambda_n^{-\frac{1}{2}} \int_T X(t) \phi_n(t)\, dt$; $f_n = \lambda_n^{-\frac{1}{2}} \int_T f(t) \phi_n(t)\, dt, f \in M_P$, where M_P is the set of admissible means of X. This problem gives (good) sufficient conditions in order that M_P is a positive cone and even linear, exemplifying the nontriviality of this property. If P_n is the n-dimensional distribution of $\{X_i, 1 \leq i \leq n\}$ with (Lebesgue) densities $p_n, n \geq 1$, suppose the following conditions hold: (i) $p_n > 0$ a.e., $n \geq 1$, (ii) $\lim_{|t_i| \to \infty} p_n(t_1, \ldots, t_i, \ldots, t_n) = 0, n \geq 1$, and (iii) $\frac{\partial p_n}{\partial t_j}, 1 \leq j \leq n$, exist, with $\sum_{j=1}^{n} \int_{\mathbb{R}^n} \left(\frac{\partial \log p_n}{\partial t_j} \right)^2 dP_n \leq K_0 < \infty, n \geq 1$. Show $M_1 = \{f \mid M_P : f = \{f_n, n \geq 1\} \in \ell^2\}$ is a positive cone. Also $M_1 = M_P$ if $\sum_{n \geq 1} \lambda_n^{-\frac{1}{2}} < \infty$. If each p_n is symmetric about the origin of $\mathbb{R}^n$, then the M_1 is also linear. In general M_P need not even be convex as the preceding problem implies. [This result is also due to Pitcher.]

4. In contrast to the preceding two problems, here we present a result that isolates (linear) admissible subspaces (of means) of M_P. For this, we need to recall a few (nontrivial and somewhat advanced) properties of (nonnegative) convex functions. Let $\Phi \colon \mathbb{R} \to \mathbb{R}^+, \Phi(0) = 0, \Phi(-x) = \Phi(x)$ convex, $|x \Phi''(x)| \leq c < \infty$ for all $x \geq 0$, and $\Phi'(x) \uparrow \infty$ as $x \uparrow \infty$. For such a convex Φ, there is (uniquely) another convex function Ψ with similar properties, which is given by $\Psi(x) = \sup\{|x|y - \Phi(y) : y \geq 0\}$. Such a Ψ is often called the *complementary function* to Φ, and they obviously satisfy the (Young) inequality, $|xy| \leq \Phi(x) + \Psi(y)$. Suppose also that $E(\Psi(\beta h_m^*(\cdot, a))) \leq C_1 < \infty$ all $\beta > 0$, where

$$h_a^*(\omega, a) = (\nabla f_n(\pi_n \omega), \pi_n a) / \tilde{f}(\pi_n \omega), \omega \in \Omega, a \in M_n$$

satisfies $\int_0^t h_n^*((\omega - sa), a)\, ds = -\log h_n(\omega, a)$. The following condition ensures that $h_n \to h$ a.e. with the desired properties. Thus let (Φ, Ψ) be the complementary convex pair as defined above with $|x\Phi''(x)| \leq C < \infty$, $x > 0$ but $\Phi'(x) \uparrow \infty$ with the complementary function $\Psi : x \mapsto \sup\{|xy| - \Phi(y) : y \geq 0\}$ and that the h^* satisfies

$$\sup_n \int_\Omega \Psi(\beta h_n^*(\omega, a))\, dP(\omega) \leq C_1 < \infty, \quad a \in M_P \qquad (\dagger)$$

for some $\beta > 0$. Then $ta \in M_P, t \in \mathbb{R}$, (so M_P is linear) and $h_n^* \to h^*$ a.e., $t \in \mathbb{R}$. The density $h(\cdot, ta) = \frac{dQ'}{dP}$ is given by

$$h(\omega, ta) = \exp\left\{-\int_0^t h^*(\omega - sa, a)\, ds\right\}, a.e.\ [P] \qquad (\ddagger)$$

implying that P_{ta} is equivalent to P.
[If $\Phi(x) = |x| \log^+ |x|$, so $\Psi(x) = e^{+|x|} - |x| - 1$, this result is given by A. V. Skorokhod (1970). If $\Psi(x) = e^{x^2} - 1$, then the corresponding Φ cannot be written explicitly, but is also covered now. It is interesting to note that the above conditions yield a result of interest in the analysis which depends on some properties of Orlicz spaces. This and related details are given in the author's paper (Rao (1977), 311–324) which may be of interest to the reader.]

5. This problem presents conditions in order that the set of all admissible means is a linear space. The result again uses a few properties of the Orlicz spaces for which a convenient reference is the book by Rao and Ren (1991). Let $\Phi: \mathbb{R} \to \mathbb{R}^+$ be a twice differentiable symmetric convex function $\Phi(0) = 0$, $|x\Phi''(x)|$ is bounded, $x \geq 0$ (and $\Phi'(x) \uparrow \infty$) with $\Psi : \mathbb{R} \to \mathbb{R}^+$ as its conjugate function, i.e., $\Psi(x) = \sup\{|x|y - \Phi(y) : y \geq 0\}$. Let P_n, Q_n be the finite dimensional distributions under the hypothesis and its alternative with densities f_n and g_n and the likelihood ratio be denoted as $h_n^*(= \frac{g_n}{f_n})$. If now $\sup_n \int_\Omega \Psi(\beta h_n^*(\omega, a))\, dP(\omega) < \infty$, for each $a \in M_P$ and some $\beta > 0$, then $ta \in M_P$ for all $t \in \mathbb{R}$ so that M_P is linear. The density $h(\cdot, ta) = \frac{dQ'}{dP}$ is given by

$$h(\omega, ta) = \exp\left\{-\int_0^t h^*(\omega - sa, a)ds\right\}, a.e.\ [P].$$

Hence P_{ta} is equivalent to P. [There are many details to fill, and the reader is referred to Rao (1977); this extends Skorokhod's (1970) work.]

6. Here we consider a discrete parameter harmonizable process that satisfies a finite difference equation subject to (or disturbed by a perhaps) different (harmonizable) error process, opening up new avenues. Let $L_n(\cdot) = \sum_{n=0}^{k} a_n(\cdot)_{m-n}$, be a kth order difference map with a_n as constant coefficients. If $\{Y_n, \in \mathbb{Z}\}$ is a (weakly) harmonizable process (observed as an output) given by the operator L_n called a *filter*, so that $Y_k = L_n X_k = \sum_{i=0}^{k} a_i X_i$, and the Y_n-process is observed and the filter L_n is assumed given, we desire to find the process $\{X_t, t \geq 0\} \subset L^2(P)$, that is a unique harmonizable one which when the X-process is strongly (or weakly) harmonizable so is the Y_n-sequence and conversely. (This problem was considered by K. Nagabhushanam (1951) if harmonizability is replaced by weak stationarity. Here we present its extension to strong and weak harmonizable processes, and indicate its use.)

To utilize the fact that L_n is an nth order difference operator acting on the (weak) harmonizable series, giving a similar process $\{Y_n, n \geq 1\}$ one uses its spectral properties (representations) and that the Y_n-process is (centered) harmonizable and hence so is the X_n-process. Thus

$$X_n = \int_T e^{in\lambda} dZ_X(\lambda), \quad E(X_n \bar{X}_m) = \int \int_{T \times T} e^{i(m\lambda - m\lambda')} d\mu(\lambda\lambda')$$

hold, with $Z_X(\cdot)$, as the stochastic and $\mu(\cdot, \cdot)$, the (scalar) spectral measures of X_n's. Given a (weakly or strongly) harmonizable process $\{Y_n, n \in \mathbb{Z}\}$ and a filter L_n such that $L_n X_k = Y_k, k \in \mathbb{Z}$, there exists a *unique* process $\{X_n, n \in \mathbb{Z}\}$ satisfying the equation above, iff (i) $|\nu|(Q \times Q) = 0$ and (ii) $\int_Q \int_{Q^C} \left| \frac{1}{F(u)\bar{F}(v)} \right| d|\nu|(u, v) < \infty$, where $\nu(\cdot, \cdot)$ is the bimeasure governing the Y_n-process and $F(\cdot)$ is the (polynomial) filter determined by $L_n(\cdot) = \sum_{j=0}^{k} a'_j(\cdot)_{n-j}$ and $F(e^{it}) = P(e^{-it}), Q = \{t \in T : F(t) = 0\}$.

[This was first obtained by K. Nagabhushanam (1951) for the weakly stationary case, extended by J. P. Kelsh (1978) for the strongly harmonizable case, and finally by D. K. Chang (1998) for the weakly harmonizable case.]

7. (a) A process $\{X_t, t \in \mathbb{R}\} \subset L^2_0(P)$ with covariance $K(s, t) = E(X_s \bar{X}_t)$ is of *class* (KF) if $K(\cdot, \cdot)$ satisfies $r(h) = \lim_{T \to \infty} \frac{1}{T} \int_0^{T-|h|} K(s, s + |h|)ds, h \in \mathbb{R}$ exists. It is seen that $r(\cdot)$ is positive definite. Verify that all stationary and even all strongly harmonizable

processes are in class (KF). [Even some (but *not* all) weakly harmonizable ones are also in it.]

(b) A process $X = \{X_t, t \in G\} \subset L_0^2(P), G$ an LCA group, is called *almost harmonizable* if there is a complex valued measure μ on $\mathcal{B}(\hat{G}) \times \mathcal{B}(\hat{G})$ and a set $\{g(\cdot, \gamma), \gamma \in \hat{G}\}$, continuous complex functions $g(\cdot, \gamma)$ being almost periodic in G, such that

(i) $\int_{\hat{G} \times \hat{G}} \|g_\gamma\|_G \|g_{\gamma'}\|_G \, d(\lambda, \lambda') < \infty$, and

(ii) $K(t, s - t) = (X_s, X_t) = \int_{\hat{G} \times \hat{G}} g_\gamma(s) \bar{g}_{\gamma^{-1}}(t) d\mu(\gamma, \gamma')$ exists.

If the process X is almost harmonizable it is asymptotically stationary as well as almost periodically correlated.

[For more on this application, see B. H. Schreiber (2004), showing how the second order processes extend with Hilbertian methods and analysis.]

8. With ideas similar to the above, and using the structural analysis of the Karhunen processes detailed in the last part of this chapter establish the following pair of properties:

(a) Every weakly harmonizable process $\{X_t, t \in \mathbb{R}\} \subset L_0^2(P)$, is a Karhunen process relative to some Borel family $\{f_t, t \in \mathbb{R}\}$ and a stochastic measure $Z: \mathcal{B}(\mathbb{R}) \to L_0^2(\tilde{P})$ with orthogonal values, where $L_0^2(\tilde{P}) \supset L_0^2(P)$ can be taken. [Hint: If $Q: L_0^2(\tilde{P}) \to L_0^2(P)$ is the projection, so that we have $X(t) = QY(t) = \int_{\mathbb{R}} \pi(e^{it(\cdot)})(\lambda)\tilde{Z}(d\lambda)$, let $f_t = \pi(e^{it(\cdot)})$, and complete the argument.]

(b) If $X: \mathbb{R} \to L^2(P)$ is a mean continuous process, then on each compact interval it coincides a.e., with a Karhunen process. [*Hint:* Using the eigenfunction expansion of the covariance on each compact interval, using Mercer's theorem, apply the preceding part to deduce this assertion.]

3.7 Bibliographical Notes

This chapter is devoted to some useful applications of harmonizable processes which often subsume the classical (weakly) stationary processes and fields; the fields part will be treated in depth in the next chapters. Here we treated both oscillatory and periodic classes including an extended analysis of both Cramér and Karhunen classes; especially their structural aspects are emphasized and detailed. Also, applications to signal extraction from noise and related problems are included.

In this analysis, some extensions related to slowly varying fields are treated. An important aspect is to obtain the (unique) input from the output if the processes are (weakly) stationary, and this was first detailed in his thesis by Nagabhushanam (1951). Conditions for uniqueness are important. This motivated an extension naturally to harmonizable classes. A basic question is the unicity of the solution. It was also given there, but in reviewing the paper, for Math. Reviews, J. L. Doob has constructed an auxiliary process of the same difference equation with zero output, so that the sum of the solutions will satisfy the equation to contradict the key unicity requirement. But the sum of these solutions is not generally stationary, as required, and this was noted by J. Kelsh who then extended Nagabhushanam's work in his thesis (1978) for strongly harmonizable classes which requires a different set of conditions. It needed a further analysis, and also different techniques to the weakly harmonizable case. This was done in his thesis (1983) by D. K. Chang. Some of these works are included in this chapter and others will be given in the next one.

It may be noted that the 'reviews' in these review publications are not generally refereed, and contains tentative opinions and views. So they should be considered as a type of mathematical "news reports" and not to be taken generally as refereed facts. [See R. H. Bing's (1981) advice, and practices on all such works and quick conclusions.]

As the reader may have noticed, most of the analysis is usually centered on the covariance properties of the second order processes. But the mean values are important as well. This was analyzed in detail by Pitcher and a few key problems on the nontriviality of the structure of the mean functions was also noted by Skorokhod. We have considered these aspects in this chapter. This shows how simple looking problems on (second order) processes are not to be taken by hunch. Even for the popular Gaussian processes, these are nontrivial but useful.

The multiple indexed classes are termed random fields and they will be discussed continuing a new concept called *isotropy* that enhances the key analysis with further problems and consequences. It will be taken up in the next chapters and the general analysis prompted by it will be continued in the rest of this volume. Thus the analysis of the current problems serves as a concrete illustration and motivation for the rest of the treatment to follow which forms the other half of the projected work.

4

Isotropic Harmonizable Fields and Applications

We have treated so far various properties of harmonizable random processes and of their structural analyses. But some key applications show that we need to study also various aspects of some related problems if the index is not a linear set as time axis, but is multidimensional such as the one corresponding to the space-time problems in evolution, as in Physics and elsewhere. This, therefore, takes us to considerations of random fields whose index is just a directed set, as in space-time problems mentioned above. It was also found out in early analysis that a stationary field $\{X_t, t \in \mathbb{R}^n, n > 1\} \subset L_0^2(P)$ which satisfies an isotropy condition is the trivial one (i.e., a constant field, with probability one). Such facts lead us to analyze and establish their nontrivial structured and related properties to be used, e.g., in isotropy for harmonizable fields.

4.1 Harmonizability for Multiple Indexed Random Classes

If the indexing of random classes $\{X_t, t \in G\}$ where $G \subset \mathbb{R}^n, n > 1$, and more generally (including $G = \mathbb{N}^n$ or $= \mathbb{Z}^n$), with $X_t \in L_0^2(P)$, then the structural analysis of the X_t-family presents new nontrivial problems, and their families are called *random fields*. Generally, these G's do not have linear ordering, and could be semi-groups. The following example motivates our possible (applicable) analysis to continue.

Recall that an orthonormal set $\{X_n, -\infty < n < \infty\} \subset L_0^2(P)$, gives a new sequence $Y_n = \Pi X_n$ where $\Pi : L_0^2(P) \to \mathcal{M} = \bar{sp}\{X_n, n \geq 0\}$ is an orthogonal projection onto $\mathcal{M}$, so that $Y_n = \{X_n, n \geq 0\}$, or $= 0, n < 0$. Thus even though the X_n-set is trivially (weakly) stationary, the Y_n-set is not stationary and not even strongly harmonizable! This is a consequence of a nontrivial extension of a theorem of F. and M. Riesz, by S. Bochner on the vanishing of Fourier coefficients on the half-line.

But it (Y_n series) is weakly harmonizable, since a weakly harmonizable sequence (or process) remains in the same family under bounded linear transformations. So we need to go beyond stationary classes and thus want to study the structural aspects of such classes again following Bochner, after introducing a new concept called *isotopy* here.

As a motivation, let $X : G \to L_0^2(P), G = \mathbb{R}^n$, be a mapping with covariance $r(s,t) = E(X_s \bar{X}_t), r(s,t) = \tilde{r}(s-t) = \bar{r}(|s-t|)$, so that it is invariant under translations and rotations. A random field $\{X_t, t \in \mathbb{R}^n\}$ with this property $(\subset L_0^2(P))$ is called *isotropic and homogeneous*. So what is the structure of $\tilde{r}$? Again Bochner has characterized such $\tilde{r}(\cdot)$, and given the fundamental integral representation as:

$$\tilde{r}(s - t) = a_n \int_0^\infty \frac{J_{\frac{n-2}{2}}(\mu\tau)}{(\tau\mu)^{\frac{n-2}{2}}} dF(\mu), \quad \tau = |s - t|, s, t \in \mathbb{R}^n, \quad (1)$$

where $a_n = 2^{\frac{n-2}{2}} \Gamma(\frac{n}{2})$, and $F(\cdot)$ is bounded and it is defined by:

$$F(u) = \int \cdots \int_{\{y:|y|<u\}} dG_0(t_1, \ldots, t_n), t = (t_1, \ldots, t_n) \in \mathbb{R}^n. \quad (2)$$

Here J in (1) is a Bessel function of the first kind, of order $\nu = \frac{n-2}{2} \geq 0$, $G_0(\cdot) \uparrow$ on $\mathbb{R}^n$.

With this as the background we now proceed ahead and introduce the corresponding concept of harmonizable isotropics based on Bochner's original representation, and discuss some developments:

Definition 4.1.1 *The covariance function r of a centered random field $X : \underset{\sim}{t} \to X_{\underset{\sim}{t}} \in L_0^2(P), \underset{\sim}{t} \in \mathbb{R}^n$, is weakly harmonizable isotropic if it is representable as:*

$$r(\underset{\sim}{s}, \underset{\sim}{t}) = \alpha_n^2 \sum_{m=0}^\infty \sum_{t=1}^{h(m,n)} S_m^l(\underset{\sim}{u}) S_m^l(\underset{\sim}{v})$$
$$\times \int_0^\infty \int_0^\infty \frac{J_{m+\nu}(\lambda s) J_{m+\nu}(\lambda' t)}{(\lambda s)(\lambda' t)} d\mu(\lambda, \lambda') \quad (3)$$

where

(i) $\underset{\sim}{s} = (s, u), \underset{\sim}{t} = (t, v)$ are spherical (polar) coordinate of $\underset{\sim}{s}, \underset{\sim}{t} \in \mathbb{R}^n$,

(ii) the spherical harmonics of mth order $S_n(\cdot)$ are given by $(S_0^l(\cdot) = 1)$:

$$S_m^l(\cdot)(1 \le l \le h(m, n)) = \frac{2(m + \nu)(m + 2\nu - 1)!}{(2\nu)!(m)!} m \ge 1,$$

are the spherical harmonics on the unit sphere of $\mathbb{R}^n$, of order m,

(iii) $\alpha_n > 0, \alpha_n^2 = 2^{2\nu} \Gamma(n/2)\pi^{\frac{n}{2}}, \nu = \frac{n-2}{2}$, and

(iv) $\mu : \mathcal{B}(\mathbb{R}^+) \times \mathcal{B}(\mathbb{R}^+) \to \mathbb{C}$ is a positive definite bimeasure of finite Fréchet (Vitali) variation, the integral for $r(s, t)$ above is in the strict Morse-Transue (Lebesgue) sense, the series for $r(\cdot, \cdot)$ converging unconditionally.

It should be observed here that, to obtain nontrivial solutions of the Laplacian $\Delta X = 0$, it is necessary *not* to restrict to the simple and easy looking stationary isotropic fields, because it was found that in this restriction only trivial isotropic fields are found as solutions (i.e., the constants) which will not conform with real applications. (See Yadrenko (1983) on this negative result.) This has been analyzed by Swift (1994) later and found that the harmonizable processes have nontrivial and meaningful solutions.

Using the addition formula for Bessel functions, one can simplify (3) with some routine (but not entirely simple) analysis to obtain an equivalent form (cf. Swift (1994) for details) as:

$$r(s, t) = \alpha_n \int_0^\infty \int_0^\infty \frac{J(|\lambda s - \lambda' t|)}{|\lambda s - \lambda' t|} d\mu(\lambda, \lambda') \tag{4}$$

where μ has finite Vitali variation iff X is strongly harmonizable isotropic. The equivalence asserted above needs a nontrivial proof. For details of this assertion, we refer the reader to Swift (1994) and the properties of the classical special functions (the Bessel class) will be needed.

For a clear understanding, it is useful to have a general characterization of weakly as well as strongly harmonizable isotropic covariances. This will also show the fundamental nature of Bochner's representation (1) above, and its extension in Definition 4.1.1 and Swift's formula (4).

Theorem 4.1.2 *Let X be a weakly harmonizable random field in $\mathbb{R}^n$, $n \ge 1$, with covariance $r : \mathbb{R}^n \times \mathbb{R}^n \to \mathbb{C}$. Then it is isotropic iff $r(\cdot, \cdot)$ is representable as:*

$$r(s, t) = 2^\nu \Gamma \left(\frac{n}{2} \right) \int_{\mathbb{R}^+} \int_{\mathbb{R}^+}^* \frac{J_\nu(|\lambda s - \lambda' t|)}{|\lambda s - \lambda' t|^\nu} \Phi(d\lambda, d\lambda'), \tag{5}$$

where Φ is a positive definite function of bounded Fréchet variation, $J_\nu(\cdot)$ is a Bessel function (of the first kind) of index $\nu \left(= \frac{n-2}{2}\right)$ and the integral is in the strict Morse-Transue sense. The corresponding strongly harmonizable characterization is obtained if Φ of (5) has finite Vitali variation and then the integral is in the standard Lebesgue sense.

Thus (4) is an aspect (modification) of this result, due to Swift (1994) and it will also be presented after this proof is detailed.

Proof. The direct part is simple since $r(gs, gt) = r(s, t)$, for all $g \in SO(n)$, the orthogonal group of matrices on $\mathbb{R}^n$, as $|g(\lambda s - \lambda' t)| = |\lambda s - \lambda' t|$ in (5), g being an isometry. Thus only the converse is to be established. This is done in four steps as follows:

I. Let $X : \mathbb{R}^n \to L_0^2(P)$ be (weakly) harmonizable. Then by the dilation, there is a larger $(\tilde{\Omega}, \tilde{\Sigma}, \tilde{P})$ containing the given space so that $L_0^2(\tilde{P}) \supseteq L_0^2(P)$, a stationary field $Y : \mathbb{R}^n \to L_0^2(\tilde{P})$ and an orthogonal projection Q onto $L_0^2(P)$, such that $X_t = QY_t, t \in \mathbb{R}^n$. Let ρ be the covariance function of Y, so $\rho(s, t) = \tilde{\rho}(s - t)$. We assert that when X is isotropic we can take $\rho(s, t)$ as $\tilde{\rho}(s - t)$. We assert that when X is isotropic we can take ρ as $\tilde{\rho}$ to be also isotropic. Now Y in $L^2(\tilde{P}) = \mathcal{H} \otimes \mathcal{H}_0$, is representable as $Y = X + X_1$, where X_1 is stationary, valued in $\mathcal{H}_0$. Although $\mathcal{H}_0$ can be realized as $L^2(\mu)$ where μ is a finite (Grothendieck) measure, we can also replace it by a standard Gaussian product measure $\tilde{\mu}$, on $\tilde{\Sigma} = \otimes_\alpha \mathcal{B}_\alpha$, as in the proof of Theorem 2.3.1 (on stationary dilations) and so can omit the detail here.

We may arrange things such that $\tilde{\mu}(gA) = \tilde{\mu}(A)$, which is true for cylinder sets A and $g \in SO(n)$, the orthogonal group. Thus the Grothendieck measure generated Hilbert space can be replaced by an isometric Hilbert space based on a Gaussian measure after the initial step.

Then the components of $X_1 = (X_{1t}, t \in \mathbb{R}^n)$ will be invariant under g so X_{1s} and X_{1t} will be identically (Gaussian) distributed. Hence $X_1(\perp X)$ is stationary and isotropic. [The actual construction of this is based on P and $\tilde{\mu}$.] That the bimeasure F of X is invariant can be verified as follows:

$$r(gs, gt) = \int_{\mathbb{R}^n} \int_{\mathbb{R}^n}^* e^{i(s \cdot (g^* \lambda) - t \cdot (g^* \lambda'))} dF(\lambda, \lambda')$$

$$= \int_{\mathbb{R}^n} \int_{\mathbb{R}^n}^* e^{i(s \cdot \tilde{\lambda} - t \cdot \tilde{\lambda}')} dF(g\tilde{\lambda}, g\tilde{\lambda}')$$

$$= r(s, t) \tag{6}$$

$$= \int_{\mathbb{R}^n} \int_{\mathbb{R}^n}^* e^{i(s \cdot \lambda - t \cdot \lambda')} dF(\lambda, \lambda'). \tag{7}$$

Since $r(s, t) = r(gs, gt)$, $\forall g \in SO(n)$, $s, t \in \mathbb{R}^n$ (by hypothesis), $F(\lambda, \lambda') = F(g\lambda, g\lambda')$ implying invariance so that Y_s- satisfies $E(Y_s \bar{Y}_t) = \rho(s, t) = \tilde{\rho}(s - t)$. This is the key refinement of the dilation $(X = QY)$ that is needed here.

II. Now consider Bochner's stationary isotropic covariance representation of Y constructed above, and let ρ be its covariance. Then following Yaglom ((1987), p. 353) one has

$$\rho(t) = C_n \int_{\mathbb{R}^+} \frac{J_v(\lambda \tau)}{(\lambda \tau)^\nu} d\Phi(\lambda) \tag{8}$$

where $t = (\tau, u)$, $|t| = \tau$, u representing the spherical polar of t, J_0 the Bessel function of order $\nu = \frac{n-2}{2}$, $\Phi(\cdot) \uparrow$ bounded and $C_n = 2^\nu \Gamma\left(\frac{n}{2}\right)$. Using the addition formula for Bessel functions (8) can be expressed as

$$\rho(s, t) = C_n^2 \sum_{m=0}^{\infty} \sum_{l=1}^{h(m,n)} S_m^l(u) S_m^l(v) \cdot \int_{\mathbb{R}^+} \frac{J_{m+\nu}(r_1 \lambda) J_{m+\nu}(r_2 \lambda)}{(r_1 r_2)^\nu \lambda^{2\nu}} d\Phi(\lambda),$$

$$\tag{9}$$

where $s = (r_1, u)$, $t = (r_2, v)$, $h(m, n) = (2m + 2\nu)\frac{(m+2\nu-1)!}{\nu! m!}$, $m \geq 1$, and $m = 0$, $S_m^l(n)$ being the spherical harmonics, orthogonal on unit sphere S_n, relative to the surface measure. For each m, there are a total of $h(m, n)$ of them. Using (9), an integral form of Y_t can be obtained. For this, we conveniently express (9) as a triangular covariance and apply Karhunen's result.

Again since only the second order properties are considered we may identify the process with a centered stationary Gaussian field with covariance S for (2nd order) computational manipulation. Thus, let

$$\tilde{\mathbb{N}} = \{(n, l) \in \mathbb{N} \times \mathbb{N} : 1 \leq l \leq h(m, n), m \geq 0\}$$

and $\mathcal{P}$ be its power set. Let $\tilde{F} : \mathcal{P} \times \mathcal{P} \times g(\mathbb{R}^+) \to \mathbb{R}^+$, be given as

$$\tilde{F}(A_1, A_2; B_1) = J(A_1 \cap A_2)\Phi(B_1), \tag{10}$$

where $\Phi(A) = \int_A d\Phi(\lambda)$, the positive bounded measure by Φ of (9) and $\zeta(\cdot)$ as the counting measure on $\mathcal{P}$, so $\tilde{F}$ extends to a σ-finite measure, and if $\tilde{g} : \tilde{\mathbb{N}} \times \mathbb{R}^n \times \mathbb{R}^+ \to \mathbb{R}$ is given by

$$\tilde{g}((m, l); t, \lambda) = S_m^l(n)\frac{J_{m+\nu}(t\lambda)}{(\lambda\tau)^\nu},$$

where $t = (\tau, u)$ is the spherical polar representation, and $\nu, S_m^l(\cdot), J_\nu$ are defined earlier. Then $\tilde{g}$ is square integrable for $\tilde{F}$ and one has the formula, with $\Lambda = \tilde{\mathbb{N}} \times \mathbb{R}^+$,

$$r(s, t) = C_n \int_\Lambda \int_\Lambda \tilde{g}((m, l); s, \lambda)\bar{\tilde{g}}((m, l); t, \lambda)d\tilde{F}. \tag{11}$$

Now (11) implies that $r(\cdot, \cdot)$ is a triangular covariance on $\mathbb{R}^n \times \mathbb{R}^n$ relative to $\tilde{F}$ (cf. Rao (1982), p. 313), and so there exists a Gaussian measure $Z : \mathcal{P} \otimes \mathcal{B}(\mathbb{R}^+) \to L_0^2(P)$, giving Y_t as:

$$Y_t = C_n \int_\Lambda \tilde{g}((m, l); t, \lambda)dZ((m, l); \lambda)$$

$$= C_n \int_{\tilde{\mathbb{N}}} \int_{\mathbb{R}^+} \tilde{g}((m, l); t, \lambda)dZ((n, l); \lambda), \tag{12}$$

where $E(Z(A, B)\bar{Z}(A_2, B_2)) = \tilde{F}(A_1, A_2; B_1, B_2) = \zeta(A_1 \cap A_2) \cdot \Phi(B_1 \cap B_2)$. If now $A_1 = \{(m, l)\}, A_2 = \{(m', l')\}$, singletons, then writing $Z(A_1, B_1) = Z_m^l(B)$ and similarly for A_2, B_2, we get for the above $E(\tilde{Z}_m(B_1)\bar{\tilde{Z}}_{m'}(B_2)) = \delta_{mm'}\delta_{ll'}\Phi(B_1 \cap B_2)$. Hence letting $B = (0, \lambda)$, the associated processes $\{Z_m^l(\lambda), 0 \le p \le h(m, n)\}$, $m \ge 0$, are orthogonal each with orthogonal increments. Hence in the case of Gaussians, the $Z_m^l(\cdot)$ become independent, with independent increments. Thus (12) becomes

$$Y_t = Y(\tau, u) = C_n \sum_{m=0}^\infty \sum_{l=1}^{h(m,n)} S_m^l(u) \int_{\mathbb{R}^+} \frac{J_{m+\nu}(\tau\lambda)}{(\tau\lambda)^\nu} Z_m^l(d\lambda), \tag{13}$$

the series converging in $L^2(\tilde{p})$-mean, the Y_t is isotropic and stationary.

III. Now let $X_t = QY_t$, with $Q : L^2(\tilde{P}) \to L^2(P)$, the orthogonal projection, in the isotropic dilated (bigger) space, and we have

$$X_t = X(\tau, u) = (QY)(\tau, u)$$

$$= C_n \sum_{m=0}^{\infty} \sum_{l=1}^{h(m,n)} S_m^l(u) \int_{\mathbb{R}^+} \frac{J_{m+\nu}(\tau'\lambda)}{(\tau\lambda)^\nu} (QZ_n^l)(d\lambda) \quad (14)$$

the interchange of the integral and the bounded Q is justified by a classical theorem of Hille, and setting $\tilde{Z}_n^l(\lambda) = (QZ_n^l)(\lambda)$, ($Q$ being linear), $\tilde{Z}_n^l(\cdot)$ are independent identically distributed (Gaussian) we have

$$E(\tilde{Z}_n^l(A)\bar{\tilde{Z}}_m^l(B)) = \delta_{ll'}\delta_{mm'}F(A, B), \quad (15)$$

where F is the bimeasure determined by the d.f.'s of $\tilde{Z}_m^l$, which will not have orthogonal increments now, but F is independent of n, and is the same, and is a bimeasure of finite Fréchet variation. Thus (13) is the desired integral representation of the field $\{X_t, t = (\tau, u)\}$.

We now compute the covariance of X_s, X_t, where $s = (\tau_1, u)$, $t = (\tau_2, v)$.

$$r(s, t) = E(X_s \bar{X}_t) = C_n^2 \sum_{m=0}^{\infty} \sum_{l=1}^{h(m,n)} S_m^l(u)S_m^l(v) \times$$

$$\int_{\mathbb{R}^+} \int_{\mathbb{R}^+}^* \frac{J_{m+\nu}(\tau_1\lambda)J_{m+\nu}(\tau_2\lambda')}{(\tau_1\lambda)^\nu(\tau_2\lambda)^\nu} F(d\lambda, d\lambda'), \quad (16)$$

the right side double integral being in strict MT-sense.

IV. Using the properties of spherical harmonics $S_m^l(u)S_m^l(v)$ we can sum the series (16) to obtain

$$\sum_{l=1}^{h(m,n)} S_m^l(u)S_m^l(v) = \frac{h(m,n)C_m^\nu(\cos(u, v))}{\omega_n C_n^\nu(1)}, \quad (17)$$

where ω_n is the surface area of the unit sphere $S_n \subset \mathbb{R}^n$, and $C_m^\nu(\cdot)$, $m \geq 0$ are the Gegenbauer or ultraspherical polynomials of order ν, for each $m \geq 0$. With (16) and (17) and using the addition formula for Bessel functions, indicated below, one gets the asserted representation (4). This gives the converse, hence the result. $\square$

The formula that we employed above is obtained on using some properties of spherical harmonics, detailed in Müller (1966). The version we need is as follows (using the same notation as above):

$$\alpha_n^2 \sum_{m=0}^{\infty} \sum_{l=1}^{h(m,n)} S_m^l(u) S_m^l(v) \frac{J_{m+\nu}(\lambda r_1) J_{m+\nu}(\lambda' r_2)}{(\lambda r_1)^\nu (\lambda' r_2)^\nu}$$

$$= 2^\nu \frac{\Gamma\left(\frac{n}{2}\right) J_\nu(\lambda R(\lambda, \lambda'))}{(\lambda R(\lambda, \lambda'))^\nu}$$

where $R(\lambda, \lambda') = \left(r_1^2 + \left(\frac{\lambda'}{\lambda}\right)^2 r_2^2 - 2\left(\frac{\lambda'}{\lambda}\right) r_1 r_2 \cos\theta\right)^{\frac{1}{2}}$, and $\cos\theta = u \cdot v$, the angle between the unit vectors u, v.

This formula can be verified using the known properties of the "classical" spherical harmonics. This was given in detail in the paper by R. J. Swift (1994) as Lemma 2.1, and we leave it to the reader, for an independent try. In any case, the simplification is needed above.

We now record some equivalent forms of entropy useful in the context of applications.

Theorem 4.1.3 *Let $X = \{X_t, t \in \mathbb{R}^n\}$ be a centered weakly harmonizable random field. Then the following are equivalent:*

(i) *X is isotropic so that the covariance $r(s, t) = r(gs, gt), g \in SO(n)$;*

(ii) *the covariance r of X admits the representation (5);*

(iii) *the covariance r of X admits a series representation (16) with the MT-integration relative to a bimeasure F of finite Fréchet variation;*

(iv) *X is representable as an $L^2(P)$-convergent series (32) with the stochastic integral representation in the standard vector (or Dunford-Schwartz) sense.*

The following approximation of a weakly harmonizable field on $\mathbb{R}^n$ with the one whose spectral measure lives on bounded Borel sets of $\mathbb{R}^n$.

Proposition 4.1.4 *Let $X : \mathbb{R}^n \to L_0^2(P)$ be weakly harmonizable (perhaps not isotropic), with $\mu(\cdot, \cdot)$ as its spectral measure that is essentially contained in a bounded Borel set, i.e., for an $\varepsilon > 0$, there is a bounded Borel set $A_\varepsilon \subset \mathbb{R}^n$ such that $\|\mu\|(\mathbb{R}^n \times \mathbb{R}^n - A_\varepsilon) = 0$. Then there is a weakly harmonizable field $X_\varepsilon : \mathbb{R}^n \to L_0^2(P)$ with spectrum in $A_\varepsilon \times A_\varepsilon$, and*

$$\|X(t) - X_\varepsilon(t)\|_2 < \varepsilon, \text{ where } t \in \mathbb{R}^n. \tag{18}$$

Proof Sketch. Since $X(t)$ has the integral representation with $Z(\cdot)$, as its stochastic measure, and $E(Z(A)\bar{Z}(B)) = \mu(A, B), \|Z\|^2(A) = \|\mu\|(A, A)$, choose Borel A_ε satisfying $\|Z\|(A_\varepsilon) < \varepsilon$, and define Z_1, Z_2 as $Z_1(\cdot) = Z(A_\varepsilon \cap \cdot)$ and $Z_2(\cdot) = Z(A_\varepsilon^c \cap \cdot)$ so that

$$X(t) = \int_{\mathbb{R}} e^{it\lambda} Z_1(d\lambda) + \int_{\mathbb{R}} e^{it\lambda} Z_2(d\lambda) = X_1(t) + X_2(t).$$

Then letting $X_\varepsilon(t) = X_1(t)$, one has

$$\|X(t) - X_\varepsilon(t)\|_2 = \|X_2(t)\|_2 \leq \|Z_2\|(\mathbb{R}) = \|Z_2\|(A_\varepsilon^c) < \varepsilon. \qquad \square$$

4.2 A Classification of Isotropic Covariances

It is interesting to consider classes of random fields relative to their second order structures in the analysis of their isotropic covariances. From our preceding work, it is clear that the following proper inclusions between the stated classes obtain:

Karhunen fields $\supset$ isotropic fields $\supset$ weakly harmonizable and isotropic

fields $\supset$ strongly harmonizable isotropic ones $\supset$ stationary isotropic fields

Our analysis also gives the following integral representation in a series from:

Theorem 4.2.1 *Let $X : \mathbb{R}^n \to L_0^2(P)$ be a random field with a continuous covariance r. Then we have:*
   (i) X is also isometric implies

$$X(\tau, u) = \sum_{m=0}^{\infty} \sum_{l=1}^{h(m,n)} \xi_m^l(\tau) S_m^l(u), \qquad (19)$$

where $\{S_m^l : 1 \leq l \leq h(m, n), m \geq 0\}$ are the spherical harmonics on the unit sphere $S_n \subset \mathbb{R}^n$, that are orthonormal for the normalized surface measure μ_m of S_n, and $E(\xi_m^l(\tau)\bar{\xi}_m^{l'}(\tau')) = \delta_{m,m'}\delta_{ll'}b_m(\tau, \tau')$ with $\sum_{n=0}^{\infty} b_m(\tau, \tau')h(m, n) < \infty$, all the other symbols as defined before.
   (ii) X is isotropic and harmonizable implies that the series (19) holds in which the variables ξ_m^l can be represented as:

$$\xi_m^l(\tau) = c_\nu \int_{\mathbb{R}^+} \frac{J_{m+\nu}(\tau\lambda)}{(\tau\lambda)^\nu} d\tilde{Z}_m^\rho(\lambda), \qquad (20)$$

*where $\nu = \frac{n-2}{2}$ and $\tilde{Z}_m^l : \mathbb{R}^+ \to L_0^2(P)$ is a stochastic measure satisfy-
ing $E(\tilde{Z}_m^l(d\lambda)\bar{\tilde{Z}}_{m'}^{l'}(d\lambda')) = \tilde{F}(d\lambda, d\lambda')\delta_{mm'}\delta_{ll'}$ with $\tilde{F}$ as a bimeasure,
$C_\nu = 2^\nu \nu \Gamma(\nu) i^m$, and the integral in (20) is in the Dunford-Schwartz
sense. Further $b_m(\tau, \tau')$ is given as:*

$$b_m(\tau, \tau') = |C_\nu|^2 \int_{\mathbb{R}^+} \int_{\mathbb{R}^+}^* \frac{J_{m+\nu}(\tau\lambda) J_{m+\nu}(\tau'\lambda')}{(\tau\lambda)^\nu (\tau'\lambda')^\nu} \tilde{F}(d\lambda, d\lambda') \quad (21)$$

*the integral now being in the strict MT-sense. The random field is
strongly harmonizable iff $\tilde{F}$ in (21) has finite Vitali variation whence
the integral in (21) becomes one in the Lebesgue-Stieltjes sense.*

*(iii) X is isotropic and stationary implies $\tilde{Z}_m^l(\cdot)$ are orthogonally
valued for all $1 \leq l \leq h(m,n), m \geq 0$ so $\tilde{F}$ of (21) reduces to a
bounded Baire measure, say Φ, and we have:*

$$b_m(\tau, \tau') = |C_\nu|^2 \int_{\mathbb{R}^+} \frac{J_{m+\nu}(\tau\lambda) J_{m+\nu}(\tau'\lambda)}{(\lambda)^{2\nu}(\tau\tau')^\nu} d\Phi(\lambda), \quad (22)$$

and the integral now is in the standard Lebesgue sense.

*In the converse direction each of the stated representations, gives the
corresponding random field.*

Proof. (i) This is a slightly revised version of Yadrenko's (1983) analysis. By the classical results, the isotropic covariance $r(\cdot, \cdot)$ of X can
be expressed as $r(s,t) = \tilde{r}(\tau_1, \tau_2, \cos(u,v))$ with $s = (\tau_1, u)$, and
$t = (\tau_2, v)$ in spherical polar coordinates (cf., e.g., Vilenkin (1968),
Ch. XI). Thus $\tilde{r}$ can be expressed as:

$$\tilde{r}(\tau_1, \tau_2, \cos(u,v)) = \sum_{m=0}^\infty \sum_{k=1}^{h(m,n)} b_m(\tau_1, \tau_2) S_m^l(u) S_m^l(v), \quad (23)$$

where $b_m(\tau_1, \tau_2) \geq 0$ are eigenvalues of $\tilde{r}$ so that by the well-known
Funk-Hecke formula (C. S. Müller (1964), p. 20), we have:

$$\int_{S_n} \tilde{r}(\tau_1, \tau_2; \cos(u,v)) S_m^l(u) d\mu_n(u) = b_m(\tau_1, \tau_2) S_m^l(v) \quad (24)$$

where μ_n is the surface measure of the unit sphere S_n of $\mathbb{R}^n$. [Recall that
this is, $\mu_n : S_n \to \frac{2\pi}{n-1}\mu_{n-2}(S_{n-2}), n \geq 2, \mu_2(S_2) = 4\pi, \mu_1(S_1) = 2\pi$.]

Letting $\xi_m^l : \tau \mapsto \int_{S_n} X(\tau, \mu) S_n^l(u) d\mu_n(u)$, we get for each $\tau \in \mathbb{R}^+$,
the r.v.'s $\xi_m^l(\tau)$ are orthogonal with the asserted properties. The series

(19) converges in mean, and then (23) follows from the addition formula for spherical harmonics.

(ii) Replacing the general isotropic fields with the weakly harmonizable class, the processes ξ_m^l just defined, can be given a sharper representation. Thus comparing the two forms of $X(\tau, u)$ given above, one has

$$\sum_{m=0}^{\infty} \sum_{l=1}^{h(m,n)} S_m^l(u)\xi_m^l(\tau) = c_n \sum_{m=0}^{\infty} \sum_{l=1}^{h(m,n)} S_m^l(u) \int_{\mathbb{R}^+} \frac{J_{m+\nu}(\tau\lambda)}{(\tau\lambda)^\nu} \tilde{Z}_m(d\lambda).$$

Multiply both sides by $S_{m'}^{l'}$ and integrate with the surface measure μ_m, using the orthogonality relations, then we find

$$\xi_m^l(\tau) = c_n \int_{\mathbb{R}^+} \frac{J_{m+\nu}(\tau\lambda)}{(\tau\lambda)^\nu} \tilde{Z}_m^l(d\lambda); \tau \in \mathbb{R}^+ \tag{25}$$

which is (20), and (21) follows from the fact that $E(\xi_m^l(\tau_1)\bar{\xi}_{m'}^{l'}(t_2)) = \delta_{ll'}\delta_{mm'} \cdot b_m(\tau_1, \tau_2)$, since $\tilde{Z}_m^l$ are orthogonal in $\mathcal{M}$, and have the same second moments in general. This will be more difficult to show generally.

(iii) This part is immediate now since each Z_m^l has orthogonal increments in (l, m) so $F(A, B) = \Phi(A \cap B)$ in the above work, and (22) obtains, the integral now is in the usual Lebesgue sense.

In the converse direction, if the process (or field) satisfies the conditions (i)–(iii) one can construct a process (or field) by Kolomogorov's basic existence theorem, and then by Karhunen's representation (on a probability space) it is isotropic as well as harmonizable or stationary. This construction is standard and can be left to the reader (it is also in Chapter 1) of the first volume of this trilogy). $\square$

A weaker form of the preceding result, admitting a series representation, has still a useful application potential, and we include the result.

Proposition 4.2.2 *Let* $X : \mathbb{R}^n \to L_0^2(P)$ *be a weakly harmonizable isotropic field. Then it is representable* ($t = (\tau, u)$) *as a series:*

$$X(\tau, u)$$

$$= \Gamma(\nu)\omega_n 2^\nu \nu \sum_{m=0}^{\infty} i^m \sum_{l=n}^{h(m,n)} S_m^l(u) \int_{\mathbb{R}^+} \int_{S_*} \frac{J_{m-\nu}(\tau\lambda)}{(\tau\lambda)^\nu} S_m^l(v) Z(d\lambda, dv) \tag{26}$$

ω_n *being the surface area of the unit sphere* S_n.

Proof. Given that X is weakly harmonizable, we can express it as

$$X(t) = X(\tau, \mu) \left[= \int_{\mathbb{R}^n} e^{its} Z(ds) \right]$$

$$= \int_{\mathbb{R}^+} \int_{S_n} e^{it\lambda \cos(u,v)} Z(d\lambda, dv), \qquad (27)$$

with $s = (\lambda, v), t = (\tau, u)$ as polar representations. Now using the known form (cf. Vilenkin (1968), p. 957) and expanding e^{itx} in a series with orthogonal ultraspherical polynomials $c_n^\nu(\cdot)$ on S_n from the above Vilenkin volume, we have for $|x| \leq 1, t = (\tau, u)$,

$$e^{itx} = \Gamma(\nu) \sum_{m=0}^{\infty} i^m (m + \nu) \frac{J_{m+\nu}(\tau)}{(\tau/2)^\nu} c_m^\nu(x). \qquad (28)$$

Putting (28) into (27), letting $x = \cos(u, v)$, and using the formula,

$$c_m^\nu(\cos(u, v)) = \sum_{l=1}^{h(m,n)} S_m^l(u) S_m^l(v) \frac{w_n c_m^\nu(1)}{h(m, n)},$$

we get

$$X(\tau, u)$$

$$= \Gamma(\nu) w_n 2^\nu \nu \sum_{m=0}^{\infty} i^m \sum_{l=1}^{h(m,n)} S_m^l(u) \int_{\mathbb{R}^+} \int_{S_n} \frac{J_{m-\nu}(\tau\lambda)}{(\tau\lambda)^\nu} S_m^l(v) Z(d\lambda, dv).$$

$$(29)$$

Since the series in (28) is $L^2(S_n, \mu_n)$-convergent, the valid substitution of c_m^ω in e^{itx} series is valued and use it in (27), to deduce (26), as desired. $\square$

We now present a characterization of isotropic weakly harmonizable fields with a parameter set in a Hilbert space. This will give a kind of completeness of the problem considered mainly here.

Proposition 4.2.3 *Let $X : \mathcal{H} \to L_0^2(P)$ be an isotropic weakly harmonizable random field, $\mathcal{H}$ being a separable Hilbert space. Then the covariance $r : (x, y) \mapsto E(X_x \bar{X}_y), x, y \in \mathcal{H}$, admits the representation (as a Morse-Transue integral) as:*

$$r(x,y) = \int_{\mathbb{R}^+} \int_{\mathbb{R}^+}^{*} \exp\{-(\lambda x - \lambda' y, \lambda x - \lambda' y)\} F(d\lambda, d\lambda'), x, y \in \mathcal{H},$$

(30)

where $F : \mathbb{R}^+ \times \mathbb{R}^+ \to \mathbb{C}$ *is a positive definite bimeasure of finite Fréchet variation, with* $(\cdot, \cdot)$ *as a scalar product in* $\mathcal{H}$.

Proof. We start with the stationary case, so $r(s,t) = \tilde{r}(s-t)$, and by isotropy $\tilde{r}$ is invariant under the rotation group so that $\tilde{r}(x) = \tilde{r}(\|x\|)$, which depends only on the length of x. Then by a key representation theorem of Schoenberg ((1958), Thm. 2) one has

$$E(X_x \bar{X}_y) = \tilde{r}(\|x-y\|) = \int_{\mathbb{R}^+} \exp\{-\lambda(x-y, x-y)\} d\Phi(\lambda), \quad (31)$$

for a bounded nondecreasing (left continuous) $\Phi : \mathbb{R}^+ \to \mathbb{R}^+$. We now replace $\mathcal{H}$ by its isometric image l^2, the square summable sequence space with each $x \in \mathcal{H}$. Let $t = t_x = \tau(x) \in l^2$, and consider

$$\exp\{-\lambda(s-t, s-t)\} = \exp\{-\lambda[(s,s) + (t,t) - 2(s,t)]\}$$
$$= \exp\{-\lambda(\|s\|^2 + \|t\|^2)\}$$
$$\times \sum_{m=0}^{\infty} \sum_{j \in l_m} \frac{(2\lambda)^m}{k_1! \cdots k_m!} (s_{i_1}, t_1)^{k_1} \cdots (s_{i_m}, t_m)^{k_m},$$

where $l_m = \{j = [(i_m, k_m)], i_m \geq 0, k_m \geq 0, k_1 + \cdots + k_m = m\}$.
    Consider ψ_j defined by

$$\psi_j(s, \lambda, m) = e^{-\|x\|^2} \frac{(2\lambda)^{\frac{m}{2}}}{(k_1! \ldots k_m!)^{\frac{1}{2}}} s_1^{k_1} \ldots s_m^{k_m},$$

with $s = (s_{i_1}, \ldots, s_{i_m})$ as a vector of reals, associated with j, so that (13) gives

$$r(s,t) = \tilde{r}(s-t) = \sum_{m=0}^{\infty} \sum_{j \in l_m} \int_{\mathbb{R}^+} \psi_j(s, \lambda, m) \psi_j(t, \lambda, n) d\Phi(\lambda). \quad (32)$$

This is a triangular covariance relative to Φ, if we set

$$\psi(x, \lambda) = (\psi_j(x, \lambda, m), j \in l_m, m \geq 0)$$

so that $r(s,t) = \int_{\mathbb{R}^+} \psi(s, \lambda) \psi(t, \lambda)^* d\Phi(\lambda)$, $\psi(\cdot, \lambda)^*$ is the adjoint $\psi(\cdot, \lambda)$. Hence there exists a stochastic measure (can be taken Gaussian) so that we have the representation:

$$X_t = \int_{\mathbb{R}^+} \psi(t,\lambda) dZ(\lambda) = \sum_{m=0}^{\infty} \sum_{j \in l_m} \int_{\mathbb{R}^+} \psi_j(t,\lambda,m) dZ_m^j(\lambda), \qquad (33)$$

where $Z_m^j(\lambda) \in L_0^2(P)$ and satisfying $E(Z_m^j(A)\bar{Z}_n^{j'}(B)) = \delta_{jj'}\delta_{mn}\Phi(A \cap B)$.

The procedure now is similar to the earlier cases. We proceed to the harmonizable case use the dilation which is valid in this case also. Thus there exists $L_0^2(\tilde{P}) \supset L_0^2(P)$ and a stationary $Y \cdot \mathcal{H} \to L_0^2(\tilde{P})$ such that $X_t = QY_t, t \in \mathcal{H}(\equiv l^2)$, Q an orthogonal projection onto $L_0^2(P)$. With the familiar procedure, we get

$$X_t = QY_t = \sum_{m=0}^{\infty} \sum_{j \in l_m} \int_{\mathbb{R}^*} \psi_j(t,\lambda,m) Q Z_m^j(d\lambda), \qquad (34)$$

where $QZ_m^j = \tilde{Z}_{m'}^j$ is the centered stochastic measure such that

$$E(\tilde{Z}_m^j(A)\bar{\tilde{Z}}_{m'}^{j'}(B)) = \delta_{jj'}\delta_{mm'}F(A,B)$$

the same $F(\cdot,\cdot)$ for all $\tilde{Z}_m^j$-processes, and $r(s,t) = r(gs,gt)$ for all relations g, and then

$$r(s,t) = \sum_{m=0}^{\infty} \sum_{j \in l_m} \int_{\mathbb{R}^+} \psi_j(s,\lambda,m)\psi_j(t,\lambda',m)F(d\lambda,d\lambda'). \qquad (35)$$

Interchanging the (legitimate) sum and the (strict) MT-integral, we get

$$r(s,t) = \int_{\mathbb{R}^+} \int_{\mathbb{R}^+}^{*} e^{-\lambda(s,s)-\lambda'(t,t)+2\lambda\lambda'(s,t)} F(d\lambda,d\lambda')$$
$$= \int_{\mathbb{R}^+} \int_{\mathbb{R}^+} e^{-(\lambda s - \lambda' t, \lambda s - \lambda' t)} F(d\lambda,d\lambda')$$

giving (30). $\square$

The preceding argument (proof) also implies the following consequence.

Corollary 4.2.4 *If $X : l^2 \to L_0^2(P)$ is an isotropic harmonizable random field, then it admits the representation as:*

$$X_t = \sum_{m=0}^{\infty} \sum_{j \in l_m} \int_{\mathbb{R}^+} \psi_j(t,\lambda,m)\tilde{Z}_m^j(d\lambda), \qquad (36)$$

where $E(\tilde{Z}_m^j)$ satisfies (34), the series converging in $L^2(P)$-mean.

Thus we have a complete extension of stationary isotropic fields to the (weakly) harmonizable class, using the (strict) Morse-Transue integration in place of the Lebesgue method. Thus much of the classical stationary field analysis can be extended. There is, however, considerable noncommutational analysis of the stationary fields given by A. M. Yaglom (1961) that needs to be and can be extended (see the next chapter) following the above work.

There are many specializations and corresponding structural analysis of the forms, most of which depend on properties of spherical harmonics and various classes of spherical functions. A considerable part of the resulting analysis was given by R. J. Swift (1994) and later, and much of it includes Yadrenko's (1983) basic (stationary) work. We present a few of the resulting properties to show how many of the familiar questions have been solved and useful conclusions drawn.

The following is a first extension to isotropic fields that can be considered obtained from the preceding work, and the Karhunen series representations. This is stated for comparison as follows.

Theorem 4.2.5 *A random field* $X : \mathbb{R}^n \to L_0^2(P)$ *is weakly harmonizable isotropic iff it is representable as*

$$X(t) = \alpha_n \sum_{m=0}^{\infty} \sum_{l=1}^{h(m,n)} S_m^l(u) \int_0^{\infty} \frac{J_{m+\nu}(\lambda r)}{(\lambda r)^{\nu}} dZ_m^l(\lambda), \qquad (37)$$

where $E(Z_m^l(B_1)\bar{Z}_m^{l'}(B_2)) = \delta_{mm'}\delta_{ll'}F(B_1 B_2), B_i \subset \mathbb{R}^n$ *Borel and* $F(\cdot, \cdot)$ *is a function of bounded Fréchet variation,* $r = \|t\|, \nu = \frac{t}{r}, \delta_{mn}$ *is the Kronecker delta with the Dunford-Schwartz integral in (37),* $\alpha_n^2 = \pi^{\frac{n}{2}} \Gamma\left(\frac{\pi}{2}\right) 2^{2\nu+1}$.

The preceding representation can be given a different formulation which leads to various applications, noted by Swift (1994). This can be presented as follows, along with some consequences.

Theorem 4.2.6 *A random field* $Y : \mathbb{R}^n \to L_0^2(P)$ *is weakly harmonizable as well as isotropic if it is representable as*

$$X(\underset{\sim}{t}) = a_n \sum_{m=0}^{\infty} \sum_{l=1}^{h(m,n)} S_m^l(u) \int_0^{\infty} \frac{J_{m+\nu}(\lambda r)}{(\lambda r)^{\nu}} dZ_m^l(\lambda), \underset{\sim}{t} \in \mathbb{R}^n, \quad (38)$$

where $E(Z_n^l(B_1)\bar{Z}_m^{l'}(B_2)) = \delta_{mm'}\delta_{ll'}F(B_1, B_2), r = \|t\|, \underset{\sim}{u} = \underset{\sim}{t}/r, a_n = 2^{\nu+} \cdot \sqrt{\Gamma(\frac{n}{2})} \cdot \pi^{\frac{n}{4}}, \delta_{mn}$ *being the Kronecker delta with* $J_m(\cdot)$ *as the Bessel* *function of order m and the vector integral is in the Dunford-Schwartz sense, with $F(\cdot, \cdot)$ having finite Fréchet variation, and the functions $J(\cdot)$ being Bessel's.*

The proof is somewhat similar to the analysis used above and we shall omit it here. The reader can reconstruct with the same procedures as before, or can also refer to Swift's (1990) paper. If $Y_m^l(r) = \alpha_m \int_0^\infty \frac{J_{m+\nu}(\lambda r)}{(\lambda r)^\nu} dZ_m^l(\lambda)$, so $E(Y_m^l(r)) = 0$, $Y_m^l(r) \perp$, and $E(Y_m^l(r)\bar{Y}_n^l(r)) = F(r, r)$, the integral is a standard D-S symbol. The point is that the above result implies the following key representation obtained by M. I. Yadrenko (1985) differently based on several other results. It is a consequence of the preceding result, and the random field is just isotropic, but not necessarily harmonizable. This was also in Swift (1994), which is of interest and so we include it.

Theorem 4.2.7 *(i) A random field $X : \mathbb{R}^n \to L_0^2(P)$ is just isotropic if it admits the (spectral) representation (n-fixed)*

$$X(\underset{\sim}{t}) = \sum_{m=0}^\infty \sum_{l=1}^{h(m,n)} S_m^l(u) Y_m^l, \quad Y_m^l \in L_0^2(P), \qquad (39)$$

and the Y_m^l are orthogonal with $E((Y_m^l)^2) = b_m, l \geq 1, m \geq 1$.

(ii) Also, $r(\cdot, \cdot)$ is the covariance of an isotropic field on S^n (unit sphere on $\mathbb{R}^n$) iff it is representable as

$$r(\underset{\sim}{s}, \underset{\sim}{t}) = \tilde{r}(\cos\theta) = \frac{1}{\omega_n} \sum_{m=0}^\infty b_m \frac{C_\infty^\nu(\cos\theta)}{C_m^\nu(1)} h(m, n)$$

with $b_m \geq 0$, $\sum_{m=0}^\infty b_m h(m, n) < \infty$.

A consequence of this complicated looking format is the next representation that is of real interest in many applications of the present analysis:

Theorem 4.2.8 *A random field $X : \mathbb{R}^n \to L_0^2(P)$ is isotropic (not necessarily harmonizable) iff its covariance is representable as:*

$$r(\underset{\sim}{s}, \underset{\sim}{t}) = \frac{1}{\omega_n} \sum_{m=0}^\infty h(m, n) \frac{C_m^\nu(\cos\theta)}{C_m^\nu(1)} b_m(r_1, r_2) \qquad (40)$$

where $b_n(0, r) = 0, m \neq 0$, and $\sum_{m=0}^\infty h(m, n) b_m(r_1, r_2) < \infty$, with $C_m^\nu(\cdot)$ being the ultraspherical polynomials. With this set up we have

$$X(\underset{\sim}{t}) = \sum_{m=0}^{\infty} \sum_{l=1}^{h(m,n)} S_m^l(\underset{\sim}{u})Y_m^l(\tau)$$

where

$$Y_m^l(\cdot), m = 0, 1, \ldots, h(m, n) \text{ and } E(Y_m^l(r)\bar{Y}_{m'}^{l'}(r)) = 0$$

except for $m = m'$, and $E(Y_m^l(\tau)\bar{Y}_m^l(\tau)) = b_m(\tau, \tau)$, $\sum_{m=0}^{\infty} b_m(\tau, \tau) \cdot h(m, n)$ is finite for each $l \geq 1$. Also $r(\cdot, \cdot)$ is the isotropic covariance of $X(\cdot)$ iff it is given by

$$r(\underset{\sim}{s}, \underset{\sim}{t}) = \tilde{r}(\tau_1, \tau_2, \cos\theta) = \frac{1}{\omega_n} \sum_{m=0}^{\infty} h(m, n)\frac{C_m^\nu(\cos\theta)}{C_m^\nu(1)}b_m(\tau_1, \tau_2),$$
(41)

where b_m satisfies $\sum_{m=0}^{\infty} h(m, n)b_m(\tau_1, \tau_2) < \infty$, $b_m(0, \tau) = 0$ if $m \neq 0$. Here $C_m^\nu(\cdot)$ are the ultraspherical polynomials.

This result was obtained by Yadrenko (cf. his book 1983). It is given here for understanding the involved nature of the isotropy property. But it is also true for harmonizable (weakly) when the $b_m(\cdot, \cdot)$ above are restricted there.

Theorem 4.2.9 *A random field $X(t), t \in \mathbb{R}^n$ is isotropic and weakly harmonizable iff the $b_n, n \geq 1$, of (40), (or (41)) admit the integral representation as:*

$$b_m(r_1, r_2) = \alpha_n^2 \int_0^\infty \int_0^\infty \frac{J_{m+\nu}(\lambda r_1)J_{m+\nu}(\lambda' r_2)}{(\lambda r_1)^\nu(\lambda' r_2)^\nu}d^2t(\lambda, \lambda')$$

where F is a function of bounded Fréchet variation, $(2\pi)^{-p/2}\alpha_n^2 = 2^{2\nu\mu}\Gamma\left(\frac{m}{n}\right)$.

Proof. If $X(t)$ is weakly harmonizable isotropic, we have

$$r(\underset{\sim}{s}, \underset{\sim}{t})$$

$$= \alpha_n^2 \sum_{m=0}^{\infty} \sum_{l=1}^{h(m,n)} S_m^l(u)S_m^l(v) \int_0^\infty \int_0^\infty \frac{J_{m+\nu}(\lambda r_1)J_{m+\nu}(\lambda' r_2)}{(\lambda r_1)^\nu(\lambda' r_2)^\nu}dF(\lambda, \lambda')$$

which on comparison with the earlier result on isotropic fields gives

$$b_m(r_1, r_2) = \alpha_n^2 \int_0^\infty \int_0^\infty \frac{J_{m+\nu}(\lambda r_1) J_{m+\nu}(\lambda' r_2)}{(\lambda r_1)^\nu (\lambda' r_2)^\nu} dF(\lambda, \lambda').$$

Conversely, if this formula holds for $b_m(r_1, r_2)$, then $r(\underset{\sim}{s}, \underset{\sim}{t})$ is obtainable as above, with Theorem 4.2.3, so that $X(t)$ obeys the weakly harmonizable isotropic condition, which completes the sketch. □

Remark 23. There are a few other forms that one can obtain with the above work, and we shall also sketch a few more properties, as complements since they can be completed by the reader with analogous computation and include a few other properties that augment the above work.

We conclude this section with an extension of the harmonizable isotropy to the vector case whose consequences may be studied by researchers in generalizing this work and applying it.

Definition 4.2.10 *A random vector field* $X_t = (X_{t_1}, \ldots, X_{t_m}) \in L_0^2(P, \mathbb{C}^n)$, *is weakly (or strongly) harmonizable isotropic if for each complex m-vector $\alpha = (\alpha_1, \ldots, \alpha_m)$ the scalar field*

$$Y_\alpha = \alpha \cdot X = \sum_{i=1}^m \alpha_i X_{t_i}, (\mathbb{R}^n \to L_0^2(P))$$

is a weakly (respectively strongly) harmonizable random field as defined.

One can see that this extension implies that the various covariance sets $B_{ij} : (s, t) \to E(X(s)\bar{X}_j(t)), 1 < i, j \le n$ are harmonizable and isotropic, and the matrix increment $(\Delta B_{ij}(s, t), 1 \le i, j \le n)$ has the earlier studied integral representation relative to (matrix) bimeasures. Thus the vector-valued case is direct, but not entirely simple.

It is also of interest to note another extension suggested by Yaglom in his research monograph (Yaglom (1987)), as follows. Namely, each permutation matrix T of order m, acting on the random field X, should have the same type of harmonizable isotropic covariance and study its structure. The work needed for its analysis is more intricate as it depends on some aspects of group representations, to be discussed in the next chapter, that will show the deeper analysis of isotropic fields (extending the locally harmonizable isotropic class) and is also of interest in this analysis. We note that some (early) aspects of this problem restricted to stationary classes had already been pioneered by Yaglom (1987). This leads to an interesting analysis. Thus the next section will be devoted to an outline of this extension.

4.3 Representations of Multiple Generalized Random Fields

The primary task here is to obtain integral representations of Cramér class [or class (C)] random fields through the analysis of generalized random functions, using the Schwartz theory of such classes. [On the latter subject see Schwartz (1957) or Gel'fand and Vilenkin (1964) for basic results with details.]

There are at least three ways of introducing such fields and it will be useful to use any one of them for analysis, since they can be shown to be *equivalent*, although this fact needs some further analysis. In the following, K denotes the Schwartz space of infinitely many times differentiable scalar functions on $\mathbb{R}^n$, vanishing off compact sets, varying with each function. Let $\tilde{C}$ be the space of complex random variables on (Ω, Σ, P). If $F : K \to \tilde{C}$ is linear it is termed as before a *generalized random field* (g.r.f.), provided for $f_n \in \mathcal{K} . f_n \to 0$ in $\mathcal{K}$ (so all the f_n live on a fixed compact set and $f_n \to 0, n \to \infty$, uniformly) then the random variables $F(f_n) \to 0$ in probability, [or $f_n \to 0$ in $\mathcal{K}$, and $F(f_n) \to 0$ in $L^2(P)$] or if $F \in \mathcal{K}^*$ (adjoint of $\mathcal{K}$), and $F(f)$ is measurable. In all the different definitions, useful in computations, F is termed a g.r.f. and their equivalence is verified by G. Y. H. Chi (1969), and we use some of these properties according to convenience below. Let us formally introduce:

Definition 4.3.1 *On the Borel σ-field $\mathcal{B}(= \mathcal{B}(\mathbb{R}^n))$ of $\mathbb{R}^n$, a mapping $Z : \mathcal{B} \to L_0^2(P)$ is called a generalized random measure if (i) $E(Z(A)) = 0, A \in \mathcal{B}$, (ii) $A, A_n \in \mathcal{B}, A = \bigcup_{n=1}^{\infty} A_n, A_n$ disjoint $\Rightarrow Z(A) = \sum_{n=1}^{\infty} Z(A_n)$, series existing in $L_0^2(P)$, and (iii) there is a tempered measure $\rho : \mathcal{B} \times \mathcal{B} \to \mathbb{C}$ such that for $A, B \in \mathcal{B}$,*

$$E(Z(A)\bar{Z}(B)) = \int_A \int_B d^2\rho(x, y). \tag{42}$$

The measure $Z(\cdot)$ so defined has orthogonal values if $\rho(\cdot, \cdot)$ concentrates on the diagonal $x = y$ whence $E(Z(A)\bar{Z}(B)) = \sigma(A \cap B); \sigma(\cdot)$ is an $\mathbb{R}^+$ valued measure.

The following standard property of $\sigma(\cdot)$, defined above, is used below.

Lemma 4.3.2 *If $\sigma : \mathcal{B} \to \mathbb{R}^+$ is a tempered measure then it is bounded on bounded sets. If $\sigma(\cdot)$ has bounded variation on each bounded set of*

$\mathbb{R}^n$, *and has a polynomial growth (i.e.,* $\sigma((0,t))/|t|^k \to 0$, *as* $|t| \to \infty$
for some $k \geq 0$), *then* $\sigma(\cdot)$ *induces a tempered measure on* $\mathcal{B}$.

Proof. If $\sigma(\cdot)$ is a tempered measure, then by hypothesis there is a $k > 0$
so that $\mu : A \mapsto \int_A \frac{d\sigma(t)}{(1+|t|^2)^k}$ defines a finite measure on $\mathcal{B}$. This shows
that $\frac{d\mu}{d\sigma}(t) = (1+|t|^2)^{-k} \leq 1$. Also $\frac{d\sigma}{d\mu}(k) = (1+|t|^2)^k$ on A, so that
$\sigma(A) = \int_A (1+|t|^2)^{-k} d\mu \leq \text{const} \cdot \mu(A) < \infty$.

Also $\sigma(t)/|t|^k \to 0(|t| > 1)$ for some $0 < k < \infty \Rightarrow \int_{\mathbb{R}^n} \frac{d\sigma(t)}{(1+|t|^2)^{1/p}} <$
∞, for some $p > 0$, so that $\sigma(\cdot)$ determines a tempered measure on $\mathcal{B}$.
$\square$

This shows that the tempered measure here also depends on $k > 0$
and hence the g.r.f. of a Cramér class depends on such a $k \geq 0$ (or ≤ 0).
Here we shall obtain an integral representation of a Cramér class which
is the most general one used, and it includes the harmonizable classes.
We then present a simple extension to the isotropic class also here.

Proposition 4.3.3 *Let* $\mathcal{K}(N)$ *be the Schwartz space of infinitely differentiable functions on a compact rectangle* $N \subset \mathbb{R}^n$ *and* F *be a g.r.f.
on* $\mathcal{K}(N)$, *centered, with covariance* $B(\cdot,\cdot)$ *and* $F : \mathcal{K}(N) \to L_0^2(P)$
*be a random field. Then there is a continuous positive definite hermitian
function* $h : N \times N \to \mathbb{C}$ *and an integer* $m > 0$ *such that*

$$B(f,g) = \int_N \int_N h(x,y)(D^\alpha g)(x)\overline{(D^\alpha g)(y)} \, dxdy, \, f,g \in \mathcal{K}(N)$$
$$< \infty, \tag{43}$$

where $D^\alpha = \frac{\delta^{|\alpha|}}{\delta x_1^{\alpha_1}...\delta x_n^{\alpha_n}}, \alpha_i \geq 0, \sum_{i=1}^m \alpha_i \leq m$ *and* $\alpha = (\alpha_n, \ldots, \alpha_m)$.

Proof. Since $B(\cdot,\cdot)$ is, by definition, a continuous bilinear functional
on $\mathcal{K}(N) \times \mathcal{K}(N)$, by the Kernel theorem of Schwartz (cf. Gel'fand-
Vilenkin (1964)) there exists a continuous linear G on $\mathcal{K}(N \times N)$ satisfying $B(f,g) = G(f\bar{g}) [(fg)(x,y) = f(x)g(y)]$ so that $fg \in \mathcal{K}(N \times N)$,
and linear combinations of such sums being dense. But then it is known
(cf. Friedman (1963)) that there is a bounded measurable $h : N \times N \to$
$\mathbb{C}$ and $m \geq 0$ such that (43) holds. By increasing m, slightly if necessary, or integrating by parts again, $h(\cdot,\cdot)$ can also be taken as continuous.
Since $B(\cdot,\cdot)$ is Hermitian positive definite (and letting f,g polynomials), it follows that $h(\cdot,\cdot)$ is a positive definite function, as needed. $\square$

Remark 24. This is essentially known but motivates the general case of class (C) to be considered.

Theorem 4.3.4 *Let Φ be a test space on $\mathbb{R}^n$ continuous, each compactly supported, real functions, and $F : \Phi \to L^2(P)$ be a centered grf of class (C), with covariance $B(\cdot, \cdot)$. Suppose F is of class (C) relative to $g(\cdot, \cdot)$ and a tempered covariance ρ. Let the g-transform of all elements of Φ exist (i.e. $f \in \Phi \Rightarrow \tilde{f}(t) = \int_{\mathbb{R}^n} g(t, x)f(x)dx$ holds). Then there is a random measure Z relative to ρ, such that we have the representation:*

$$F(f) = \int_{\mathbb{R}^n} \tilde{f}(t)dZ(t), \qquad f \in \Phi \tag{44}$$

where $\tilde{f}$ is the g-transform of f and

$$B(u, v) = \int_{\mathbb{R}^n} \int_{\mathbb{R}^n} \tilde{u}(t)\bar{\tilde{v}}(s)d^2\rho(t, s), u, v \in \Phi. \tag{45}$$

The stochastic integral in (44) is defined in the mean square sense. On the other hand if $g(\cdot, \cdot)$ and ρ are given with properties noted earlier and $Z(\cdot)$ is a random measure, on $\mathcal{B}$ (of $\mathbb{R}^n$), and $\mathcal{K} \subset \Phi$ is dense, then $F(\cdot)$ of (44) defines a g.r.f. of class (C) on $\mathcal{K}$ and has a unique extension to Φ.

Proof. We present the argument in steps, shortening later some (similar) arguments.

I. Let $F: \Phi \to L_0^2(P)$ be of class (C), so $E(F(f)) = 0$, and $B(f_1, f_2) = E(F(f_1)\bar{F}(f_2))$. We assume, for convenience, that $B(\cdot, \cdot)$ is strictly positive definite, and introduce the inner product in Φ as:

$$(f_1, f_2) = B(f_1, f_2) = \int_{\mathbb{R}^n} \int_{\mathbb{R}^n} f_1(t)\bar{f}_2(t)d^2\rho(t, s), f_1, f_2 \in \Phi. \tag{46}$$

II. Let $\mathcal{K}_0 = \text{sp}\{F(f) : f \in \Phi\} \subset L^2(P)$ and let $\mathcal{H}$ be its completion. It follows that

$$(F(f_1), F(f_2)) = B(f_1, f_2) = (f_1, f_2), \tag{47}$$

so that $f_i \mapsto F(f_i)$ gives an isometry of Φ onto $\mathcal{K}_0$. If $L^2(\rho)$ is the completion of Φ in the norm induced by (46) then the map $f \to F(f)$ has an extension uniquely form $L^2(\rho)$ onto $\mathcal{H}$. The correspondence $\chi_A \to F(\chi_A)(= Z(A))$ gives $Z : \mathcal{B} \to \mathcal{H}$ as a random measure relative to ρ, using the standard simplification

$$\left\| Z(A) - \sum_{i=1}^{n} Z(A_i) \right\| = \left\| \chi_A - \sum_{i=1}^{n} \chi_{A_i} \right\| = \left\| \chi_{\cup_{i>n} A_i} \right\| \leq \rho(A \times A) < \infty$$

where $\tilde{P}$ is the measure determined by P. Letting $n \to \infty$, we get $Z(A) = \sum_{i=1}^{\infty} Z(A)$. This easily implies that $Z(\cdot)$ is a random measure relative to $\rho(\cdot)$.

III. If $\tilde{\Phi} = \{\tilde{f} : f \in \Phi\}$, then $\tilde{\Phi} \subset L^2(\rho)$ and is dense in it since

$$(\tilde{f_1}, \tilde{f_2}) = \int_{\mathbb{R}_n} \int_{\mathbb{R}^n} \tilde{f}(t)\bar{\tilde{f}}_2(s) d^2 \rho(t, s) = B(f_1, f_2) = (f_1, f_2),$$

by the preceding computations showing $f \to \tilde{f}$ to be an isometry. The density of Φ in $L^2(\rho)$ implies that of $\tilde{\Phi}$ in $L^2(P)$. It also follows in a standard way that $\tilde{\Phi}$ in $L^2(\rho)$ has the same property. Hence

$$F(f_n) = \sum_{i=1}^{m} a_{m_i} Z(A_i) = \int_{\mathbb{R}^n} f_n(t) dZ(t),$$

and $\|f - f_n\| \to 0$ so that $\{F(f_n), n \geq 1\} \subset \mathcal{K}$ is a Cauchy sequence in the latter, and we get with the usual notations:

$$F(f_n) = \int_{\mathbb{R}^n} \hat{f}_n(t) dZ(t), \hat{f}_n \in L^2(\rho),$$

for simple and then for all functions in Φ, so that

$$F(f) = \int_{\mathbb{R}^n} \hat{f}(t) dZ(t), \qquad f \in \Phi,$$

giving (44).

IV. For the opposite direction, only the continuity is to be verified to see that it defines a class (C) random field. For this it suffices to show the continuity, since then $B(f, g) = \int_{\mathbb{R}^n} \int_{\mathbb{R}^n} \tilde{f}(s)\bar{\tilde{g}}(t) d^2 \rho(t, s)$ will be of class (C).

But $g(\cdot, \cdot)$ is bounded on bounded sets, it follows that $B(f_n, f_n) \to 0$ as $f_n \to 0$ a.e. also, boundedly, $B(f_n, f_n) \to 0$ in $\mathcal{K}$, so that F is continuous on $\mathcal{K}$. But $\mathcal{K} \subset \Phi$ with a stronger topology, hence F is continuous in the topology of Φ in which $\mathcal{K}$ is dense. This implies easily that F on Φ is of class (C), as asserted. $\square$

This analysis implies the following consequence:

Corollary 4.3.5 *Let $\mathcal{K} \subset \Phi \subset \mathcal{L}$ be a test space, the inclusions being also topological. If $F : \Phi \to L^2(P)$ is a g.r.f. which is harmonizable w.r.t. a tempered covariance ρ, then there is a random measure $Z : \mathcal{B}(\mathbb{R}^n) \to L^2(P)$ such that*

$$F(f) = \int_{\mathbb{R}^n} \hat{f}(t)dZ(t) = \int_{\mathbb{R}^n} \int_{\mathbb{R}^n} e^{itx} f(x)dxdZ(t), f \in \Phi, \quad (48)$$

uniquely. Conversely, F given by (48) with $Z(\cdot)$ and ρ, is a generalized harmonizable random field relative to ρ.

Since $\Phi \subset \mathcal{L}$ and the Fourier transform on $\mathcal{L}$ is isomorphism onto, the hypothesis holds and the result is a consequence of the theorem.

We now present the Cramér classes with conditions imposed on their local behavior. This was shown by A. M. Yaglom (1987) for vector-valued stationary random fields, and he appreciated (in a letter) the author's extension to the class (C) g.r.f. We thus present the result here.

Theorem 4.3.6 *On the test space Φ, let $F : \Phi \to L^2(P)$ be a grf, locally class (C), relative to $g(\cdot, \cdot)$, so that the g-transforms exist. Suppose $g(t, x - y) = g(t, x)\overline{g(t, x)}$, for all t, x, y in $\mathbb{R}^n$, and that (i) $\frac{\partial g}{\partial x}(s, 0) \neq 0$ for $s \neq 0$, (ii) $\frac{\partial g}{\partial s_k}(s, r) = \alpha_k(s, r)r_k$ and $\alpha_k(0, r) = \alpha_k \neq 0$ the α_k being independent of r. [s_k, r_k are kth elements of vectors s, r.] Then there is a tempered covariance ρ on $\mathcal{B}(\mathbb{R}^k - 0)^2$ such that for some p (the index of temperedness of ρ) one has:*

$$\int_{\mathbb{R}^n - \{a\}} \int_{\mathbb{R}^n - \{a\}} \frac{|x||y||d^2\rho(x, y)|}{[(1 + |x|^2)(1 + |y|^2)]^{\frac{p+1}{2}}} < \infty \quad (49)$$

and there is a random measure Z on $\mathcal{B}(\mathbb{R}^n - \{0\})$ relative to ρ and a random vector $\alpha = (\alpha_1, \ldots, \alpha_n)$ centered, uncorrelated with Z, so that

$$F(f) = \int_{\mathbb{R}^n - \{0\}} \tilde{f}(t)dZ(t) + (\alpha, \tilde{\nabla}\tilde{f}(0)), f \in \Phi, \quad (50)$$

where $(\cdot, \cdot)$ is the inner product, $\tilde{\nabla} = -(-\frac{1}{\alpha_k}\frac{\partial}{\partial x_k}, k = 1, \ldots, x)$ and $\tilde{f}$ is the g-transform. On the other hand, a map F given by (50) is a grf, locally of class (C) on Φ, relative to g.

The restrictions are satisfied if $g(t, x) = e^{i(tx)}$, and $\alpha_k = i$ for all k, where $\mathcal{K} \subset \Phi \subset \mathcal{S}$, in the test space notation for the stationary case. For a proof of the above general case, the following special form is useful.

Proposition 4.3.7 *Let F be a g.r.f. on Φ as in the above theorem with B as its covariance functional. Then there is a tempered covariance $\rho : \mathcal{B}(\mathbb{R}^n - \{0\}) \times \mathcal{B}(\mathbb{R}^n - \{0\}) \to \mathbb{C}$, satisfying (49) and a positive definite matrix A such that for $f_i \in \Phi, i = 1, 2$:*

$$B(f_1, f_2) = \int_{\mathbb{R}^n - \{0\}} \int_{\mathbb{R}^n - \{0\}} \tilde{f}_1(t) \bar{\tilde{f}}_2(s) d^2 \rho(t, s) + (A\tilde{\nabla} f_1 \tilde{\nabla} f_2)(0),$$

(51)

where $\tilde{\nabla}$ is given in (50). Conversely (51) defines a covariance functional of some locally class (C) generalized random field on Φ_1 as in the preceding result.

Proof. Let $F = (F_{r_1}, \dots, F_{r_k})$ where $F_r(f) = F(T_r f), f \in \Phi$ and $T_r = I - \tau_r$, τ_r being the translation operator, $r \in \mathbb{R}^n$. Then F is a multidimensional g.r.f. of class (C) for each k, and hence its covariance functional $B = (B_{ij})$ is given by

$$B_{ij}(f_1, f_2) = \int_{\mathbb{R}^n} \int_{\mathbb{R}^n} \widetilde{(T_{r_i} f_1)}(t) \overline{\widetilde{(T_{r_j} f_2)}}(s) d^2 \rho_{r_i r_j}(t, s), \qquad (52)$$

where $(\rho_{r_i r_j})$ is a positive definite matrix of tempered covariances. Since the τ_{r_i} commute it follows for $f \in \Phi$ that

$$F_{r_1' + r_1''}(f) = F_{r_i'}(f) + F_{r_1''}(\tau_{r_1'} f) = F_{r_1''}(f) + F_{r_1'}(\tau_{r_1''} f). \qquad (53)$$

Setting (53) into (52) with $r_1 = r_1' + r_1''$ and simplifying

$$\int_{\mathbb{R}^n} \int_{\mathbb{R}^n} \widetilde{T_{r_1''} f_1}(t) \bar{\tilde{f}}_2(s) \rho(dt, ds; r_1', r_2)$$

$$= \int_{\mathbb{R}^n} \int_{\mathbb{R}^n} \widetilde{T_{r_1'} f_1}(t) \bar{\tilde{f}}_2(s) \rho(dt, ds; r_1'', r_2). \qquad (54)$$

Similarly for $r_2 = r_2' + r_2''$, and simplifying one gets:

$$\int_{\mathbb{R}^n} \int_{\mathbb{R}^n} \tilde{f}_1(t) \overline{\widetilde{T_{r_2''} f_2}}(s) \rho(dt, ds; r_1, r_2')$$

$$= \int_{\mathbb{R}^n} \int_{\mathbb{R}^n} \tilde{f}_1(t) \overline{\widetilde{T_{r_2'} f_2}}(s) \rho(dt, ds'; r_1, r_2'').$$

For bounded Borel $A, B \subset \mathbb{R}^n - \{0\}$, we can define $\tilde{\rho}$ by,

$$\tilde{\rho}(A \times B) = \int_A \int_B \frac{\tilde{f}_1(t) \bar{\tilde{f}}_2(s)}{(T_{r_1} f_1)(\tilde{t})(T_{r_2} f_2)(s)} \rho(dtds; r_1, r_2), f_i \in \Phi. \quad (55)$$

Using the hypothesis on g one shows that $\tilde{\rho}$ is independent of g, and we may conclude that it gives the required tempered ρ satisfying (51).

Similar routine but tedious competitions (detailed in the author's paper [Rao (1969), Section 4]) show that the statement holds as given. □

Using this result the general Theorem 4.3.6 is obtained which is also proved in the same paper and the detail is omitted here. The interested reader will find that extending the method from the stationary case to the Cramér class is not quite straightforward, but working out the corresponding treatment is *instructive* to use in the general results for applications. Readers will find this paper (Rao (1969)) useful in understanding the generalized case better.

We now conclude this work by indicating the generalized isotropy classes that have been studied above for the harmonizable fields. This extension gives a further idea that these extended classes may be used in applied works.

4.4 Remarks on Harmonizability and Isotropy for Generalized Fields

Here we consider briefly the random fields that are defined on a Schwartz space Φ. It is the class of infinitely differentiable, (real) functions on $\mathbb{R}^n$ living on compact sets, as defined early. A mapping $F : \Phi \to L^2(P)$ is thus a grf (generalized random field) if $F(f_n) \xrightarrow{P} 0$ whenever $f_n \in \Phi$, $f_n \to 0$ in the (Schwartz) topology of Φ. If $\mathcal{G}$ denotes the space of all orthogonal matrices on $\mathbb{R}^n$, and $F : \Phi \to L^2(P)$ is a generalized random field (g.r.f.) then it is *isotropic* whenever, with $m(f) = E(F(f))$, the covariance $B(f_1, f_2) = E[(F(f_1) - m(f_1))\overline{(F(f_2) - m(f_2))}]$ and $m(\sigma_u f) = m(f)$ satisfy $B(\sigma_u f_1, \sigma_u f_2) = B(f_1, f_2)$, for all $u \in \mathcal{G}$, where $(\sigma_u f)(x) = f(ux), f \in \Phi$. Thus the means and covariances functions of the random field are invariant under the action of orthogonal transformation, and this can be characterized relatively easily, as compared to the isotropy considered early.

Proposition 4.4.1 *Let $F : \Phi \to L^2(P)$ be a grf of class (C) relative to a $g(\cdot, \cdot)$-family and a tempered covariance ρ, the Φ being our test space on $\mathbb{R}^n$ (with the Schwartz topology), for which the g-transformation exists, and $g(ut, x) = g(t, u^*x)$ for all $u \in \mathcal{G}$, the group of orthogonal matrices,* *u* being the adjoint of u. Then F is isotropic if and only if $\rho(ux, uy) = \rho(x, y)$ for $x, y \in \mathbb{R}^n$ and $u \in \mathcal{G}$.*

Proof. The covariance $B(\cdot, \cdot)$ of F can be given, from definition

$$B(f_1, f_2) = \int_{\mathbb{R}^n} \int_{\mathbb{R}^n} \tilde{f}_1(s) \bar{\tilde{f}}_2(t) d^2 \rho(s, t) \tag{56}$$

and for $u \in \mathcal{G}, i = 1, 2$ we have $[(\sigma_u f)(x) = f(ux), f \in \Phi]$,

$$\widetilde{(\sigma_u f_i)}(s) = \int_{\mathbb{R}^n} f_i(ux) g(s, x) dx$$

$$= \int_{\mathbb{R}^n} f_i(x) g(s, u^*x) dx,$$

since $u^{-1} = u^*$. Hence $B(\sigma_u f_1, \sigma_u f_2) = B(f_1, f_2)$ giving

$$\int_{\mathbb{R}^n} \int_{\mathbb{R}^n} \widetilde{\sigma_u f_1}(s) \overline{\widetilde{\sigma_u f_2}}(t) d^2 \rho(s, t)$$

$$= \int_{\mathbb{R}^n} \int_{\mathbb{R}^n} f_1(x) \bar{f}_2(y) \left[\int_{\mathbb{R}^n} \int_{\mathbb{R}^n} g(us, x) \overline{g(ut, y)} d^2 \rho(s, t) \right] dx dy$$

$$= \int_{\mathbb{R}^n} \int_{\mathbb{R}^n} \tilde{f}_1(s) \bar{\tilde{f}}_2(t) d^2 \rho(us, ut) = B(f_1, f_2). \tag{57}$$

But $\tilde{\Phi} = \{\tilde{f} : f \in \Phi\} \subset L^2(\rho)$ is dense so that from the representation of $B(f_1, f_2)$ above, and earlier analysis it follows that $\rho(us, ut) = \rho(s, t)$, as desired. The converse is immediate from definition. $\square$

With the specialization $g(t, x) = e^{i(t,x)}$, in the above result, we have the following consequence.

Corollary 4.4.2 *If $\mathcal{K} \subset \Phi \subset \mathcal{L}$, and F is a generalized harmonizable random field on Φ relative to a tempered covariance ρ, then F is isotropic if and only if $\rho(\cdot, \cdot)$ is invariant under the action of the orthogonal group $\mathcal{G}$.*

A multidimensional extension of the proceeding statements is straightforward and we present it (for reference) as follows:

Proposition 4.4.3 *Let $\Phi = \mathcal{K}$ and suppose $F : \Phi \to L^2(P)$ be a g.r.f., centered and with covariance functional $B(\cdot, \cdot)$ of function space type. Then $B(\cdot, \cdot)$ is harmonizable relative to a tempered covariance ρ iff there is a sequence $\{B_n(\cdot, \cdot), n \geq 1\}$ of covariances such that the following conditions (i)–(iii) hold:*

(i) $B_m(f_1, f_2) \to B(f_1, f_2)$ uniformly on $\Phi \times \Phi$ as $m \to \infty$,

(ii) Support $(B_n) \subset E_n \uparrow, \bigcup_{n=1}^{\infty} E_m = \mathbb{R}^n \times \mathbb{R}^n$, E_n compact, rectangle,

(iii) h_n is the continuous covariance giving $B_2 : (f, g) \to \int_{\mathbb{R}^n} \int_{\mathbb{R}^n} h(x, y) \times f^n(x) g^{\alpha}(y) dx dy$, with supports in compact increasing sets.

Proof Sketch. If $B(\cdot, \cdot)$ is harmonizable, then there is a tempered covariance ρ such that $B(\cdot, \cdot)$ is representable relative to ρ:

$$B(f_1, f_2) = \int_{\mathbb{R}^n} \int_{\mathbb{R}^n} \tilde{f}_1(t) \bar{\tilde{f}}_2(s) d^2 \rho(t, s), f_1, f_2 \in \Phi. \qquad (58)$$

If $\rho_m = \rho | E_m$ and B_m is the corresponding covariance functional, then $B_n \to B$ uniformly in $\Phi \times \Phi$ (B_m has compact support). So (i) and (ii) hold. For (iii), since B, hence B_n is of function type, it is representable with a continuous positive definite $h_m(\cdot, \cdot)$ as

$$B_m(f, g) = \int \int_{F_m} h_n(x, y) f(x) \overline{g(y)} dx dy. \qquad (59)$$

Hence (58) and (59) imply for B_m

$$\int \int_{E_m} f(x) \overline{g(y)} [h_m(x, y) - \int_{\mathbb{R}^n} \int_{\mathbb{R}^n} e^{i((t,x) - (s,y))} d^2 \rho(t, s)] dx dy = 0.$$
$$(60)$$

Since $f, g \in \Phi$ are arbitrary, it follows from (59) that $h(\cdot, \cdot)$ is harmonizable relative to ρ. It then is seen that h_m satisfies condition (iii) of the Proposition.

The converse follows from a similar argument, and uses a well-known theorem of Alexandroff's (cf. Dunford-Schwartz (1958), p. 316) so that $\rho_m \to \rho$ on compact sets of $\mathbb{R}^n \times \mathbb{R}^n$, one gets that ρ is a covariance of bounded variation on each such set and then B is harmonizable. $\qquad \square$

4.5 Summability Methods for Second Order Random Processes

In this section we enlarge the preceding analysis from stationary and harmonizable classes to include some more processes that are not necessarily harmonizable, using certain summability methods pioneered by

J. Kampé de Fériet (and Frankiel). These ideas enlarge the scope of applications of several nonstationary processes that are of great interest in extending some areas of stochastic analysis.

Definition 4.5.1 *Let* $X : \mathbb{R} \to L_0^2(P)$ *be a (centered) second order process whose covariance* $K : (s,t) :\to E(X_s \bar{X}_t)$ *is said to be Kampé de Fériet and Frankiel (to be abbreviated as* KF*) class if the following limit exists for all* $h \in \mathbb{R}$*:*

$$r(h) = \lim_{T \to \infty} \frac{1}{T} \int_0^{T-|h|} K(s, s+h)ds (= \lim_{T \to \infty} r_T(h)), \qquad (61)$$

giving a positive definite function $r(\cdot)$ *when* X *is measurable.*

This (positive definite measurable) $r(\cdot)$ then is expressible as:

$$r(h) = \int_{\mathbb{R}} e^{ith} F(dt), \qquad (62)$$

for almost all $h \in \mathbb{R}$ relative to Lebesgue measure. The positive bounded function $F(\cdot)$ is termed an *associated spectral function* of the process X.

It is clear that the stationary processes are in class (KF). This class and a closely related "asymptotically stationary" one introduced by Parzen (1962) will be of interest in our analysis. It will be seen that these include strongly (but not necessarily all weakly) harmonizable processes. To unify and extend further using (e.g. Hardy's) summability methods, Swift (1997) introduced the following (more) *general class* and analyzed it.

Definition 4.5.2 *A process* $X : \mathbb{R} \to L_0^2(P)$ *with a continuous covariance* r *is of* class $KF(p), p \geq 1$, *if for each* $h \in \mathbb{R}$,

$$\tilde{r}(h) = \lim_{T \to \infty} r^{(p)}(h) = \lim_{T \to \infty} \begin{cases} \frac{1}{T} \int_0^T r^{(p-1)}(h)dh, & p > 1, \\ \frac{1}{T} \int_0^{T-|h|} r(s, s+|h|)dh, & p = 1, \end{cases} \qquad (63)$$

exists, so that if $p = 1$ *it becomes the original class* (KF).

The early study for $p = 1$, has been developed by several people and the generalized class $KF(p), p > 1$, due to Swift (1992) does unify and extend the previous work.

The point of studying the above classes, with summability methods, is to analyze (in a general way) the processes introduced by Karhunen and Cramér subsuming the harmonizable families employing the same (summability) analysis which already became standard. This enhances the basic ideas of nonstationary analysis bringing it into the Fourier fold in general, and especially the applicational potential is enhanced.

The following representation leads to the generalized analysis:

Theorem 4.5.3 *Let* $X : T \to L_0^2(P)$ *be a process* $(T \subset \mathbb{R})$ *of weak Cramér class relative to* $\{g(t, \cdot), t \in T\}$ *real Borel functions so that the covariance* $r(\cdot, \cdot)$ *of* X *is given by*

$$r(t_1, t_2) = \int_S \int_S g(t_1, \lambda) \overline{g(t_2, \lambda')} d^2 F(\lambda, \lambda'), \tag{64}$$

with respect to a positive definite function $F(\cdot, \cdot)$ *of locally bounded variation on* $S \times S, (S, \mathcal{S})$ *being a measurable set and* $g(t, \cdot)$ *satisfies*

$$0 \le \int_S \int_S g(t, \lambda) \overline{g(t_2 \, \lambda_1')} d^2 F(\lambda, \lambda') < \infty, t_1 \in T. \tag{65}$$

Then there exists a vector measure $Z : \mathcal{S} \to L_0^2(P)$ *such that*

$$X(t) = \int_S g(t, \lambda) dZ(\lambda), t \in T, \tag{66}$$

$$E(Z(A)\bar{Z}(B)) = F(A, B), A, B \in \mathcal{B} = \sigma(S). \tag{67}$$

Conversely, if $X(\cdot)$ *is defined by* (66) *with* (67) *holding, then it is of weak class* (C).

The proof is quite standard and we leave it as an exercise for the reader. [The details can also be found in the paper, Chang and Rao (1986, Thm. 1 in Sec. 7). They are indeed standard using the MT-integrals in place of the Lebesgue integrals.] Motivated by these ideas of summability analysis Swift (1997) introduced an extension of class (KF) as follows.

Definition 4.5.4 *A process* $X : \mathbb{R} \to L_0^2(P)$ *with a continuous covariance* r *is termed of class* $\{(KF, p), p \ge 1\}$ *if for* $h \in \mathbb{R}$, *we have*

$$\bar{r}(h) = \lim_{T \to \infty} r_T^{(p)}(h),$$

existing for each h, *where*

$$r_T^{(p)}(h) = \begin{cases} \frac{1}{T}\int_0^T r^{(p,1)}(t)dh, & p > 1 \\ \frac{1}{T}\int_0^T r(s,s+|h|)dh, & p = 1. \end{cases}$$

[Thus $(KF,1)$ is the original class (KF).]

The classical summability analysis implies that these are inclusive with p i.e., class $(KF) \subset$ class $(KF,p) \subset$ class $(KF,p+1)$. Also, we have the following result obtained by Swift (1997).

Theorem 4.5.5 *The (C,p)-summable Cramér processes are contained in the class $(KF,p), p \geq 1$, relative to a class $\{g(t,\cdot), t \in \mathbb{R}\}$ of Borel functions with $g(t,\cdot)$ as in Theorem 4.5.3. If $Q : L_0^2(P) \to L_0^2(P)$ is a bounded linear operator, then $Y(t) = QX(t), t \in \mathbb{R}$ is (C,p) summable $(p \geq 1)$ weak class (C) relative to the functions $g(t,\cdot)$ and $F(\cdot,\cdot)$ as in (64), (65).*

There are numerous extensions and alterations of the above classes with the corresponding analysis, nontrivially. We invite the interested workers in the subject to study, Swift's extended analysis with details given as noted above. The profitable applications of the well-developed summability analysis results, in the stochastic theory, are of immense use in extensions as well as in applications. It is possible to go further and use almost periodic analysis in this context and advance the subject. This will be indicated in the complements section below. We treat another aspect now.

4.6 Prediction Problems for Stochastic Flows

There is always a deep interest to study the structure of stochastic flows governed by certain (stochastic) differential equations with orthogonal or (naturally) harmonizable noise processes. An early work by Dolph and Woodbury (1953) and the latter one extended by Dym (1966) are useful contributions for analysis and discussion here.

Consider a linear differential operator L_t defined by:

$$(L_t g)(t) = \sum_{i=0}^{n} \alpha(t)\frac{d^{n-1}g(t)}{dt^{n-2}}, \alpha_0(t) \neq 0, t \geq t_0. \tag{68}$$

Then the process $\{Y(t), t \in T\}$ satisfying

$$(L_t Y)(t) = Z(t), t \in T = [t_0, t], \tag{69}$$

is termed a *stochastic flow* driven by the (harmonizable) noise process, $\{Z(t), t \in T\}$, where $\alpha(t)$'s are $(n-1)$ times continuously differentiable real functions. Here (69) is "interpreted" as satisfying (formally let $dZ(t) = Z'(t)dt$)

$$\int_T Q(t)(L_t Y)(t)dt$$

$$= \int_T Q(t)Z'(t)dt \left(= \int_T Q(t)dZ(t) \right), \tag{70}$$

for all $Q(\cdot)$ with compact supports and the right side is the Bochner (or a D-S) integral.

Theorem 4.6.1 *Let $\{Y(t), t \geq t_0\}$ be a process defined by*

$$Y(t) = \int_{t_0}^t R(s,t)dZ(s)$$

where $Z(t)$ is a centered weakly harmonizable (noise) process as in (69). Then it belongs to a weak Cramér class so that its covariance r satisfies

$$r(s,t)$$
$$= (X(s), Y(t))$$
$$= \int_{\tilde{\mathbb{R}}} \int_{\tilde{\mathbb{R}}}^* g(s,\lambda)g(t,\lambda')dF(\lambda,\lambda') \tag{71}$$

for $(s,t) \in \mathbb{R} \times \mathbb{R}$, where $\tilde{\mathbb{R}}$ is the dual group of $\mathbb{R}$, and r is $2n-2$ times continuously differentiable with $(2n-1)$st derivative having a (possible) jump at $t = s$, given by:

$$\lim_{\varepsilon \downarrow 0} [(D_1^{2n-1}r)(t, s+\varepsilon) - (D_1^{2n-1}r)(t, s-\varepsilon)]$$
$$= (-1)^n(\alpha(t))^{-2}. \tag{72}$$

Proof. Since $Z(\cdot)$ and $Z'(\cdot)$ are centered, one has for $r(s,t)$:

$$r(s,t) = E(Y(s)\bar{Y}(t))$$

$$= \int_{t_0}^{s} \int_{t_0}^{t} R_s(w)\bar{R}_t(v)E[Z'(w)\bar{Z}'(v)]dudv,$$

the interchange of $E(\cdot)$ and integral being valid,

$$= \int_{t_0}^{s} \int_{t_0}^{t} R_t(w)\bar{R}_t(v) \int_{\mathbb{R}} \int_{\mathbb{R}}^{*} e^{iu\lambda - iv\lambda'} dG(\lambda\lambda')dudv,$$

with $dG(\lambda, \lambda') = \lambda\lambda' dF(\lambda, \lambda')$, F being the spectral bimeasure of $Z(\cdot)$ and $Z'(\cdot)$, which is weakly harmonizable,

$$= \int_{\mathbb{R}} \int^{*} \left[\int_{\mathbb{R}} e^{iu\lambda}\bar{R}(u)du \right] \left[\int_{\mathbb{R}} e^{iv\lambda'}\bar{R}(v)dv \right] dG(\lambda, \lambda'),$$

where $\tilde{R}(u)(\chi_{(t,s)}, R_s)(u)$ and the integral interchange being justifiable,

$$= \int_{\mathbb{R}} \int_{\mathbb{R}}^{*} \hat{R}(\lambda)\hat{\bar{R}}(\lambda')dG(\lambda, \lambda'). \tag{73}$$

Here $\hat{R}_s$ is the Fourier transform of $\tilde{R}_s$, well-defined by the properties of R_s. Since G is a bimeasure, (73) implies that the (covariance and hence) Y-process is of (weak) Cramér class. The other differentiability properties are deduced from (73) and the fact that L_t and the integrals in (73) commute, so that $(L_t r_s)(t_0) = 0$ and $(L_t r_s)(t) = 0$ as well as the remaining assertions are verified by simple computations.

Since $(L_t R_t)(t_0) = 0$, one has $(D_t^n R)(s,t) = -\sum_{t=1}^{n} \alpha(t)(D_1^{n-k}R) \times (s,t)$, so that $D_1^k r$ exists, $0 \le k \le n - 1$, and then $D_2^k r$, $0 \le k \le n$ also exist. Finally

$$\lim_{\varepsilon \downarrow 0}(D_1^k r)(s - t, s) = \int_{t_0}^{s} \int_{t_0}^{s} D_1^k R(s, u)\bar{R}(s, v)dudv + (-1)^{n-1}\frac{\delta_k^{2n-1}}{\alpha_0^2(t)},$$

since $(D_t^{n-1}R)(s,t) = \alpha_0(t)^{-1}$ where δ_k is the Kronecker symbol above. If the $\alpha_i(\cdot)$ are C^∞-smooth, then $r(\cdot)$ will be also C^∞-smooth. This property plays a basic role in Dym's (1911) analysis of stationary measures, in differential equations governed by white noises. This completes the essentials of proof of this result. $\square$

The following result extends a basic fact on stochastic flows observed by Dolph and Woodbury (1952) which played a key motivational role in the theory of unbiased linear least squares prediction:

Theorem 4.6.2 *Let* $X(t) = Y(t) + Z(t), a \leq t \leq b$ *be a signal plus noise model, with the* Y, Z *processes having two moments. Suppose the covariances and cross covariance of* Y, Z *are known and are smooth in that they are* n-*times continuously differentiable. If the process* $\{X(t), a \leq t \leq b\}$ *is observed, an unbiased linear weighted least-squares predictor* $\hat{X}(t_0)$ *of* $Y(t_0)$ *of the form*

$$\hat{X}(t_0) = \sum_{k=0}^{n} \int_a^b X^{(k)}(t) dp_k(t, t_0) \tag{74}$$

relative to a set of complex weight functions $p_k(\cdot, t_0), k = 0, 1, \ldots, n$, *of bounded variation is possible if and only if the* p_k *minimize* $J_0(p)$:

$$0 \leq J_0(p) = \sum_{k,l=0}^{n} \int_a^b \int_a^b [D_1^k D_2^l K(s,t) dp_k(s, t_0) + r_k(t_0, t_0)]$$

$$+ 2 \sum_{j=0}^{n} \lambda(t_0) g_j(t_0) - 2 \sum_{k=0}^{n_0} \int_a^b [\mathrm{Re}\,(D_1^k K_1)(t, t_0)]$$

$$+ \sum_{j=1}^{m} \lambda_j(t_0) g_j^{(k)}(t)] dp_k(t, t_0)$$

with λ_j *as Lagrange multipliers. Thus the minimum of* J_0 *yields the best solution.*

The many details, some of which are technical, are given in the author's paper (1994) which improves and generalizes the basic work by Dolph and Woodbury (1952). The reader is advised to read through both these papers to understand the nontrivial mathematical analysis that is basic to this type of applications. Some related extensions are also given there and should be of interest to the workers now.

An extended form of the preceding result, useful is a number of applications, given in a Karhunen form that contains both the stationary and harmonizable classes, is defined by:

$$Y(t) = \int_{\mathbb{R}} \tilde{R}(t, u) dZ(u) \tag{75}$$

for a suitable kernel $R(\cdot, \cdot)$ so that we have (with centered Z)

$$r(s,t) = E(Y(s)\bar{Y}(t)) = \int^{s \wedge t} R(s,u)\bar{R}(t,u)dH(u)$$

$$= \int_{\mathbb{R}} \tilde{R}(s,u)\bar{\tilde{R}}(t,u)dH(u) \tag{76}$$

where $\tilde{R}(s,u) = (\chi_{(-\infty,u)}R(s,\cdot))(u)$. This is Karhunen's form. For such processes we have the following general result of interest.

Theorem 4.6.3 *Let $X(t) = Y(t) + Z(t), a \leq t \leq b$, be a second order process with $Y(t)$-process as signal and $Z(\cdot)$, the noise, with orthogonal increments having $H(\cdot)$ as the variance function, and $E(Y(t)) = \sum_{i=1}^{m} a_i g_i(t), a_i \in \mathbb{R}$. Suppose that the noise process $Z(\cdot)$ also has orthogonal increments with the variance function $H(\cdot)$. Assume that the covariance function of $Y(\cdot)$ and $Z(\cdot)$-processes are $r_y(s,t), r_{yz}(s,t)(= \mathrm{cov}\,(Y(s), Z(t)))$ as given. Then an unbiased linear weighted least-squares predictor $\hat{X}(t_0)$ of $Y(t_0)$ of the form*

$$\hat{X}(t_0) = \int_a^{t_0} X(t)dp(t,t_0) \tag{77}$$

relative to $p(\cdot, t_0), t_0 > b$ of bounded variation exists when $p(\cdot, t_0)$ is a solution of the integral equation

$$K_1(t,t_0) + \sum_{j=1}^{m} \lambda_i(t_0)g_j(t) = \int_a^b K(s,t)dp(s,t_0) \tag{78}$$

with $K_1(t,t_0) = (r_y + r_{yz})(t,t_0), K(s,t) = r_y(s,t) + H(s \wedge t) + 2r_{yz}(s,t)$, the $\lambda_i(t_0)$ being Lagrange multipliers and $\int_a^b g_i(t)dp(t,t_0) = g_j(t_0), i = 1, \ldots, m$. The minimum mean-squared error $\sigma^2 = E(\hat{X}(t_0) - Y(t_0))^2$ is given by:

$$\sigma^2 = r_y(t_0,t_0) + \sum_{i=1}^{m} \lambda_j(t_0)g_j(t_0) - \int_a^b K_1(t,t_0)dp(t,t_0). \tag{79}$$

Remark 25. In the case that the signal and noise are uncorrelated, whence $r_{yz} = 0$ and so $K = r_y$ and $K = r_y + H$, the above equations simplify with $K = r_y + H$. If Y is nonrandom, then $r_y = 0 = r_{yz}$ and (79) reduces to $\sum_{j=1}^{m} \lambda_i(t_0, g_j(t_0)) = \int_a^b H(s \wedge t)dp(s,t_0)$. But if no signal is sent (so $Y = 0$), then with $p'(\cdot, t_0)$ existing, we have

$\int_a^b dp(t, t_0) = 0$, and other simplifications result. If $n = 0$ and $m = 2, r(s, t) = \exp[-\beta(t - s)]$ with $\beta > 0$ Z-Gaussian, the result was detailed in Dolph and Woodbury (1952). Specializations for Cramér and Karhunen classes can now be obtained and they are useful.

The preceding analysis motivates the following extension of interest and the related analysis for people desiring to go further.

Proposition 4.6.4 *Consider a second order process* $X(t) = Y(t) + Z(t)$ *on the interval of observations* $a \leq t \leq b$ *where the signal process* Y *is satisfying:* $E(Y(t)) = \sum_{i=1}^m a_i g_i(t), a_i \in \mathbb{R}$, *the noise* $Z(\cdot)$ *has orthogonal increments with variance function* $H(\cdot)$ *and the covariances of* Y *and* Z *are given, so that* $r_y : (s, t) \rightarrow \mathrm{cov}\,(Y(s), Y(t)), r_{yz} : (s, t) \rightarrow \mathrm{cov}\,(Y(s), Z(t))$ *are also given along with* $a_j, 1 \leq j \leq n$. *Then an unbiased weighted linear least squares predictor* $\tilde{X}(t_0)$ *of* $Y(t_0)$ *of the form:*

$$\tilde{X}(t_0) = \int_a^b X(t) dp(t, t_0) \tag{80}$$

relative to a real weight $p(\cdot, t_0)$ *for* $t_0 > b$ *exists provided* p *satisfies:*

$$K_1(t, t_0) + \sum_{i=1}^n \lambda_i(t_0) a_i(t) = \int_a^b K(s, t) dp(s, t_0) \tag{81}$$

with $K_1(t, t_0) = (r_y + r_{yz})(t, t_0), K(s, t) = r_y(s, t) + H(s \wedge t) + 2r_{yz}(s, t)$, *the* $\lambda_i(t_0)$ *being the Lagrange multipliers and*

$$\int_a^b g_j(t) dp(t, t_0) = a_j(t_0), j = 1, \ldots, m. \tag{82}$$

The minimum mean-squares error σ^2 *is given by*

$$\sigma^2 = E(\hat{X}(t_0) - Y(t_0))^2 = r_y(t_0, t_s) + \sum_{j=1}^m \lambda_i(t_0) g_0(t_0)$$

$$- \int_a^b K(t, t_0) dp(t, t_0). \tag{83}$$

Remark 26. If the signal and noise are uncorrelated, so $r_{yz} = 0$ and $K = r_y$ and if the signal $Y(\cdot)$ is also deterministic so $r_y = r_{yz} = 0$ then (83) reduces to

$$\sum_{j=1}^{m} \lambda_j(t_0) g_j(t_0) = \int_a^b H(\min(s,t)) dp(s,t_0). \qquad (84)$$

If no signal is sent, so $Y = 0$, one gets $\int_a^b dp(s, t_0) = 0$ and $\sigma_2 = 0$.

If $m = 2$, $n = \infty$ and $r(s,t) = \exp[-\beta|(t - s)|]\beta > 0$ with Z-Gaussian, an example is detailed in Dolph-Woodbury (1952), which may be of interest to the reader who wants to study with more details, in order to appreciate the problems in applications.

We now discuss (briefly) the spectral function estimation of a (weakly) harmonizable process. Thus if a process $\{Y_t, -T \le t \le T\}$ is assumed weakly harmonizable, centered, with a covariance kernel R, so that

$$Y_t = \int_{t_0}^t R_t(s) dZ(s) \qquad (85)$$

whose covariance $r_y(\cdot, \cdot)$ is represented by

$$r_y(s,t) = E(Y_s \bar{Y}_t) = \int_{\mathbb{R}} \int_{\mathbb{R}} R_y(u) \bar{R}(v) dG(u,v) \qquad (86)$$

where $dG(\lambda, \lambda') = \lambda \lambda' dF(\lambda, \lambda')$, the $F(\cdot, \cdot)$ being the spectral function of Z in (85) and $dG(\lambda, \lambda') = \lambda \lambda' dF(\lambda, \lambda')$. We now give an estimator $\hat{F}_T$ of F (so it gives a $\hat{G}_1$ by the above) when the process is observed on $-T \le t \le T$, with the following properties:

(i) $\hat{F}_T$ is unbiased in limit as $T \to \infty$ (i.e., asymptotically unbiased).
(ii) $\hat{F}_T$ is consistent, i.e., $(\hat{F}_T - F_T) \to 0$ in probability as $T \to \infty$.
(iii) $g(T)^{-1}(\hat{F}_T - F_T) \to$ a random variable (in distribution), as $T \to \infty$.
(iv) The speed of convergence in these limits is given.

The following brief discussion on the spectral estimation may be of interest for research workers in this area. It is a consequence of Theorem 4.6.3 above and is given to explain the subject for applications.

Proposition 4.6.5 *In a signal plus noise model $X(t) = Y(t) + Z(t)$, $a \le t \le b$ where the observed process $X(\cdot)$ is a signal process $Y(\cdot)$, disturbed by the noise $Z(\cdot)$, a process of orthogonal increments, $a_j \in \mathbb{R}$, with a variance function $H(\cdot)$, and covariance $r_{yz}(s,t)$ be that of Y, Z processes, and similarly r_{yy}, r_{yz} are defined for the Y and Y, Z*

and $a_i, i \leq j \leq m$ given. Then a weighted linear unbiased least squares predictor $\hat{X}(t_0)$ of $X(t_0)$ of the form

$$\hat{X}(t_0) = \int_a^b X(t)dp(t, t_0)$$

relative to a weight function $p(\cdot, t_0)$ for $t_0 > b$ exists when $p(\cdot, t_0)$ solves

$$K_1(t, t_0) + \sum_{j=1}^m \lambda_i(t_0)a_j(t) = \int_a^b K(s, t)dp(s, t_0) \qquad (87)$$

where K_1 and K are given by $K_1(t_1 t_0) = (r_{yy} + r_{yz})(t, t_0)$, $\lambda_j(t_0)$ being Lagrange multipliers, $K(s, t) = r_y(s, t) + H(s \wedge t) + 2r_{yz}(s, t)$, and

$$\int_a^b g_j(t)dp(t, t_0) = g_j(t_0), j = 1, \ldots, m. \qquad (88)$$

The minimum mean squared error σ^2 is given by

$$\sigma^2 = E(\hat{X}(t_0) - Y(t_0))^2$$
$$= r_y(t_0, t_0) + \sum_{i=1}^m \lambda_i(t_0) - \int_a^b K_1(t, t_0)dp(t, t_0). \qquad (89)$$

Note: If the noise and signal are uncorrelated so that $r_{yz} = 0$ and $K_1 = r_y, K = r_y + H$, the above equations slightly simplify. When Y is deterministic, so that $r_y = r_{yz} = 0$, then we have

$$\sum_{j=1}^m \lambda_i(t_0)g(t) = \int_a^b H(s \wedge t)dp(s, t_0). \qquad (90)$$

Thus this reduces for $m = 1$, $Y = a$ (constant) so $g = 1$ and (90) becomes with the unbiasedness constraint

$$\lambda_1 = \int_a^b H(s \wedge t)dp(s, t_0), \int_a^b dp(s, t_0) = 1. \qquad (91)$$

Thus, in this case, the minimum value of the estimator is λ_1 and the values of λ_1 and p are obtainable from the equations (91).

If $Y = 0$ (no signal is sent), then assuming for $p(\cdot, t_0) = \int_a^b dp(s, t_0)$ $= 0$, and

$$0 = \int_a^b H(s)p'(s,t_0)ds \to H(t)\int_a^b p'(s,t_0)ds,$$

one concludes that $p(\cdot) = 0$ and $\tilde{X}(t_0) = 0$ so $\sigma^2 = 0$ the error variance.

We conclude this section with a few results on generalized harmonizable, and also of Cramér (Karhunen) type, random fields. The following technical result will clarify in studying the structural properties of these classes.

Theorem 4.6.6 *Let $r : \mathbb{R}^n \times \mathbb{R}^n \to \mathbb{C}$ be a bounded continuous covariance function. Then it is (weakly) harmonizable if and only if there exist a sequence $r_m, m \geq 1$ of uniformly bounded continuous covariances such that (i) $r_m \to r$ uniformly in $\mathbb{R}^n \times \mathbb{R}^n$ as $m \to \infty$; (ii) the support E_m of r_m is compact, $E_m \subset E_{m+1}$, and $\cup_{m \geq 1} E_m = \mathbb{R}^n \times \mathbb{R}^m$ and (iii) for each m, if $\{\phi_k^m, k \geq 1\}$ denotes a complete set of eigenfunctions of r_m for some $\{G_k^m\}$ of bounded variation on $\mathbb{R}^n$.*

Remark 27. If $r(s,t) = \tilde{r}(s-t)$ then we can take $r_m = \phi_k^m, k \geq 1$, when the r_m are suitable restrictions of r and the G_k^m are then positive finite measures as in Bochner's classical characterization of covariance generally. We sketch a proof of the result depending on a theorem of A. D. Alexandriff (cf. Dunford-Schwartz (1958) p. 366) and present its extension which is of interest here.

Proof. In the forward direction, if $r(\cdot, \cdot)$ is harmonizable then there is a covariance function of bounded variation, ρ, such that

$$r(s,t) = \int_{\mathbb{R}^n}\int_{\mathbb{R}^n} e^{((i(s,x)-i(t,y))}d^2\rho(x,y). \tag{92}$$

If now $E_n \subset E_{n+1} \subset \mathbb{R}^n \times \mathbb{R}^n$, is a sequence of compact rectangles covering the whole space let $\rho_n = \rho|E_n$ and r_n be the resulting integral of (92) with ρ_n for ρ there. That the sequence $\{r_n, n \geq 1\}$ of (92) satisfies (ii) and (i) follows from the classical A. D. Alexandroff's result (cf., Dunford-Schwartz (1958, p. 316). About (iii) let $\rho_n = \alpha, r_m = h$ and by Mercer's theorem

$$h(x,y) = \sum_{i=1}^{\infty} \mu_i \phi_i(x)\bar{\phi}_i(y), \quad \mu_i \geq 0, \tag{93}$$

where $\{\mu_i\}, \{\phi_i\}$ are its eigenvalues and the normalized complete set of eigenfunctions, the series (93) converging uniformly and absolutely. Also if $\mu_j\phi_j(x) = \int_{\mathbb{R}^n} h(x,y)\phi_j(y)dy$ so that (92) gives (for h):

$$\phi_j(x) = \int_{\mathbb{R}^n} e^{i(t,x)} dG_j(t)$$

where $G_j(t) = \frac{1}{\mu_n} \int_{\mathbb{R}^n} \hat{\phi}_j(s) d_s \alpha(t,s)$, $\hat{\phi}_j$ being the Fourier transform of ϕ_j. So G_j is of bounded variation, whence (iii) holds.

For the converse, let (i)–(iii) hold and consider an r_m. Then an expansion of the type (93) holds for r_n with some G_j^m. If $F_j^m(s,t) = G_j^m(t)G_j^m(s)$, then from the (uniform) convergence of $F^n(s,t) = \sum_{j=1}^{\infty} \mu_j^m F_j^m(s,t)$ one easily obtains, with the convergence of (93),

$$r_m(x,y) = \int_{\mathbb{R}^m} \sum_{j=1}^{\infty} u_j^m \int_{\mathbb{R}^n} e^{i[(s,x)-(t,y)]} d^2 F_j^m(s,t)$$

$$= \int_{\mathbb{R}^m} \int_{\mathbb{R}^n} e^{i[(s,x)-(t,y)]} d^2 F^n(s,t) \qquad (94)$$

and that the F^m is a covariance of bounded variation, so that r_m is a harmonizable covariance, and (iii) already characterizes continuous harmonizable covariance on a compact set. Now conditions (i) and (iii) can be used to infer harmonizability of $r(\cdot,\cdot)$. Indeed, since F^m determines a finite positive regular measure on $\mathbb{R}^n \times \mathbb{R}^n$, (i) implies that the integrals in (94) converge when F_m is considered as determining a regular measure and then invoke the converse part of Alexandroff's theorem as used above, so that $F^m \to \rho$, a covariance of bounded variation on $\mathbb{R}^n \times \mathbb{R}^n$. This gives the converse and the theorem follows. $\square$

This result admits extensions to Cramér and Karhunen processes. We include that result, of interest here, to complete this set of ideas. It also admits an easy characterization of generalized harmonizable fields, such an extension is of interest for this analysis as well as applications.

Recall that $\mathcal{K}$ is the L. Schwartz space of infinitely differentiable compactly supported functions (complex) on $\mathbb{R}^n$, defined and discussed in Chapter 2 (and used there for the Gel'fand representation theorem on local functions). We use some properties here without repeating the concepts and related details. It will be motivational but also useful to present the development of the result and summarize the work in a proposition as it also gives the motivation. Thus $X : \mathbb{R} \to L_0^2(P)$ is a (measurable) mapping, centered, and its covariance $r : (s,t) \to E(X(s)\bar{X}(t))$ is of *Karhunen class* where we set $r(\cdot,\cdot) : (s,t) \to E(X(s)\bar{X}(t)) = (X(s), X(t))$, with inner product notations. It admits an (LS)-representation as:

$$r(s,t) = \int_{\mathbb{R}} g(s,\lambda)\bar{g}(t,\lambda)F(d\lambda), \quad s,t \in \mathbb{R}, \tag{95}$$

relative to a class of Borel functions $\{g(s,\cdot), s \in \mathbb{R}\}$ and a σ-finite Borel measure $F : \mathcal{B}(\mathbb{R}) \to \mathbb{R}^*$. Then the process is shown to be representable as

$$X(t) = \int_{\mathbb{R}} g(t,\lambda)Z(d\lambda), \quad t \in \mathbb{R}, \tag{96}$$

where $Z : \mathcal{B}_0(\mathbb{R}) \to L_0^2(P)$, such that $(Z(A), Z(B)) = F(A \cap B)$, for $A, B \in \mathcal{B}_0(R)$ of bounded Borel sets of $\mathbb{R}$, so that $Z(\cdot)$ has orthogonal increments, and this has an extension, to be called a *Cramér process* if $Z(\cdot)$ has non-orthogonal increments, but covariance is to be representable as

$$r(s,t) = \int_{\mathbb{R}} \int_{\mathbb{R}} g(s,\lambda)\overline{g(t,\lambda')}\tilde{F}(d\lambda, d\lambda'), s,t \in \mathbb{R}, \tag{97}$$

relative to Borel functions $\{g(s,\cdot), s \in \mathbb{R}\}$ and a given covariance bimeasure F, satisfying

$$0 \leq \int_{\mathbb{R}} \int_{\mathbb{R}} g(s,\lambda)\overline{g(t,\lambda')}F(d\lambda, d\lambda') < \infty, \tag{98}$$

of finite Fréchet variation on $\mathcal{B}_0(\mathbb{R}) \times \mathcal{B}_0(\mathbb{R})$. Evidently, the Cramér class includes the Karhunen family, somewhat analogous to the harmonizable class includes the stationary family. Here MT integration is used. [A complete account of the (restricted) MT-integration is detailed in a chapter by D. K. Chang and the author (1986) and will be used here. The reader can fill in the omitted details easily from it.] Using this we now discuss here how the Cramér and Karhunen classes of process that extend the stationary and harmonizable classes to some extent, and the limitations of the method as compared with the harmonizable case will be made explicit. Our guidance (and real motivation) here is that, on an extended $L^2(P)$-space, we consider a (weakly) stationary process whose orthogonal projection gives a (weakly) harmonizable process, and we wish to extend these ideas to some nonstationary classes, especially Karhunen and Cramér classes on suitably enlarged $L^2(P)$-spaces. We want to make this idea precise and get a possible extension of classes involving these two families which also show the limitations of this method.

The motivation here is the classical de la Valée Pussion's criterion of uniform (Lebesgue) integrability of a bounded set of (real) functions on the line. This is needed for a vector measure $\nu : \Sigma \to \mathcal{X}$ relative to a σ-finite measure $\mu : \Sigma \to \bar{\mathbb{R}}^+$ so that we have

$$\| \int_\Omega f d\nu \|_{\mathcal{X}} \leq \|f\|_{f,\mu}, \quad f \in L^f(\mu) \tag{99}$$

for some $1 \leq f < \infty$, and then ν is termed dominated by μ. A solution of this problem is needed for us in our ensuring analysis. Here we need to recall a generalized variation of a vector measure and obtain a solution, and the $L^1(\mu)$ spaces should be replaced by more general Orlicz spaces to understand the problem. Thus if $\phi : \mathbb{R} \to \mathbb{R}^*$ is a symmetric convex function, $\phi(0) = 0$, its conjugate $\psi : \mathbb{R} \to \mathbb{R}^*$ is defined as $\psi(x) = sup\{|x|y - \phi(y) : y \geq 0\}$, so that $\psi(\cdot)$ is also a symmetric convex function, called the conjugate of ϕ and satisfies $|xy| \leq \phi(x) + \psi(y), x, y \geq 0$, and the norm, $(L^\phi(\mu), \| \cdot \|_\phi)$ is given as

$$\|f\|_{\psi,\mu} = \inf \left\{ \alpha > 0 : \int_\Omega \psi \left(\frac{f}{\alpha} \right) d\mu \leq 1 \right\}. \tag{100}$$

With this preparation, the ϕ semi-variation of $\nu : \Sigma \to \mathcal{X}$, a Banach space, is defined as:

$$\|\nu\|_\phi(A) = \sup\{\| \int_A f(\omega)\nu(d\omega)\| : \|f\|_{\psi,\mu} \leq 1\}, \tag{101}$$

where the norm of f is given by $\|f\|_{\psi,\mu} = \inf\{\alpha > 0 : \int_\Omega \psi \left(\frac{s}{\alpha} \right) d\mu \leq 1\}$ which is the usual Lebesgue integral ($\|f\|_{\psi,\mu}$) is the norm of $f \in L^\psi(\mu)$, $\psi(\cdot)$ a convex Young function, called the *conjugate of the convex* ϕ. In the classical case when $\phi(x) = |x|^p/p, p \geq 1$, then the conjugate of ϕ is $\psi(\cdot)$ given by $\psi(y) = (y)^q/q, p^{-1} + q^{-1} = 1$. For us, the pair (ϕ, ψ) is quite general. With this introduction we have the following:

Theorem 4.6.7 *Let (Ω, Σ) be a measurable space, $\mathcal{X}$ is a Banach space with $\nu : \Sigma \to \mathcal{X}, \sigma$-additive. Then there is a measure $\mu : \Sigma \to \mathbb{R}^+$, a Young function $\phi : \mathbb{R} \to \mathbb{R}^+, \frac{\phi(x)}{x} \uparrow \infty$ as $x \uparrow \infty$, with $\|\nu\|_\phi(\Omega) < \infty$ in (101) and so ν is dominated by the couple (ϕ, μ) [μ is often termed a control measure of ν. If $\phi(x) = \frac{x^2}{2}$, so $\psi(x) = \frac{x^2}{2}$ also, then ν is said to have 2 semi-variation finite].*

Proof. Since the weak and strong σ-additives are known to be the same, for a disjoint $\{A_n, n \geq 1\} \subset \Sigma$, and $x^* \in \mathcal{X}^*, \|x^*\| \leq 1$, we have

$$0 = \lim_{n \to \infty} \|\nu(\bigcup_{n=1}^{\infty} A_n) - \sum_{m=1}^{n} \nu(A_m)\|$$

$$= \limsup_{n \to \infty} \{|x^*(\nu)[\bigcup_{n \geq 1} A_n] - \sum_{K=1}^{n} (x^*(\nu))(A_k)| : \|x^*\| \leq 1\} \quad (102)$$

and thus the scalar measures $\{x^*(\nu_n), \|x^*\| \leq 1\}$ are uniformly σ-additive on Σ. Then by a well-known result (Dunford-Schwartz (1958), IV.16.5) there is a finite positive dominating measure $\mu : \Sigma \to \mathbb{R}^+$ such that $x^* \circ \nu \ll \mu, \|x^*\| \leq 1$. Hence by the classical Radon-Nikódym theorem $g_{x^*} = \frac{dx^*(\nu)}{d\mu)}$ exists and (102) implies:

$$0 = \lim_{\mu(A) \to 0} |x^* \circ \nu(A)| = \lim_{\mu(A) \to 0} \int_A g_{x^*}(t)\mu(dt), |x^*| \leq 1. \quad (103)$$

Thus $\{g_x, x^* \in S^*\} \subset L^1(\mu)$ is bounded and uniformly integrable, since $\mu(r) < \infty$, by the classical de la Vallée Poussin's theorem there exists a convex ϕ of the desired type such that

$$\int_{\Omega} \phi(|g_{x^*}(t)|)\mu(dt) \leq K_0 < \infty, \quad \|x^*\| \leq 1. \quad (104)$$

To simplify further, let $\psi : \mathbb{R} \to \mathbb{R}^*$ be the conjugate of ϕ so that (104) gives:

$$\|\nu\|_{\phi}(\Omega) = \sup\{\|\int_{\Omega} f(\omega)\nu(d\omega)\|; \|f\|_{\psi,\mu} \leq 1\}$$

$$= \sup\left\{ \sup\left[\left|\int_{\Omega} f(\omega)g_{x^*}(\omega)\mu(d\omega)\right|\right] : x^*\varepsilon\mathfrak{X}^*, \right.$$

$$\left. \|f\|_{\psi,\mu} \leq 1\right\}$$

$$\leq 2\sup\{\sup \|g_{x^*}\|_{\phi,\mu}\|f\|_{\psi,\mu}; x^* \in S^*, \|f\|_{\psi,\mu} \leq 1\},$$

by Hölder's inequality for the L_{μ}^{ϕ}-spaces,

$$\leq 2\{\|g_{x^*}\|_{\phi,\mu} : x^*\varepsilon S^*\} < 2k_0 < \infty, \quad \text{by (104)}.$$

This completes the proof.

Remark 28. Since $\|\nu\|(\Omega) = \|\nu\|_1(\Omega)$ in this notation with $\phi(x) = |x|$, and $\mu(\Omega) < \infty$, it is clear that $L^{\phi}(\mu) \subset L^1(\mu)$ and then $\|\nu\|_{\phi}(\Omega) \leq$

$C_0\|\nu\|(\Omega) < \infty$, and $\phi(\cdot)$ can grow exponentially. Thus if $p > 1$, $\phi(x) = |x|^p$, it is nontrivial to find a good $0 < C < \infty$, with $\|\nu\|_\phi(\Omega)$ bounded by a simple verifiable bound C in terms of $\|\cdot\|_p$ for immediate applications. The following special result, established by a particular method, due to J. Lindenstrauss and A. Pelczyński (1968), will be given for some immediate applications towards Cramér and Karhunen processes, showing open-ended extension of our theory. The special result originally obtained by these authors is as follows.

Theorem 4.6.8 *[Lindenstrauss and Pelczyński] Let (Ω, Σ) be a given measurable couple with $B(\Omega, \Sigma)$ as the Banach space of scalar bounded measurable functions and $\mathcal{Y}$ be an L_p-space, $1 \leq p \leq 2$ (i.e. an L_ϕ with $\phi(x) = x^p$ in the above). Then a vector measure $\nu : \Sigma \to \mathcal{Y}$ is $(2, \mu)$-dominated, so that there is a finite positive μ on Σ such that*

$$\left\| \int_\Omega f(\omega)\nu(d\omega) \right\|_y \leq \|f\|_{2,\mu}, \quad f \in B(\Omega, \Sigma), \tag{105}$$

whenever ν has 2-semi-variation relative to μ finite.

The proof of this result with full details is in the author's graduate text (cf. Rao (2004)), *Measure Theory and Integration* pp. 527–529) and need not be reproduced here. The interested reader may study there the details and applications as well as some extensions. Here we consider only applications to Cramér and Karhunen processes. Recall that a centered second order process $\{X_t, t \in \mathbb{R}\}$ with covariance $r(s,t)$ is of *Cramér class* if $r(\cdot, \cdot)$ has the Morse-Transue integral representation:

$$r(s,t) = \int_\mathbb{R} \int_\mathbb{R} g(s, \lambda)\bar{g}(t, \lambda')F(d\lambda, d\lambda') \tag{106}$$

relative to $\{g(s, \cdot), s \in \mathbb{R}\}$, a Borel class, and F of finite Vitali variation, and it is of *Karhunen class* if $F(\cdot)$ concentrates on the diagonal $s = t$ which is σ-finite (on the diagonal). These are extensions of stationary and harmonizable processes studied earlier. We now discuss how the earlier analysis extends to this generalized class taking over the stationary and harmonizable families in which the Fourier analysis is desired to be extracted and find its possible limitations. Thus let $X : \mathbb{R} \to L_0^2(P)$ be a Karhunen process relative to a σ-finite F on $\mathcal{B}(\mathbb{R})$, as in (95) above and if $T : L_0^2(P) \to L_0^2(P)$ is a bounded linear operator, consider $Y(t) = TX(t), t \in \mathbb{R}$ so that we have

$$Y(t) = T \int_{\mathbb{R}} g(t, \lambda) Z(d\lambda) = \int_{\mathbb{R}} g(t, \lambda)(T \circ Z)(d\lambda), \qquad (107)$$

by classical vector analysis, as extended by Thomas ((1970), p. 79) where $\tilde{Z} = T \circ Z : \mathcal{B}(\mathbb{R}) \to L_0^2(P)$, and the Y-process becomes a Cramér-process relative to a bimeasure $\tilde{F} : (A, B) \mapsto (\tilde{Z}(A), \tilde{Z}(B))$, $A, B \in \mathcal{B}(\mathbb{R})$, so that $Y(t)$ is indeed of the Cramér class. We next discuss the opposite (a kind of converse) inclusion.

Thus consider $\{X(t), t \in \mathbb{R}\}$, a Cramér process, so that

$$X(t) = \int_{\mathbb{R}} g(t, \lambda) \tilde{Z}(d\lambda), \quad t \in \mathbb{R} \qquad (108)$$

relative to $\tilde{Z}$-integrable $\{g(t, \cdot), t \in \mathbb{R}\}$, and $\tilde{F} : (A, B) \mapsto (\tilde{Z}(A), \tilde{Z}(B))$ and that it is well-defined with $\tilde{Z} : \mathcal{B}_0(\mathbb{R}) \to L_0^2(P)$ as a vector measure. We need to use (a form of) the Lindenstrauss-Pelczyński theorem specialized from the above and proceed to get:

Theorem 4.6.9 *Let $X : \mathbb{R} \to L_0^2(P)$ be a process and $\{g(t, \cdot), t \in \mathbb{R}\}$ be a family of Borel functions such that X is a Karhunen process relative to the $g(t, \cdot)$-family and a σ-finite F on $\mathcal{B}(\mathbb{R})$, and $T : L_0^2(P) \to L_0^2(P)$ be a bounded linear mapping. Then $\{Y(t) = TX(t), t \in \mathbb{R}\}$ is a Cramér process relative to the $g(t)$-family and a suitable bimeasure. Conversely, if $\{g(t, \cdot), t \in \mathbb{R}\}$ is a bounded Borel set and $X : \mathbb{R} \to L_0^2(P)$ is a Cramér process relative to this family and a suitable covariant bimeasure, then there is an extension space $L_0^2(\tilde{P})(\supset L_0^2(P))$ determined by the given process, a Karhunen process $Y : \mathbb{R} \to L_0^2(\tilde{P})$ with the same g family and a suitable Borel measure on $\mathbb{R}$ such that $X(t) = QY(t), t \in \mathbb{R}$, where $Q : L_0^2(\tilde{D}) \to L_0^2(P)$ is an orthogonal projection onto the range.*

The details are extensions of the stationary to harmonizable case and they are spelled out in the author's paper (cf. Rao (1981) pp. 304–308), and they will not be reproduced here, as they are easily available. But the following consequences are interesting. We have seen earlier that each harmonizable process is obtainable as a (an orthogonal) projection of a weakly stationary process on a super Hilbert space. The work depended on the fact that the class $g(t, \lambda) = e^{it\lambda}$ so that the classical harmonic analysis results and methods are employable. But now the $\{g(t, \cdot), t \in \mathbb{R}\}$ is more general and just (uniform) boundedness is not sufficient. In fact, on attending the presentation of this paper by the author, Erik

Thomas whose (Radon) vector measure theory of the 1970s is important to our work, has constructed a counter example to show that there exist processes of Cramér class that are not projections of a Karhunen class on a larger (or super) Hilbert space so that the harmonic analysis is crucial here.

The preceding discussion and the result show that the class of Cramér processes is quite large and some of its members may not be obtained from a (fixed) Karhunen class membership. Also, the Cramér class is closed under continuous (or bounded) linear mappings, and thus this class is closed under such transformations, following from a result of E. Thomas (1970), p. 79). Note that in all the work above the index set $\mathbb{R}$ can be replaced by a locally compact group with the (spectral) measure F being Radon on these spaces.

We end this chapter with an operator *characterization of the Cramér process* as it contains many others, considered above.

Definition 4.6.10 *Let $\mathfrak{X}$ be a Banach space $A : \mathfrak{X} \to \mathfrak{X}$ be a (not necessarily bounded) linear operator with spectrum $\sigma(A) \subsetneq \mathbb{C}$. Let $\mathcal{F}(A) = \{f : \mathbb{C} \to \mathbb{C} \text{ analytic in a neighborhood of } \sigma(A) \text{ and } A\}$. The neighborhood can depend on f and not connected. For $f \in \mathcal{F}(A)$, define*

$$f(A) = f(\infty)I + \frac{1}{2\pi i} \int_\Gamma f(\lambda)R(\lambda, A)d\lambda, R(\lambda, A) = (A - \lambda I)^{-1}.$$

$$(109)$$

The operator $f(A)$ is well-defined, and assume that $g(\cdot, \cdot) = 1$ so that $r(0,0) = k < \infty$ and $F(\cdot)$ of (106), for the Karhunen process, a finite measure. We now establish the following representation:

Theorem 4.6.11 *Let $X : \mathbb{R} \to L_0^2(P)$ be a Cramér process relative to $\{g(t, \cdot), t \in \mathbb{R}\}$ of bounded Borel functions with $g(0, \cdot) = 1$. Then there is an extension space $L_0^2(P') \supset L_0^2(P)$, some $Y_0 \in L_0^2(P')$ and an unbounded linear operator $A : L_0^2(P') \to L_0^2(P)$ such that $A|_{L_0^2(P)}$ is symmetric with dense domain $\overline{\text{sp}} \{X(t), t \in \mathbb{R}\} \subset L_0^2(P)$, and $g(t, A)$'s are as above and*

$$X(t) = g(t, A)Y_0, \quad t \in \mathbb{R}. \tag{110}$$

Conversely, if A is symmetric and densely defined in $L_0^2(P)$, $X_0 \in L_0^2(P)$ and the $g(t, \cdot)$ are as above, then $Y(t) = g(t, t)X_0, t \in \mathbb{R}$ is always a Cramér process relative to the given g-family.

Proof. Let $X : \mathbb{R} \to L_0^2(P)$ be a Cramér process relative to the given g-family. Then by Theorem 4.6.9 above, there is $L_0^2(P') \supset L_0^2(P)$ and a Karhunen process $Y : \mathbb{R} \to L_0^2(P')$ with $X(t) = QY(t), t \in \mathbb{R}$ and Q as the orthogonal projection onto $L_0^2(P)$. Since $g(0, \cdot) = 1$, the representing measure is finite. However, the Karhunen process can also be given in operator theoretic form as: $Y(t) = g(t, \tilde{A})Y(0)$ for an $\tilde{A}$, with dense domain $\bar{\mathrm{sp}}\{Y(t), t \in \mathbb{R}\} \subset L_0^2(P')$. [This version has been verified by Getoor (1956), Thm. 3.4.] Hence we have

$$g(t, \tilde{A})Y(0) = \int_{\mathbb{R}} g(t, \lambda)\tilde{E}(d\lambda)Y(0)$$

with $\{\tilde{E}(t), t \in \mathbb{R}\}$ as the resolution of the identity of $\tilde{A}$. It now follows from the extended spectral calculus of Thomas (1970), that

$$X(t) = QY(t) = \int_{\mathbb{R}} g(t, \lambda)(Q\tilde{E}), (d\lambda)Y(0) = g(t, A)Y(0),$$

where $A = \int_{\mathbb{R}} \lambda E(d\lambda) \cdot [E(\lambda) = Q\tilde{E}(\lambda), \lambda \in \mathbb{R}]$ [$E(\lambda)$ is a *generalized spectral family* in that the increments are only positive]. It is verified easily that $A|_{L_0^2(P)}$ is symmetric and densely defined, $g(0, A) = Q$ and so $X(0) = QY(0)$. The converse depends on Naĭmark's theorem.

Thus, if A is symmetric and density defined, Naĭmark's theorem implies that it extends to $\tilde{A}$, a self-adjoint operator onto the extension $L_0^2(P')$ so that $A = Q(\tilde{A})$ and $g(t, A) = Qg(t, \tilde{A})$. But $Y(t) = g(t, \tilde{A})Y_0, t \in \mathbb{R}$, is a Karhunen process on $L_0^2(P')$ relative to the g-family. So by the representation $X(t) = QY(t), t \in \mathbb{R}$, is a Cramér process in $L_0^2(P)$ for the g-family, completing the proof. $\square$

Remark 29. Each vector measure on $(\mathbb{R}, \mathcal{B}(\mathbb{R}))$ into a Hilbert space is derived from a generalized spectral family.

We close this section with the following interesting result on self-adjoint dilations of certain operators useful in abstract analysis.

Theorem 4.6.12 *Let A be a densely defined symmetric operator in a Hilbert space $\mathcal{H}$ and $\{g_t, t \in \mathbb{R}\}$ be a class of bounded Borel functions with $g_0 = 1$. Then the family $\{T_t = g_t(A), t \in \mathbb{R}\}$ defines a class of bounded operators on $\tilde{\mathcal{H}} \supset \mathcal{H}$, ($\tilde{\mathcal{H}}$ a Hilbert space) on which there is a s.a. operator $\tilde{A}$ extending A such that $T_t = Qg_t(\tilde{A})$ where $Q : \tilde{\mathcal{H}} \to \mathcal{H}$ is an orthogonal projection. Conversely, every densely defined* *self-adjoint $\tilde{A}$ on $\mathcal{H}$ and a family $\{g_t, t \in \mathbb{R}\}$ defines a class of closed operators $T_t = Q g_t(\tilde{A})|_{\mathcal{H}} = g_t(A)|_{\mathcal{H}}$ where $A = Q\tilde{A}$ and $\mathcal{H} = Q(\mathcal{K})$ with Q as orthogonal projection on $\mathcal{H}$, $A = Q\tilde{A}$.*

The preceding discussion and analysis easily allow one to complete the details of proof of this result and thus will be left to the reader, thereby concluding the section. As before we include some complements and problems to round out the discussion of the section (and the chapter).

4.7 Complements and Exercises

1. Let $\mathcal{K}$ be the L. Schwartz space of compactly based real C^∞-functions on $\mathbb{R}^n$, and $F : \mathcal{K} \to C$ be a second order generalized random field with $B(\cdot, \cdot)$ as its covariance functional. Then there exists a continuous positive definite compactly supported hermitian function $h : \mathbb{R}^n \times \mathbb{R}^n \to C$, such that

$$B(f, g) = \int_{\mathbb{R}^n} \int_{\mathbb{R}^n} f(x)\overline{g(y)}h(x, y)dxdy, \quad f, g \in \mathcal{K}.$$

   [Thus $h(\cdot, \cdot)$ is the covariance function of an ordinary $L^2(P)$-field.]
2. The following class was introduced by Cramér (1964). If $\mathcal{K}$ is the L. Schwartz space on $\mathbb{R}^n$ and Φ is a separable Hausdorff space $F : \mathbb{R} \times \Phi \to \tilde{C}$ is a random functional, $F(t, f) = F_t(f) \in L^2(P)$ which is continuous in t and f separately, let $\mathcal{K}_t = \bar{sp}\{F_s(f) : s \le t\}$ and $\cap_{t \in \mathbb{R}} \mathcal{K}_t = \{0\}$. Then Cramér terms F *purely nondeterministic*. In this case one can show that $F(tf) = F_t(f) = \int_{-\infty}^{t} G(t, u)dZ(u)$ for a process $Z(\cdot)$ of orthogonal increments, and G is an operator valued function taking $\mathcal{K}_t$ into $\mathcal{K}_\infty$. [Cramér (1961) has shown this if Φ is one dimensional, and Kallianpur-Mandrekar (1963) extended this result for the general Φ as given. The reader should follow these works for related matters. See the author's (1967) PNAS note for details and related analysis.]
3. The following result was given by Sz.-Nagy, as an *appendix* to the Second Edition of the classical Riesz-Sz.-Nagy book *Functional Analysis* under the title: "Extensions of Linear Transformations in Hilbert Space Which Extend Beyond This Space". Verify that the result can be obtained from Theorem 4.6.12 (the last result of this section) which depended only on Naĭmark's theorem. The result is:

If $\{T_t, t \in \mathbb{R}\}$ is a weakly continuous positive definite set of contractive operators on a Hilbert space $\mathcal{H}$, $T_0 = $ id., then there is a super (or extension) Hilbert space $\mathcal{K} \supset \mathcal{H}$, a unitary group of operators $\{U_t, t \in \mathbb{R}\}$ on $\mathcal{K}$ such that $T_t = QU_t, t \in \mathbb{R}$, holds. On the other hand, every weakly continuous group of unitary operators $\{U_t, t \in \mathbb{R}\}$ defines a weakly continuous positive definite contractive set of operators $\{T_t = QU_t, t \in \mathbb{R}\}$ on $\mathcal{H} = Q(\mathcal{K})$ for each orthogonal projection Q on $\mathcal{K}$. [This shows that the Naĭmark and Sz.-Nagy extensions are abstractly equivalent. Here $Q : \mathcal{K} \to \mathcal{H}$ is $\perp$ projection.]

4. Using the concepts of Exercise 1 above, let $F : \mathcal{K} \to \mathbb{C}$ be a generalized random field of class (C) with orthogonal values for functions of disjoint supports in $\mathcal{K}$. Then there is a random measure $Z : \mathcal{B}(\mathbb{R}^n) \to L^2(P)$, on the Borel field of $\mathbb{R}^n$, with orthogonal values such that, if $\tilde{f}(t) = \int_{\mathbb{R}^n} f(x)g(t,x)dx$, we have

$$F(f) = \int_{\mathbb{R}^n} \tilde{f}(t)dZ(t), \quad f \in \mathcal{K},$$

and the covariance $B : (y, y) \mapsto \int_{\mathbb{R}^n} g(t)\bar{y}(t)d\sigma(t)$, a tempered measure σ in $\mathbb{R}^n$ is the spectral measure of F with $g(\cdot)$, related to class (C). On the other hand (i.e., conversely) if $Z(\cdot)$ is an orthogonally valued random measure relative to a tempered $\sigma(\cdot)$, and F is given by the integral above, then the resulting $F(\cdot)$ is a generalized random field on $\mathcal{K}$ of class (C) with orthogonal values. [If $g(t, x) = e^{itx}$ here, the representation gives the earlier formulas of A. M. Yaglom (noted in Gel'fand-Vilenkin (1964)), and the first case of stationary fields of K. Ito (1954).]

5. This problem gives an extension of Bochner's characterization of the classical continuous stationary covariance to a larger class. Thus $r(\cdot, \cdot)$ is a harmonizable covariance on $\mathbb{R}^n \times \mathbb{R}^n \to \mathbb{C}$ if and only if there exist uniformly bounded covariances $r_m : \mathbb{R}^n \times \mathbb{R}^n \to \mathbb{C}$ such that

(i) $r_m(s,t) \to r(s,t)$, uniformly in $(s,t) \in \mathbb{R}^n \times \mathbb{R}^n$ as $m \to \infty$,

(ii) r_m is supported on a compact rectangle $E_m \subset E_{m+1}, \bigcup_{m=1}^{\infty} E_n = \mathbb{R}^n$,

(iii) if $\{\phi_k^m, k \geq 1\}$ is a complete set of eigenfunctions of r_m and for each m, $\{Q_k^m\}$ is the set of eigenfunctions of r_m, then ϕ_k^m is the Fourier transform of some G_k^m of bounded variation in $\mathbb{R}^n$, for

$m, k \geq 1$. [If $r(s, t) = \tilde{r}(s - t)$, the result reduces to Bochner's classical theorem.]

6. A weakly stationary centered random field $X : \mathbb{R}^n \to L_0^2(P)$ is *isotropic* if its covariance $r : (s, t) \to \mathbb{C}$ is invariant under (not only translations but also) relations of the (s, t)-coordinates. In this case, Bochner (introduced this concept and) characterized the resulting representation of $r(\cdot, \cdot)$. The following formula extends his representation to weakly harmonizable classes. Thus Bochner's characterization of *isotopy* of a *(weak) stationary random field* $X : \mathbb{R}^n \to L_0^2(P)$ with covariance $r(s, t) = \tilde{r}(s - t)$ is given by (cf. Gikhman-Skorokhod (1969), 37–39) the representation:

$$r(s, t) = \tilde{r}(s - t) = 2^\nu \Gamma\left(\frac{n}{2}\right) \int_{\mathbb{R}^n} \frac{J_\nu(\lambda|s - t|)}{(\lambda|s - t|)^n} dG(\lambda), s, t \in \mathbb{R}^n,$$

with $\nu = \frac{n-2}{2}, G : \mathbb{R}^{-1} \to \mathbb{R}^{-1}$ is a bounded Borel measure and $|s - t|^2 = (s - t, s - t), n \geq 1$ the (squared) Euclidean distance. Here J_ν is a Bessel function (of the first kind) of order ν. Extending this, we say that X is *weakly harmonizable isotropic* (centered) if there is a positive definite bimeasure $\mu : \mathcal{B}(\mathbb{R}^n) \times \mathcal{B}(\mathbb{R}^+) \to \mathbb{C}$ and n-variate covariance $r(\cdot, \cdot)$ given by

$$r(\underset{\sim}{s}, \underset{\sim}{s}) = \alpha_n^2 \sum_{m=0}^{\infty} \sum_{l=1}^{h(m,n)} S_m^l(\underset{\sim}{u}) S_m^l(\underset{\sim}{v})$$
$$\times \int_{\mathbb{R}^+} \int_{\mathbb{R}^+} \frac{J_{m+\nu}(\lambda s) J_{m+\nu}(\lambda' r)}{(\lambda s)^\nu (\lambda' r)^\nu} d\mu(\lambda, \lambda')$$

where (i) $g = (s, \underset{\sim}{u}), t = (r, \underset{\sim}{v})$ are the spherical polar coordinates of $\underset{\sim}{s}, \underset{\sim}{t}$ in $\mathbb{R}^n$, (ii) $s_m^l(\cdot), 1 \leq l \leq h(m, n) = (lm + 2\nu)(m + 2\nu - 1)![(2\nu)!m!]^{-1}, 1 \leq m$, are the spherical harmonics on the unit sphere of $\mathbb{R}^m$, (iii) $\alpha_n > 0 \, \alpha_n^2 = 2^{2\nu} \Gamma\left(\frac{n}{2}\right) \pi^{\frac{n}{2}}$. The integrals are in the (strict) Morse-Transue sense. If the integrals are strongly MT-sense, then the integrals here can be strengthened to Lebesgue's. With this explanation, we can assert the following integral representation of isotropic harmonizable random fields extending the stationary case. Thus let $X : \mathbb{R}^n \to L_0^2(P)$ be a weakly harmonizable isotropic field. Then it admits an integral representation as:

$$X(\underset{\sim}{t}) = X(r, \underset{\sim}{v}) = \alpha_n \sum_{m=0}^{\infty} \sum_{l=1}^{l} (m, \nu) S_m^l(v)$$

$$\times \int_{\mathbb{R}^+} \frac{J_{m+\nu}(\lambda r)}{(\lambda r)^\nu} Z_m^l(d\lambda)$$

with $Z_m(\cdot)$ verifying $E(Z_m^l(B_1)\bar{Z}_{m'}^{l'}(B_2)) = \delta_{mm'}\delta_{ll'}F(B_1, B_2)$, and $F(\cdot, \cdot)$ a bimeasure of $Z_n^l(\cdot)$ in that $E(Z_m^l(B_1)\bar{Z}_{m'}^{l'}(B_2)) = \delta_{mm'}\delta_{ll'}$ $\times F(B_1, B_2)$, the $F(\cdot, \cdot)$ being a bimeasure of finite Fréchet variation. [The result is important and a nontrivial extension of Bochner's classical representation of the weakly stationary and harmonizable cases that we have discussed earlier. The detailed analysis of the problem is given in the author's basic extension in (Rao (1991), Thm. 4). [This and much of the related extensions are in Swift (1997) with references to earlier works motivated by these problems.]

7. This problem presents an approximation of a general weakly harmonizable random field by one whose spectral mass is essentially on a bounded Borel rectangle of $\mathbb{R}^n \times \mathbb{R}^n$. Thus let $X : \mathbb{R}^n \to L_0^2(P)$ be weakly harmonizable (may not be isotropic) with $\mu(\cdot, \cdot)$ as its spectral bimeasure. Then for each $\varepsilon > 0$ there is a weakly harmonizable $X_\varepsilon : \mathbb{R}^n \to L_0^2(P)$, with spectrum in $A_\varepsilon \times A_\varepsilon$ satisfying $\|X(t) - X_\varepsilon(t)\|_2 < \varepsilon, t \in \mathbb{R}^n$.
[This result implies that the spectrum of the $X_\varepsilon(t)$ field is contained in a compact rectangle. The result also gives analogs of the sampling theorems of the Kotehnikov-Shannon type that are useful in applications.]

8. The dilation properties obtained earlier for harmonizable fields have analogs for isotropic harmonizable families also, as shown by the work of Swift (1994). Here is a key extension. Thus let $X : \mathbb{R}^n \to L_0^2(P)$ be a weakly harmonizable isotropic field. Then there is an extension space $L_0^2(\tilde{P}) \supset L_0^2(P)$ and a stationary isotropic random field $Y : \mathbb{R}^n \to L_0^2(\tilde{P})$ with $X = QY, Q : L_0^2(\tilde{P}) \to L_0^2(P)$ being the orthogonal projection onto.
If X is strongly harmonizable isotropic, then the covariance function is m times continuously differentiable in the representing coordinates τ_1, τ_2, θ where $m = \left[\frac{n-1}{2}\right]$, the integral part, and θ is $\arccos(\underset{\sim}{s}, \underset{\sim}{t})$. For the strongly harmonizable isotropic case [so $r(\underset{\sim}{s}, \underset{\sim}{t}) = r(g\underset{\sim}{s}, g\underset{\sim}{t})$ for any unitary mapping g on $\mathbb{R}^n$], the μ with Vitali variation

$$r(s,t) = \alpha_n \int_0^\infty \int_0^\infty \frac{J_\nu(|\lambda s - \lambda' t|)}{|\lambda s - \lambda' t|^\nu} d\mu(\lambda, \lambda').$$

[Many of these representations have analogs if we replace the index T by an LCA group G, and one has to use deeper properties of the resulting harmonic analysis (see Swift (1994) for details). Some of this extension will be discussed in the next chapter indicating the interplay of the probability and harmonic analysis at a deeper level, and showing the prospects as well as some interesting applications.]

4.8 Bibliographical Notes

This chapter contains some crucial aspects of second order random fields, their structural analysis and the impact of the new property of isotropy that arises when the index is the (vector) group including the n-dimensional space $\mathbb{R}^n, n > 1$. The structural problems are deeper but are useful for many real applications. When the index set is $\mathbb{R}^n, n > 1$, the new concept of isotropy appears and has several interesting (deep) applications but the corresponding mathematical analysis is also deeper. It is fortunate that the great harmonic analyst S. Bochner has taken interest and obtained key integral representations and also extended the Fourier analysis as needed.

The Russian Probabilists, Yaglom (1987), Yadrenko (1988) and others have used these formulas for stationary random fields. Serious limitations were noted for stationary isotropic analysis and even though there existed problems that are more general, the methods and existing analysis were not adequate. The early real extension for more general random fields, starting with the harmonizable class, Swift (1994) was able to generalize the classical stationary analysis, and made several useful applications and extensions thereafter. The problems are such that one needs to use the analysis of spherical harmonics. We have presented the main ideas without detailing the many aspects of these results but giving the necessary references.

The main thrust here is to present the analysis not only on a (weak/strong) harmonizable class that uses Fourier analysis for the most part, but we tried to show how its extensions are formulated by H. Cramér (1951) and K. Karhunen (1947) as well as the works of Bochner (1956), Loève (1947) and Rozanov (1959), among others.

In some sense, the material presented in this chapter is to be regarded as an introduction and a preview of what harmonic analysis contributes to this area of stochastic process and fields. This work will be a good motivation for random fields on LCA and related groups some to be considered in the next chapter, opening up new ways that many problems can be formulated and solved with a full use of (abstract) harmonic analysis on (LCA) groups as well as on some generalized classes, termed hypergroups, that appear to have potential both for interesting applications as well as theory. This will be discussed in the next chapter.

5

Harmonizable Fields on Groups and Hypergroups

This chapter is devoted to a comprehensive treatment of harmonizable random fields on LCA groups and some results extended to hypergroups which are of interest in applications as well as in extending the subject to some new directions. We start with a brief (but abstract) account of bimeasures to use it in the ensuing analysis freely, mostly abstracted from the basic weakening of the Lebesgue analysis by M. Morse and W. Transue rendering and enabling it for the (weaker) Fréchet integrals which play a significant role in a study of (weakly) harmonizable processes. This class lies between the Riemann and Lebesgue integral analyses. Here we include a necessary outline to appreciate the general subject. Then the work proceeds with LCA groups and hyper groups.

5.1 Bimeasures and Morse-Transue (or MT-) Integrals

A systematic study of bimeasures and their integrals started with the work of Fréchet (1915) and detailed by Morse and Transue (1949–56) as well as P. Lévy (1946). This is sketched as follows.

If $(\Omega_i, \Sigma_i), i = 1, 2$ are measurable spaces and $\beta : \Sigma_1 \times \Sigma_2 \to \mathbb{C}$ is a mapping with $\beta(A, \cdot)$ and $\beta(\cdot, B)$ as complex measures for $A \in \Sigma_1, B \in \Sigma_2$, then $\beta(\cdot, \cdot)$ is called a *bimeasure*, which may *not* have an extension to be a (complex) measure on the product σ-algebra $\Sigma_1 \otimes \Sigma_2$. We present an alternative form to explain and extend the concept so as to define integrals for applications below, and that $\beta(\cdot, \cdot)$ may have an extension to a (complex) measure on the generated product σ-algebra $\Sigma_1 \otimes \Sigma_2$, and then using it to get the desired integrals for analyzing the (stochastic) theory of processes and fields. To appreciate the subject better, we need to briefly recall tensor products of Banach spaces with their cross-norms, clarifying our general approach. A few of the results needed here will utilize some work of Voropoulos (1967), and then we apply it to the MT-integration which is prominent in our work here.

If $\mathcal{X}_1, \mathcal{X}_2$ are Banach spaces, let $\mathcal{X}_1 \otimes \mathcal{X}_2$ denote the vector space of formal sums $f \otimes g = \sum_{i=1}^n f_i \otimes g_i, f_i \in \mathcal{X}_1, g_i \in \mathcal{X}_2$. A norm α on this product space satisfying $\|f \otimes g\| = \|f\|\|g\|$, is called a *cross-norm* $\alpha(\cdot)$, and $\mathcal{X}_1 \otimes_\alpha \mathcal{X}_2$ denotes the completed space so obtained for $\alpha(\cdot)$. There (clearly) exist several (such) norms, but the following two denoted $\| \cdot \|_\gamma, \| \cdot \|_\lambda$ called the *greatest* and *least* cross-norms are needed and defined as:

$$\|f \otimes g\|_\gamma = \inf \left\{ \sum_{i=1}^n \|f_i\|_{\mathcal{X}}\|g_i\|_{\mathcal{X}} : f \otimes g = \sum_{i=1}^n f_i \otimes g_i, n \geq 1 \right\}, \quad (1)$$

and if $\mathcal{X}_i^*$ is the adjoint of $\mathcal{X}_i$, let

$$\|f \otimes g\|_\lambda = \sup \left\{ \left| \sum_{i=1}^n x_1^*(f_i)x_2^*(g_i) \right| : x_i^* \in \mathcal{X}_i^*, \|x_i^*\| \leq 1, i = 1, 2 \right\}.$$

The corresponding completed spaces are denoted by $\mathcal{X}_1 \otimes_\gamma \mathcal{X}_2, \mathcal{X}_1 \otimes_\lambda \mathcal{X}_2$ and sometimes simply as $\mathcal{X}_1 \hat{\otimes} \mathcal{X}_2$ and $\mathcal{X}_1 \overset{\wedge}{\otimes} \mathcal{X}_2$, and termed the tensor product spaces. For the spaces of continuous linear functions denoted $\mathcal{X}_i^*, i = 1, 2$, the following are true and well-known relations.

Proposition 5.1.1 *If $\mathcal{X}_1, \mathcal{X}_2$ are Banach spaces, then we have:*
 (a) $(\mathcal{X}_1 \hat{\otimes} \mathcal{X}_2)^* \cong L(\mathcal{X}_1, \mathcal{X}_2^*)(\cong L(\mathcal{X}_2, \mathcal{X}_1^*))$,
 (b) $(\mathcal{X}_1 \overset{\wedge}{\otimes} \mathcal{X}_2)^* \hookrightarrow L(\mathcal{X}_1^*, \mathcal{X}_2)$
where "$\cong$" is an isometric isomorphism in (a) and the symbol in (b) is an isometric imbedding of the first into the second space.

The point of recalling these results is that the property (a) gives an alternative definition of bimeasures and illuminates the MT-integrals. Thus if Ω_i is locally compact, $\mathcal{X}_0 = C_0(\Omega_i)$ as the continuous function space with functions vanishing at "∞", that $V(\Omega_1, \Omega_2) = C_0(\Omega_1) \otimes C_0(\Omega_2)$, then one verifies that $V(\Omega_1, \Omega_2)^* = L(C_0(\Omega_1), C(\Omega_2)^*)$ and the correspondence is given by

$$B(f, g) = F(f \otimes g) = T(f)g, \forall f \in C_0(\Omega_1), g \in C_0(\Omega_2), \quad (2)$$

with $F \in V(\Omega_1, \Omega_2)^*, T : C_0(\Omega) \to C_0(\Omega_1)^*$, bounded and linear. Now $B : C_0(\Omega_1) \times C_0(\Omega_2) \to \mathbb{C}$ is a bounded bilinear form of F. Since $C_0(\Omega_2)^* = M(\Omega_2)$, bounded regular Radon measures, we get by the Riesz-Markov theorem the representation:

$$(Tf)(\cdot) = \int_{\Omega_1} f(\omega_1)\mu(d\omega_1) \in M(\Omega_2). \qquad (3)$$

Here we use the facts that T is weakly compact, $M(\Omega_2)$ is weakly sequentially complete (cf. Dunford-Schwartz (1958), VI.7.3, IV.29). Letting $\mu^f(\cdot) = (Tf)(\cdot)$ in (3), (2) is simplified as:

$$\begin{aligned} F(f \otimes g) = B(f, g) &= (Tf)(g) \\ &= \int_{\Omega_2} g(\omega_2)\mu^f(d\omega_2) \\ &= \int_{\Omega_2} g(\omega_2) \left[\int_{\Omega_1} f(\omega_1)\mu(d\omega_1, \cdot) \right] (d\omega_2) \\ &= \int_{\Omega_2} \int_{\Omega_1} (f, g)(\omega_1, \omega_2)\mu(d\omega_1, d\omega_2). \end{aligned}$$

Thus $B(\cdot, \cdot)$ can be identified with $\mu(\cdot, \cdot)$, a bimeasure. Also

$$\begin{aligned} \|F\| = \|B\| &= \sup \{|B(f, g)| : \|f\|_\infty \leq 1, \|g\|_\infty \leq 1\} \\ &= \sup \left\{ \left| \int_{\Omega_1} \int_{\Omega_2} (f, g)(\omega_1, \omega_2)\mu(d\omega_1, d\omega_2) \right| : \right. \\ &\qquad\qquad\qquad\qquad \left. \|f\|_\infty \leq 1, \|g\|_\infty \leq 1 \right\} \\ &= \|\mu\|, \text{ the semi-variation of } \mu. \qquad (4) \end{aligned}$$

For a bimeasure μ, $\|\mu\|$ is the Fréchet variation. Since $V(\Omega_1, \Omega_2)^* \supsetneq M(\Omega_1 \times \Omega_2)(= C_0(\Omega_1 \times \Omega_2)^*)$, μ is generally not a restriction of a scalar measure on $[C_0(\Omega_1 \times \Omega_2)]^*$. Thus we need to consider a generalized (called an MT) integral for our work. This is perhaps the reason that Varopoulos calls the members of $V(\Omega_1, \Omega_2)^*$ just bimeasures. This distinction is needed.

We next sketch the bimeasure integrals with a slight update and modifications, using some works of Kluvanek (1981), and Vrem (1979), as modified and used from Morse-Transue theory. The discussion is meant to explain the Morse-Transue (vector) type integration as a Riemann-type (nonabsolute in real distinction with the Lebesgue type extensions in the literature) and the Kluvanek modification is included which still retains the nonabsolute conditions.

Thus let $f_n : \Omega = \Omega_1 \times \Omega_2 \to \mathbb{C}$ be a simple function given as $f_n = \sum_{i=1}^{k_n} \sum_{j=1}^{k_n} a_{ij}^n \chi_{A_i^n} \chi_{B_j^n}$, $A_i^n \in \Sigma_1; B_j^n \in \Sigma_2$ (disjoint) and let $\beta : \Sigma_1 \times \Sigma_2 \to \mathbb{C}$ be a measure; now set

$$\int_\Omega f_n d\beta = \sum_{i=1, j=1}^{k_n} a_{ij} \beta(A_i^n, B_j^n). \tag{5}$$

If $f_n \to h$, pointwise and $\{\int_\Omega f_n d\beta, n \geq 1\} \subset \mathbb{C}$ is Cauchy with limit α_0, define $\alpha_0 = \int_\Omega h d\beta$. It can be verified that $h \mapsto \int_\Omega h d\beta$, is well-defined and linear. [An equivalent form is given by Kluvanek (1981), valid even if $\mathbb{C}$ is replaced by Banach spaces with some extension.]

We now designate a subclass called *strict integrals* to use. If (Ω_i, Σ_i) is a measurable pair, $f_i : \Omega_i \to \mathbb{C}$ is measurable (Σ_i) and $\beta : \Sigma_1 \times \Sigma_2 \to \mathbb{C}$ is a bimeasure, let f_1 be $\beta(d\omega_1, \cdot)$ and f_2 be $\beta(\cdot, d\omega_2)$ integrable with $\beta : \Sigma_1 \times \Sigma_2 \to \mathbb{C}$ as a bimeasure, and suppose that f_1 is $\beta(d\omega_1, \cdot)$, f_2 is $\beta(\cdot, d\omega_2)$-integrable (DS) as vector measures. Then the complex measures $\tilde{\beta}_1^F, \tilde{\beta}_2^F$ given by

$$\tilde{\beta}_1^F(A) = \int_F f(\omega_2) \beta(A, d\omega_2), \quad \tilde{\beta}_2^F(B) = \int_F f_1(\omega_1) \beta(d\omega_1, B)$$

for $A \in \Sigma_1, B \in \Sigma_2$, and for each $E \in \Sigma_1, F \in \Sigma_2$ clearly define complex measures. The pair (f_1, f_2) is termed *strictly β-integrable* if one has

$$\int_E f_1(\omega_1) \tilde{\beta}_1^F(d\omega_1) = \int_F f_2(\omega_2) \tilde{\beta}_2^E(d\omega_2). \tag{6}$$

The common value is denoted $\int_E \int_F^* f_1(\omega_1) f_2(\omega_2) \beta(d\omega_1, d\omega_2)$. In the earlier case, this was demanded only for a single pair $E = \Omega_1, F = \Omega_2$. This strengthening renders the dominated convergence valid as well as the absolute continuity of the integral; but it is still weaker than the Lebesgue theory, and e.g., the Jordan decomposition is still not valid here.

It should be noted that these extended concepts are interesting even in integral representations of bilinear operators having analogs in multilinear theory. Some aspects of these results with applications to abstract analysis have been given by Dobrakov (1987). For instance, a vector bimeasure β has finite Fréchet variation if $\|\beta\|(\Omega_1, \Omega_2) < \infty$ where

$$\|\beta\|(\Omega_1, \Omega_2) = \sup \left\{ \left\| \sum_{i,j=1}^{n} a_i b_j \beta(A_i, B_j) \right\|_{\mathcal{X}} : A_i \in Z_1, B_j \in \Sigma_2, \right.$$

$$\left. \text{disjoint}, |a_i| \leq 1, |b_j| \leq 1, \text{scalars}, n > 1 \right\} < \infty. \quad (7)$$

This follows from the fact that $l \circ \beta$ is a scalar bimeasure, $l \in \mathcal{X}^*$. With a scalar extension method, Ylinen (1978) has obtained the following version of the Riesz-Markov representation stated for reference:

Theorem 5.1.2 *Let $(\Omega_i, \mathcal{B}_i), i = 1, 2$ be Borelian l.c spaces, and $C_0(\Omega_i)$ as the continuous real function space with functions vanishing at "∞". If $\mathcal{X}$ is a reflexive Banach space and $B : C_0(\Omega_1) \times C_0(\Omega_2) \to \mathcal{X}$ is a bounded bilinear form, then there is a regular (in each component) bimeasure $\beta : \mathcal{B}_1 \times \mathcal{B}_2 \to \mathcal{X}$, such that*

$$B(f_1, f_2) = \int_{\Omega_1} \int_{\Omega_2} f_1(\omega_1) f_2(\omega_2) \beta(d\omega_1, d\omega_2), f_i \in C_0(\Omega_i),$$

satisfying the bound $\|B\| = \sup \{\|B(f_1, f_2)\|_{\mathcal{X}} \cdot \|f_i\|_\infty \leq 1, i = 1, 2\}$ with $\|B\| = \|\beta\|(\Omega_1, \Omega_2)$. For nonreflexive $\mathcal{X}$, the same holds provided B maps bounded sets of $V(\Omega_1, \Omega_2) = C_0(\Omega_1) \hat{\otimes} C_0(\Omega_2)$ into relatively compact sets.

The problem receives a complete and comprehensive view when the mapping $X : \mathbb{R} \to L_0^2(P)$ with covariance $r : (s, t) \to E(X\bar{X}_t)$ is weakly stationary provided (by the classical Bochner theorem) as, $r(s, t) = \tilde{r}(s - t) = \int_{\mathbb{R}} e^{i(s-t)\lambda} \mu(d\lambda)$, for a bounded Borel μ on LCA groups, and for (weakly) harmonizable case if

$$r(g_1, g_2) = \int_{\hat{G}} \int_{\hat{G}} \langle g_1, \lambda \rangle \overline{\langle g_2, \lambda' \rangle} \beta(d\lambda, d\lambda'). \quad (8)$$

When $\beta(\cdot, \cdot)$ has finite (Fréchet) variation, then it was generalized by Cramér (and Karhunen) from the harmonizable concepts. All of this is better understood if G is an LCA group, as we now put it in perspective for harmonizability in the present context.

5.2 Harmonizability on LCA Groups

This section is denoted to some abstract extensions of the harmonizability concepts when the indexing set is a general locally compact group with some extensions in the next section to hypergroup indexing. Only then the real potential of the subject and its uses are appreciated. This study opens up essentially the full scope of the subject for analysis.

As a motivation let us restate the Cramér and Karhunen classes when they are given with index $G = \mathbb{R}^n$ or $\mathbb{Z}^n, n > 1$. Thus if $\hat{G}$ is the dual group, and $X : G \to L_0^2(P)$ is weakly (strongly) harmonizable, then the covariance $r(g_1, g_2) = E(X_{g_1} \bar{X}_{g_2})$ is representable as ($\langle g, \cdot \rangle \in \hat{G}$ is a character of G):

$$r(g_1, g_2) = \int_{\hat{G}} \int_{\hat{G}} \langle g_1, r_1 \rangle \overline{\langle g_2, r_2 \rangle} \beta(dr_1, dr_2), \qquad (9)$$

with β as a positive definite bimeasure on $\mathcal{B}(\hat{G}) \times \mathcal{B}(\hat{G})$ of Fréchet (or finite Vitali) variation of the random field $\{X_g, g \in G\}$.

Let $\{X_t, t \in T\} \subset L_0^2(P)$ be a family with $r(s, t) = E(X_s \bar{X}_t)$ as its covariance. If $\mathcal{B}$ is a σ-algebra of T, then the X_t-family is said to be of *class* (C) if there is a measurable pair $(S, \mathcal{S})$ and a positive definite bimeasure $\beta : \mathcal{S} \times \mathcal{S} \to \mathbb{C}$ with finite Vitali variation such that

$$r(s, t) = \int_S \int_S g_s(\lambda) \overline{g_t(\lambda')} \beta(d\lambda, d\lambda'), s, t \in T, \qquad (10)$$

relative to a β-integrable collection $\{g_s, s \in T\}$ of scalar functions such that $r(s, t) < \infty, s \in T$, and if β is only of finite Fréchet variation, then we have a *weak class* (C). If β in (10) concentrates on $S = T$ one has the *Karhunen class* and then (10) becomes

$$r(s, t) = \int_S g_s(\lambda) \overline{g_t(\lambda)} \mu(d\lambda), s, t \in T \qquad (11)$$

and this was introduced by Karhunen in 1947. We can have the Karhunen field, extending the original stationary and harmonizable classes:

Proposition 5.2.1 *Every weakly or strongly harmonizable random field* $X : G \to L_0^2(P)$, *G an LCA group, belongs to a Karhunen class. In fact, if the family is weakly harmonizable then there is a finite Borel measure* μ *on* $\hat{G}$ *and a family* $\{g_s, s \in G\} \subset L^2(\hat{G}, \mu)$ *such that* (11) *holds with* $T = G$ *and* $S = \hat{G}$.

We specialize the analysis now for harmonizable fields on LCA groups and also present some analogs for the nonabelian case, as this reveals an internal structure of the problem better. It will be useful to recall the D-S integral of a scalar function relative to a vector measure. Thus if (Ω, Σ) is a measurable space, $f : \Omega \to \mathbb{C}$ is (Σ) measurable, and $Z : \Sigma \to \mathcal{X}$, a Banach space, is a vector measure, then f is D-S integrable for Z if the following pair of conditions holds: (i) there is a sequence $f_n : \Omega \to \mathbb{C}$ of simple (Σ-measurable) members such that $f_n \to f$ pointwise, and (ii) if $f_n = \sum_{i=1}^{k_n} a_i^n \chi_{A_i^n}$, and $\int_C f_n dZ = \Sigma a_i^n Z(E \cap A_i^n) \in \mathcal{X} \Rightarrow \{\int_E f_n dZ, n \geq 1\}$ is Cauchy in $\mathcal{X}$ for each $E \in \Sigma$. Then the unique limit is denoted by $\int_E f dZ, E \in Z$. It may be verified that the D-S integral above is a well-defined element of $\mathcal{X}$, is linear, and the dominated convergence theorem is valid for it. However, the evaluation of $\int_E f dZ$, as a Stieltjes integral is generally false. Note that the convergence in (i) is pointwise and not uniform.

Now let $L^1(Z)$ be the space of scalar (D-S) integrable functions for Z_0, and $L_0^2(\beta)$, the collection of strictly β-integrable (MT-sense) functions $f : S \to F$ when $\beta : (A, B) \mapsto E(Z(A)\overline{Z(B)})$ is a bimeasure for Z, with $\mathcal{X} = L_0^2(P)$, on (Ω, Σ, P). Here $Z(\cdot)$ is termed a stochastic measure and β its *spectral bimeasure* of the second order process, we then have:

Theorem 5.2.2 *Let $(S, \mathcal{S})$ be a measurable space $\beta : \mathcal{S} \times \mathcal{S} \to \mathbb{C}$ be a positive definite bimeasure. Then there is a probability triple (Ω, Σ, P) and a stochastic measure $Z : \Sigma \to L_0^2(P)$ such that*
   (i) $E(Z(A)\overline{Z(B)}) = \beta(A, B), A, B \in \mathcal{S}$, and
   (ii) $L^1(Z) = \mathcal{L}_*^2(\beta)$, equality between sets of functions.

This result can be obtained quickly using the (Aronszajn) reproducing kernel Hilbert space techniques. A general form of the latter is as follows. If $(S, \mathcal{S})$ is a Borelian space, (S-topological) a bimeasure $\beta : \mathcal{S} \times \mathcal{S} \to \mathbb{C}$ is said to have locally finite Fréchet (or Vitali) variation if $\beta : \mathcal{S}(E) \times \mathcal{S}(E) \to \mathbb{C}$ has finite Fréchet (or Vitali) variation on each bounded Borel set $E \subset S$. [See Edwards (1955) on the distinctions.] With these ideas, the following general representation, to be used later, holds:

Theorem 5.2.3 *Let S be locally compact and $(S, \mathcal{S})$ be Borelian pair, and $\{X_t, t \in T\} \subset L_0^2(P)$, on a probability space (Ω, Σ, P), be a locally weakly class (C) relative to a positive definite bimeasure β :*

$S_0 \times S_0 \to \mathbb{C}$ *of locally finite Fréchet variation and a family* $g_t : S \to$ $\mathbb{C}$, *each* g_t *being locally strictly* β-*integrable where* S_0 *is the* δ-*ring of bounded sets of* S. *Then there exists a vector measure* $Z : S_0 \to L_0^2(P)$ *such that for the index set* T:

$$(i) \qquad X_t = \int_S g_t(\lambda)Z(d\lambda), t \in T, (D\text{-}S \text{ integral}) \qquad (12)$$

$$(ii) \qquad E(Z(A)\bar{Z}(B)) = \beta(A, B), A, B \in S_0. \qquad (13)$$

In the opposite direction, if $\{X_t, t \in T\}$ *is given by (i), then the process is of local weak class* (C) *for a bimeasure* β, *given by (ii) and the* g_t *of (i) being locally strictly* β-*integrable. The process is of (local) Karhunen class iff (i) and (ii) hold with* $\beta(A, B) = \mu(A \cap B)$ *for a* σ-*finite measure* μ *on* S.

Sketch of Proof. If $K \subset S$ is compact, consider $S(K)$ of S restricted to K and $\beta : S(K) \times S(K) \to \mathbb{C}$ is a positive definite bimeasure and the theorem applies for it. If now $\tilde{Z} : S(K) \to L_0^2(P)$ is the corresponding representing stochastic measure, then one will have:

$$\int_K g_t(\lambda)\tilde{Z}(d\lambda) = j\left(\int_K g_t(\lambda)\beta(d\lambda, \cdot)\right) \in L_0^2(P), \qquad (14)$$

where j is the mapping above between $\tilde{Z}$ and β. With the local compactness we can define a measure $Z : S \to L_0^2(P)$ and extend the above representation using the standard procedure (cf. Hewitt-Ross (1963), pp. 133–134). Without local compactness, this piecing together can fail. From now on we will follow the standard, but not completely trivial, and which can be given (see e.g. Chang and Rao (1986), p. 53), but the details will not be reproduced here. $\square$

To understand the type of functions g_t, it may be of interest to specialize the class for harmonizable and stationary representations. We present the special result for reference.

Theorem 5.2.4 *Let* $\{X_t, t \in G\} \subset L_0^2(P)$ *be a class with* G *as an LCA group. Then the set is weakly (or strongly) harmonizable relative to a positive definite bimeasure* β *on* $B(\hat{G}) \times B(\hat{G}) \to \mathbb{C}$ *(of finite Vitali variations) iff there is a stochastic measure* $Z : B(\hat{G}) \to L_0^2(P)$ *such that*

$$(i) X_t = \int_{\hat{G}} \langle t, \lambda \rangle Z(d\lambda), t \in G, (D\text{-}S \text{ integral}) \qquad (15)$$

$$(ii) E(Z(A)\overline{Z(B)}) = \beta(A, B), \ A, B \in \mathcal{B}(\hat{G}), \tag{16}$$

where $\langle t, \cdot \rangle$ is a character of G. When these conditions hold the mapping $t \mapsto X_t$ is strongly uniformly continuous on $L_0^2(P)$; and the random field $\{X_t, t \in G\}$ is weakly stationary iff (i), (ii) hold with $\beta(A, B) = \mu(A \cap B)$ for a bounded Borel measure μ so that $Z(\cdot)$ has orthogonal increments.

It is natural to ask if the corresponding result obtains for G that may not be abelian. The nonabelian G shows the potential on the use of ideas and methods of Fourier that are also at the base of the deep and satisfying applications of harmonic analysis. Thus we discuss the harmonizable extension of $X : T \to L_0^2(P, \mathcal{X})$ for $\mathcal{X}$-valued strongly measurable fields so that $l(X) : G \to L_0^2(P)$, is its scalar correspondent. Then $l(X) : G \to L_0^2(P), l \in \mathcal{X}^*$, is a harmonizable field so that $l(X) : G \to L_0^2(P)$ admits a representation as:

$$l(X_t) = \int_{\hat{G}} \langle t, \lambda \rangle Z_l(d\lambda), t \in G, \tag{17}$$

where Z_l is a stochastic measure. The mapping $l \mapsto Z_l$ is linear (Z_l is a regular vector measure with finite semi-variation) so $\|Z_l\|(\hat{G}) < \infty$, where

$$\|Z_l\|(\hat{G}) = \sup_t \|l(X_t)\|_2 \leq \|l\| \sup_t \|X_t\|_2 < \infty, \tag{18}$$

since $X(G)$ is bounded. By the uniform boundedness principle $\|Z_t\|(\hat{G}) \leq K$ for some $K < \infty$, and there is $\tilde{Z}$ with $Z_l = l(\tilde{Z}(\cdot)) \in \mathcal{X}^{**} = \mathcal{X}$ here.

By reflexivity of $\mathcal{X}$, there is $\tilde{Z}$ with $Z_l = l(\tilde{Z})$, also reflexivity and uniform boundedness imply $E(\tilde{Z}(A)) \in \mathcal{X}$ so that we have (for $l \in \mathcal{X}^*$)

$$l(X_t) = \int_{\hat{G}} \langle t, \lambda \rangle l(\tilde{Z})(d\lambda) = l\left(\int_{\hat{G}} \langle t, \lambda \rangle Z(d\lambda)\right).$$

It follows from this that

$$X_t = \int_{\hat{G}} \langle t, \lambda \rangle \tilde{Z}(d\lambda) \quad t \in G. \tag{19}$$

This gives us the representation as:

Theorem 5.2.5 *Let G be an LCA group, $\mathcal{X}$ a reflexive B-space and $X : G \to L_0^2(P, \mathcal{X})$, such that $X(G)$ is relatively weakly compact,* *(equivalently norm bounded). Then X is weakly harmonizable iff there is a stochastic measure $\tilde{Z}$ such that* (19) *obtains.*

This implies that it may be possible to characterize weakly harmonizable fields without using bimeasure integration. If $G = \mathbb{R}$, Niemi (1975) considered some special representations, and certain others were given in Chang and Rao (1988). The above result unifies all these cases.

To appreciate and study the potential of harmonizable random fields, it is desirable to consider these objects also on a few nonabelian groups G as well. But these involve Fourier analysis on nonabelian groups containing G, since then there is no dual group of G, one needs to invoke now some ideas and results of C^*-algebras, and this was done by Ylinen (1975, 1984, and 1987) using certain related techniques of Edwards (1964) which can be simplified if the LC group G is restricted to be separable and unimodular. We present a little of this analysis simply to show the real potential of the subject and for future extended analysis. Here we include a brief account to be used in our analysis essentially following Mautner (1955); especially Tatsuma (1967) and Segal (1950). [See also Naĭmark (1964), Chapter 8.]

(i) If $\hat{G}$ denotes the set of all irreducible strongly continuous representations of G into a Hilbert space, then we can endow $\hat{G}$ a topology with which it becomes a locally compact Hausdorff space. Thus one can endow $\hat{G}$ a topology making it also a locally compact Hausdorff space. If now μ is a Haar measure on G, then there is a unique Radon measure ν on $\hat{G}$ so that $(\hat{G}, \nu)$ becomes a dual gauge of (G, μ) and the Plancherel formula holds for it.

(ii) The representation Hilbert space $\mathcal{H}$ can be taken as $L^2(G, \mu) \times (L^2(G))$ and then $\mathcal{H} = \oplus_{y \in \hat{G}} \mathcal{H}_y$, the direct sum, where $\mathcal{H}_y$ is the representation space for y in $\hat{G}$. If $\mathcal{A}_y \subset L(\mathcal{H}_y)$ is the weakly closed self-adjoint subalgebra of $L(\mathcal{H}_y)$, generated by the strongly continuous unitary family $\{U_y(g), g \in G\}$, then $\mathcal{A}_y$ is of 'type I or II', and

$$L^2(G) = \int_{\hat{G}}^{\oplus} \mathcal{H}_y \nu(dy), \text{ direct integrals.} \tag{20}$$

Further, if $(L_a f)(x) = f(a^{-1}x), x \in G$, then the weakly closed self-adjoint algebra $\mathfrak{a}$ generated by $\{L_a, a \in G\} \subset L(\mathcal{H})$, gives a direct sum decomposition of $\mathcal{A}_y, y \in \hat{G}$, and for each $f \in L^1(G) \cap L^2(G)$, the following (Bochner) integral exists:

$$\hat{f}(y) = \int_G U_y(g)f(g)\mu(dg), U_y(g) \in \mathcal{A}_y, y \in \hat{G}, \tag{21}$$

and defines a bounded linear mapping on $\mathcal{H}$. Moreover, $\hat{f}(y)$ may be extended uniquely to a dense subspace of $\mathcal{H}$ containing $L^1(G) \cap L^2(G)$ so that it is closed and self-adjoint. This extended form $y \mapsto \hat{f}(y)$, is the *generalized Fourier transform of f*, and this general setup is needed for us here.

(iii) There is a *trace* functional $\tau_y : \mathcal{A}_y \to F$ which is positive, linear, normal, semi-finite and faithful in terms of which there is the Plancherel formula, that we need for $f_i \in L^2(G), i = 1, 2$, with $\hat{f}_i^*$ as the adjoint of $\hat{f}$, so that one has:

$$\int_G f_1(g)\overline{f_2(g)}\mu(dg) = \int_{\hat{G}} \tau_y(\hat{f}_1(y)\hat{f}_2^*(y))\nu(dy). \tag{22}$$

The desired measurability of $\hat{f}_i$ and $y \to \tau_y(\hat{f}_1(y)\hat{f}_2^*(y))$ for ν are not obvious, but can be proved in this work, and $f \to \hat{f}$ is one-to-one. These are established in Mautner (1955).

(iv) If $A(y) \subset \mathcal{A}, y \mapsto A(y)$ is measurable, norm bounded, and $\int_{\hat{G}} \tau_y(A(y)A^*(y))dy < \infty$, then there is $f \in L^2(G)$ such that $\hat{f}(y) = A(y), y \in \hat{G}$. Also if $h \in L^2(G)$ and $h = f * f$ for some $f \in L^2(G)$, then one has the inversion, ('*' being the convolution product):

$$h(g) = \int_{\hat{G}} \tau(U_g(g)^*\hat{h}(y))\nu(dy), g \in G.$$

With this abstract set up on the generalized Fourier transform we can present the needed formulation of Bochner's classical notion of V-boundedness, and the (long awaited) characterization of weakly harmonizable random fields. Set forth as follows:

Definition 5.2.6 *Let G be a separable locally compact unimodular group and $X = \{X_g : g \in G\} \subset L_0^2(P)$ be a random field. Then X is weakly harmonizable if it is weakly continuous, and*

$$\left\{ \int_G X_g\phi(g)\mu(dg) : \|\hat{\phi}\|_\infty = \sup_{y \in G} \|\hat{\phi}(y)\| \le 1, \phi \in L^1(G) \cap L^2(G) \right\}$$

is bounded in $L_0^2(P), \hat{\phi}$ being the above defined generalized Fourier transform of ϕ.

With this preparation, the general integral representation of a weakly harmonizable random field can be given as follows.

Theorem 5.2.7 *Let $X : g \to X_g \in L_0^2(P), g \in G$, be a weakly harmonizable random field. Then there exists (i) a weakly σ-additive regular operator valued measure $\mathcal{M}(dy)$ defined on $\hat{G}$, operating on $\mathcal{H}_y$ into $L_0^2(P)$, vanishing on ν-null sets, and (ii) a trace functional $\tau_y : \mathcal{A}_y \to \mathbb{C}$ such that*

$$X_g = \int_G trace(U_g(y)\underset{\sim}{\mathcal{M}}(dy)), g \in G \ (Bartle's\ integral) \qquad (23)$$

and $X_{(.)}$ is uniformly continuous in the strong topology of $L_0^2(P)$. Conversely, a weakly continuous $X : g \to X_g$ given by (23) is weakly harmonizable. Further, the covariance function r of the weakly harmonizable X, of (23), has its covariance represented as:

$$r(g_1, g_2) = \int_{\hat{G}} \int_{\hat{G}} \tau_{y_1} \otimes \tau_{y_2}(U_{g_1}(y_1) \otimes U_{g_2}(y_2))\beta(dy_1, dy_2), \qquad (24)$$

where β is an operator valued bimeasure on $B(\hat{G}) \times B(\hat{G})$, with $B(\hat{G})$ as the Borel σ-algebra of $\hat{G}$.

Proof. If $f \in L^1(G) \cap L^2(G)$, let $\hat{f}$ be defined by (21) above which is a measurable (operator) function. It is also bounded since $\mathcal{H} = \int_{\hat{G}}^{\oplus} \mathcal{H}_y \nu(dy)$ and $\mathcal{H}_y$ can be embedded in $\mathcal{H}$, and may treat it as a closed subspace. So $U_y(g) = U(g, y)$ is in $L(\mathcal{H}_y)$ and be extended as $\tilde{U} = U$ on $\mathcal{H}_y$ and as identity on $\mathcal{H}_y^{\perp}$ so that $\{\tilde{U}(g, y), g \in G\}$ again unitary in $L(\mathcal{H})$ with $\{\tilde{U}(g, \cdot) \in L(\mathcal{H}), g \in G\}$ as a unitary family in $L(\mathcal{H})$ with $\tilde{U}(g, \cdot) \in L(\mathcal{H}), g \in G$.

If the resulting operators of (21) obtained by replacing U with $\tilde{U}$, denoted by $\hat{f}$, then it is measurable and $\mathfrak{a}(\mathcal{H}) = \{\hat{f}(y) \in L(\mathcal{H}_y), y \in \hat{G}\}$ is identifiable with a subalgebra of $L(\mathcal{H})$.

If $T : f \to \hat{f}$ is the mapping, then it is one-to-one and is a contraction, the former property is a consequence of the general theory, and its contraction is verified as follows: ($\| \cdot \|_{op}$ denotes operator norm)

$$\|\hat{f}(y)\|_{\mathrm{op}} = \left\|\int_G f(g)\tilde{U}(g,y)\mu(dg)\right\|_{\mathrm{op}}$$

$$\leq \int_G |f(g)|\|\tilde{U}(g,y)\|_{\mathrm{op}}\mu(dg), \quad \text{as a vector integral,}$$

$$\leq \int_G |f(g)|\mu(g) = \|f\|.$$

Hence $\sup_{y\in\hat{G}} \|\hat{f}(y)\|_{\mathrm{op}} \leq \|f\|_1 < \infty$, so that $T : L^1(G) \cap L^2(G) \to \mathfrak{a}(\mathcal{H})$ is a contraction. Since X is weakly harmonizable, we have for each $f \in L^1(G) \cap L^2(G)$ the following:

$$T_1(f) = \int_G f(g)X_g\mu(dg) \in L^2(P). \tag{25}$$

Clearly, T_1 is bounded. Let $\tilde{T} = T_1 \circ T^{-1}$ so that

$$\tilde{T}(\hat{f}) = T_1(T^{-1}(\hat{f})) = T_1(f), \quad f \in L^1(G) \cap L^2(G),$$

and $\tilde{T}$ is well-defined. Also,

$$\|\tilde{T}(\hat{f})\|_2 \leq c\|f\|_2, \quad (c, \text{ a constant}). \tag{26}$$

Thus $\tilde{T}$ can be expressed as a direct sum of bounded operators from $\mathfrak{a}(\mathcal{H}_y)$ into $L_0^2(P), y \in \hat{G}$, by the general theory. Since the range of $\tilde{T}$ is in $L_0^2(P), \tilde{T}$ is weakly compact. Applying a form of the Riesz-Markov theorem (e.g. see, Dinculeanu (1967), p. 398, Thm. 9) and using the theory of direct integrals for which the separability conditions of G are needed, (cf., Naĭmark (1964), Ch. 8, Sec. 4), one gets a regular weakly σ-additive operator measure $\mathcal{M}$ on $\mathcal{B}(\hat{G})$ into the bounded operator class $L(\mathfrak{a}(\mathcal{H}), L^2(P))$ such that $(\text{tr} = \text{trace})$

$$\tilde{T}(\hat{f}) = \int_{\hat{G}} \text{tr}\left(\hat{f}(y)\underset{\sim}{\mathcal{M}}(dy)\right), \hat{f} \in \mathfrak{a}(\mathcal{H}), \tag{27}$$

where the integral is in the Bartle-Dunford-Schwartz sense. Here $\underset{\sim}{\mathcal{M}}(\cdot) : \mathcal{B}(\hat{G}) \to L_0^2(P)$ is σ-additive and regular for each $x \in \mathfrak{a}(\mathcal{H})$. It follows now from (18) and the preceding analysis that the ensuing computations are valid.

$$\int_G f(g)X_g\mu(dg) = \int_{\hat{G}} \mathrm{tr}\left(\hat{f}(y)\underset{\sim}{\mathcal{M}}(dy)\right)$$

$$= \int_{\hat{G}} \mathrm{tr}\left[\int_G f(g)\tilde{U}(g,y)\mu(dg)\underset{\sim}{\mathcal{M}}(dy)\right]$$

$$= \int_G f(g)\int_{\hat{G}} \mathrm{tr}\left[\tilde{U}(g,y)\underset{\sim}{\mathcal{M}}(dy)\right]\mu(dg)$$

since f is scalar and the trace is linear and commutes with the integral over G, so that Fubini's argument applies. The above can be rearranged to obtain:

$$\int_G f(g)\left[X_g - \int_{\hat{G}} \mathrm{tr}\left(\tilde{U}(g,y)\underset{\sim}{\mathcal{M}}(dy)\right)\right]\mu(dg) = 0. \qquad (28)$$

Since $f \in L^1(G) \cap L^2(G)$ is arbitrary and this product set is dense in both $L^1(G)$ and $L^2(G)$, we get that the function in $[\,] = 0$. Now replacing $\tilde{U}$ with U in (28), which is valid, (28) implies (23).

The converse is similar. In fact, if (23) holds and $\phi \in L^1(G) \cap L^2(G)$, then by the standard arguments one has

$$\int_G X_g\phi(g)\mu(dg) = \int_G \phi(g)\left[\int_{\hat{G}} \mathrm{tr}(U(g,y)\underset{\sim}{\mathcal{M}}(dy)\right]\mu(dg)$$

$$= \int_G \int_{\hat{G}} \mathrm{tr}\left[\phi(g)U(g,y)\underset{\sim}{\mathcal{M}}(dy)\right]\mu(dg)$$

$$= \int_{\hat{G}} \mathrm{tr}\left[\left(\int_G \phi(g)U(g,y)\mu(dg)\right)\right]\underset{\sim}{\mathcal{M}}(dy)$$

$$= \int_{\hat{G}} \mathrm{tr}\left[\hat{\phi}(y)\underset{\sim}{\mathcal{M}}(dy)\right].$$

From this we conclude:

$$\left\|\int_G X_g\phi(g)\mu(dy)\right\| \leq \left\|\hat{\phi}\right\|_{\mathrm{op}} \|\underset{\sim}{\mathcal{M}}\|(\hat{G}), \qquad (29)$$

where $\|\underset{\sim}{\mathcal{M}}\|(\cdot)$ is the semi-variation of $\underset{\sim}{\mathcal{M}}$ (cf. Dinculeanu (1967) Sec. 19).

Letting $C = \|\underset{\sim}{\mathcal{M}}\|(\hat{G}) < \infty$ in (29) it follows that $X(\cdot)$ is weakly harmonizable since it is clearly weakly continuous.

Finally, to obtain (24), we can calculate the covariance r of $X(g)$, using a few other properties of the MT-integral as follows:

$$r(g_1, g_2) = E(X_{g_1} \cdot \bar{X}_{g_2})$$

$$= E \left(\int_{\hat{G}} \text{tr}_{y_1}(\tilde{U}(g_1, y_1) \underset{\sim}{\mathcal{M}}(dy_1) \cdot \int_{\hat{G}} \text{tr}_{y_2}(\tilde{U}_1(g_2, y_2) \underset{\sim}{\mathcal{M}}(dy_2) \right) \quad (30)$$

$$= E \left(\int_{\hat{G} \otimes \hat{G}} \text{tr}_{y_1} \otimes \text{tr}_{y_2} \left[\tilde{U}(g_1, y_1) \otimes \tilde{U}(g_2, y_2) \underset{\sim}{\mathcal{M}}(dy_1) \otimes \underset{\sim}{\mathcal{M}}(dy_2) \right] \right)$$

using the tensor product properties of trace

functionals (cf., Hewitt-Ross (1970), Appendix D),

$$= \int_{\hat{G}} \int_{\hat{G}} \text{tr}_{y_1} \otimes \text{tr}_{y_2} [\tilde{U}(g_1, y_1) \otimes \tilde{U}(g_2, y_2)] E(\underset{\sim}{\mathcal{M}}(dy_1) \otimes \underset{\sim}{\mathcal{M}}(dy_2))$$

$$= \int_{\hat{G}} \int_{\hat{G}} \text{tr}_{y_1} \otimes \text{tr}_{y_2} [\tilde{u}(g_1, y_1) \otimes \tilde{u}(g_2, y_2)] \beta(dy_1, dy_2) \quad (31)$$

where $\beta(\cdot, \cdot)$ is the operator valued positive-definite bimeasure.

This is just (24) written in a different form. Thus the result follows. $\square$

Remarks. 1. The representation (23) can be used to solve filter equations in other applications. If G is abelian, so $\hat{G}$ is also a group, and $\mathcal{H}_y = \mathbb{C}$. Thus the result reduces to a previously known case (cf. Rao (1982)).

2. The simpler representation for stationary random fields was first obtained by Yaglom (1960, 1961). It is possible now to extend his work for homogeneous spaces as well as multidimensional fields also to the weakly harmonizable case.

3. The measure $\underset{\sim}{\mathcal{M}}(\cdot)$ in (27) need not be σ-additive in the uniform operator topology. In the LCA case, this difficulty disappears, since then $\mathcal{H}$ is $\mathbb{C}$ and the classical Pettis theorem implies stating that weak and strong σ-additives agree.

4. If G is not a separable group, the decomposition, runs into difficulties, and one may have to use the C^*-algebra approach, as was initiated by Ylinen (1975). The representation (23) is not valid in general.

5. If the group G is compact, then the Fourier transforms can be derived through the Peter-Weyl theory. Thus a representation similar to (23) can be obtained through it. (See Hewitt and Ross (1970) for the general Fourier analysis on compact groups.)

We close the main part of this section with an *application to linear filters*: $\Lambda X = Y$. Let G be an LCA group, and $X, Y : G \to L^2(P, \mathbb{C}^k)$, be k-vector random fields. Then $\Lambda : X \to Y$ is assumed to satisfy $T_h(\Lambda X)(g) = \Lambda(T_h X)(g)$ for all $g, h \in G$. Here $(T_h X)(g) =$

$X(g+h), g \in G$, (translation) and $(\Lambda X)(g) = Y(g), g \in G$, is the filter equation. Thus Λ is called a *linear filter* and we discuss its solution if Λ is of the form,

$$Y(g) = (\Lambda X)(g) = \int_G \Lambda(s)X(g-s)ds. \tag{32}$$

In the context of a harmonizable process, one has:

Theorem 5.2.8 *Let the output Y be a k-dimensional weakly harmonizable random field with β_y as its $k \times k$ matrix spectral bimeasure. For the filter equation (32) given above, there exists a weakly harmonizable solution iff*

$$(i) \quad \int_D \int_D^* (I - FF^{-1})(\lambda)\beta_y(d\lambda, d\lambda')(I - FF^{-1})^*(\lambda') = 0 \tag{33}$$

for all Borel sets $D \subset \hat{G}$ where $F = \hat{\Lambda}$, the Fourier transform of Λ, with F^{-1} denoting the generalized inverse of F and '*' for the adjoint of a matrix, the integral in (i) being in the strict MT-sense, and where the following integral similarly exists:

$$(ii) \quad \int_{\hat{G}} \int_{\hat{G}}^* F(\lambda)^{-1}\beta_y(d\lambda, d\lambda')(F(\lambda')^{-1})^*. \tag{34}$$

Under these conditions, the solution X of the problem is given as:

$$X_t = \int_{\hat{G}} \langle t, \lambda \rangle F^{-1}(\lambda)Z_y(d\lambda), \quad t \in G, \tag{35}$$

where $Z_y(\cdot)$ is the stochastic measure representing Y. The solution is unique iff $F(\lambda)$ is nonsingular for each $\lambda \in \hat{G}$.

Here $F(\lambda)$ in (34) is usually called the *spectral characteristic of the filter Λ*. Under further restrictions on $A(\cdot)$ one can obtain the following simpler result:

Proposition 5.2.9 *Let F be the spectral characteristic of the filter Λ in (32). If (i) and (ii) of the above theorem hold, and if there is a $k \times k$ matrix function f, which is Lebesgue integrable, whose Fourier transform $\hat{f}$ satisfies $\|F^{-1} - \hat{f}^*\|_{2,\beta_y} = 0$, with the norm $\|f\| = \|T^{\frac{1}{2}}f\|_2$, then the solution is given by:*

$$X_t = \int_G f(s)Y(t-s)ds, \quad t \in G. \tag{36}$$

When $G = \mathbb{R}^n$, similar problems were considered by Chang and Rao (1986), and their methods, extend to the present case of an LCA group G. The further conditions under which the solution X is stationary have been discussed by Bochner (1956). These results were extended by the author (1984) to include the above considerations, and the interested readers are referred to it.

5.3 Harmonizability on Hypergroups

In some statistical analyses, for example, sample means of stationary or harmonizable sequences, we are often, confronted with 'new' processes that are not of the original form. They are of second order classes which are not easily compared and classified with the original processes. But still many of these form a generalization which is now termed '*hypergroups*', and the subject has potential for an extended analysis. It has applicational prospect and potential for developments, as seen from the detailed and lucid exposition by Heyer (2014), we discuss harmonizability for such classes (on hypergroups) and present some results on their representation. The abstract concept is as follows:

Definition 5.3.1 *A locally compact space K is called a* hypergroup *if the following conditions are satisfied for it:*

There exists an operation, termed convolution, $* : K \times K \to M(K)$ *such that for $x, y \in K$, $(x, y) \to \delta_x * \delta_y$ (δ_x is Dirac measure) where $M(K)$ is the set (or space) of Radon probability measures on K, endowed with the weak*-(also termed vague) topology and so $M(K)$ can be regarded as the dual space of $C_0(K)$, such that:*

(i) $\delta_x * (\delta_y * \delta_z) = (\delta_x * \delta_y) * \delta_z$,

(ii) $\delta_x * \delta_y$ has compact support,

(iii) there is an involution denoted "$\sim$", on K such that $(x^\sim)^\sim = x$ and $(\delta_x * \delta_y)^\sim = \delta_{\tilde{x}} * \delta_{\tilde{y}}, x, y \in K$ where for a measure μ on K, $\tilde{\mu}(A) = \mu(\tilde{A})$ with $\tilde{A} = \{\tilde{x} : x \in A\}$, and there is a unit element 'e' in K so that $\delta_e * \delta_x = \delta_x * \delta_e = \delta_x$,

(iv) 'e' $\in \sup(\delta_x * \delta_y)$ iff $x = y$ and the mapping $(x, y) \to supp(\delta_x * \delta_y)$ is continuous when the space 2^K is given the Kuratowski topology.

Note: If (iv) is not assumed, then the space K having properties (i)–(iii) is called a *weak hypergroup*. A standard reference for hypergroups is the book by Bloom and Heyer (1995).

A hypergroup satisfying conditions (i)–(iii) is termed a *weak hypergroup*. A number of examples and details of these objects are given by Lasser (1983), and a general theory is detailed in the above noted reference volume of Bloom and Heyer (1995). Also, there exist some further extensions and recent research on the subject seen in a nice exposition by Heyer (2010).

It was shown by Spector (1978) that an abelian hypergroup K admits an invariant (or Haar) measure and if $\hat{K}$ is its dual (so $K \subset \hat{K}$) it is termed a *strong* hyper group provided $K = \hat{K}$. A considerable amount of classical analysis extends to such hypergroups. In the probabilistic context, we say that $X : K \to L_0^2(P)$, with covariance $\rho : (a, b) \to E(X_a \bar{X}_b)$, is *hyper-weakly stationary* if ρ is representable as:

$$\rho(a, b) = \int_K \rho(x, 0)(\delta_a \times \delta_b)(dx), \quad a, b \in K, \tag{37}$$

and then X is *hyper-weakly stationary* on the commutative hypergroup K. We now introduce *hyper weak harmonizability* as a useful concept.

Definition 5.3.2 *Let K be a commutative hypergroup with its dual object $\hat{K}$, and $X : K \to L_0^2(P)$ be a centered random field. If $\rho : (a, b) \mapsto E(X_a \overline{X}_b), a, b \in K$, denotes its covariance function, then X is termed hyper-weakly (strongly) harmonizable if ρ admits a representation:*

$$\rho(a, b) = \int_{\hat{K}} \int_{\hat{K}}^* \alpha_1(a)\overline{\alpha_2(b)}\beta(d\alpha_1, d\alpha_2), \tag{38}$$

where $\beta : \mathcal{B}(\hat{K}) \times \mathcal{B}(\hat{K}) \to \mathbb{C}$ is a positive definite bimeasure (of finite Vitali variation) and the integral (with 'star') is in the MT- (or Lebesgue-Stieltjes) sense.

It is now known that β has always a finite Fréchet variation on the Borel σ-algebra $\mathcal{B}(\hat{K})$. The Fourier transform is well-defined and the strict MT-integral will be used (as detailed in the article by Chang and Rao (1986)). This is one-to-one and the Fourier transform is also one-to-one. With these properties, the following result can be established.

Theorem 5.3.3 *Let $X : K \to L_0^2(P)$ be a hyper-weakly harmonizable random field defined above. Then there exists a stochastic measure $Z : \mathcal{B}(\hat{K}) \to L_0^2(P)$ such that*

$$X_a = \int_{\hat{K}} \alpha(a) Z(d\alpha), \quad a \in K \tag{39}$$

with $E(Z(A_1)\bar{Z}(A_2)) = \beta(A_1, A_2)$ defining a bimeasure β. Indeed a second order weakly continuous random field on a commutative hypergroup K admits the representation (39) so that it is hyper weakly harmonizable, iff the following set is bounded in $L^2(P)$:

$$\left\{ \int_K \phi(a)X_a d\mu(a) : \|\hat{\phi}\|_\infty \leq 1, \phi \in L^1(K, \mu) \cap L^2(K, \mu) \right\} \subset L_0^2(P)$$

(40)

with μ being a Haar measure carried on K and $\hat{\phi}$ is the Fourier transform, given as $\hat{\phi}(\alpha) = \int_K \alpha(a)\phi(a)d\mu(a), \alpha \in \hat{K}$.

The result is obtained from the works of R. Lasser and R. C. Verm noted above. There is also much activity by these authors on related results. Further discussion on the problems will be omitted here.

5.4 Remarks on Strict Harmonizability and V-Boundedness

Instead of restricting to the second moment properties, some extensions will be useful for comparison as well as for some applications. Hence we consider some related problems based on the complete distributional properties for their structural and probabilistic features when Fourier analysis is still available.

Definition 5.4.1 *On a measurable couple $(S, \mathcal{S})$ let $Z : \mathcal{S} \to L^p(P), 0 < p \leq 2$, be a mapping. Then $Z(\cdot)$ is termed an independently valued random measure of exponent p if:*

(i) $A_i \in \mathcal{S}, i = 1, \ldots, n, n \geq 1$, disjoint, $\{Z(A_i), 1 \leq i \leq n, n \geq 1\}$ is a mutually independent set of random variables,

(ii) for each $A \in \mathcal{S}, Z(A)$ is a stable random variable of exponent p_i,

(iii) $A_n \in \mathcal{S}$, disjoint, $A = \cup_{n \geq 1} A_n \Rightarrow Z(A) = \sum_{n=1}^{\infty} Z(A_n)$, the series converging in probability (hence also with probability one).

The above concept allows us to present an integral representation of certain *strictly stationary* processes with p moments, $1 \leq p \leq 2$ that admit integral representation relative to a stochastic measure, with stable independent values, opening up some new directions than before. This shows how new types of integrals arise in such classes of studies.

Theorem 5.4.2 *Let $X : \mathbb{Z} \to L^\alpha(P), 1 \leq \alpha \leq 2$ be a (discrete) process. Then there exists an independently valued stable random measure*

Z of exponent α, which is isotopic on the Borel field $\mathcal{S}$ of the interval $(-\pi, \pi)$ such that

$$X_n = \int_{-\pi}^{\pi} e^{int} Z(dt), \quad n \in \mathbb{Z}, \tag{41}$$

iff $\{X_n, n \in \mathbb{Z}\}$ is strictly stationary and V-bounded in the sense that

$$\left\| \sum_{K=1}^{n} a_k X_k \right\|_{\alpha} \leq C_0 \sup \left\{ \left| \sum_{k=1}^{n} a_k e^{-2\pi i \alpha_k} \right| : t \in (-\pi, \pi) \right\}, \tag{42}$$

where C_0 is constant, and if $\alpha = 1$, $\{\sum_{k=1}^{n} a_k X_k, n \geq 1, a_k \in \mathbb{C}\}$ is relatively weakly compact in $L^1(P)$ in addition. When these conditions hold we obtain their characteristic functions as:

$$\phi_{n_1, \ldots n_k}(u_1, \ldots u_k) = E \left(\exp \left[\sum_{j=1}^{k} i n X_{n_j} \right] \right)$$

$$= \exp \left\{ -\int_{-\pi}^{\pi} | \sum_{j=1}^{k} u_i e^{i\lambda u_i} |^{\alpha} dG(\lambda) \right\}, \tag{43}$$

for a bounded increasing function $G \geq 0$.

Some Remarks. It may be noted that the integral representation (41) is obtained using (42) and Bochner's V-boundedness condition for processes valued in Banach spaces. The calculations, giving the representation (42) are due to Hosoya (1982), who also extended it thereafter for $0 < \alpha < 1$ and, for X_t of (41). There are some related extensions of the integral formula (41) by several authors, including Combanis (1983), Weron (1985), Urbanik (1968), Kuelbs (1993), Okazaki (1979), Rosinski (1986), Marcus (1987), and perhaps some others.

We end this section with a strict sense version of the Karhunen class due to Kuelbs (1973) which is an extension of Schilder's (1970) result who considered this for a symmetric stable class. This is given for comparison and reference here.

Theorem 5.4.3 *Let $\{X_t, t \in T\}$ be a random field indexed by a separable Hausdorff space T whose finite dimensional distributions form a symmetric stable class of index α, $0 < \alpha \leq 2$. Then there exists an independently valued stable random measure $Z(\cdot)$ of index $\alpha \in [-\frac{1}{2}, \frac{1}{2}]$*

and a family $\{f_t, t \in T\}$ contained in $L^\alpha(-\frac{1}{2}, \frac{1}{2})$, the Lebesgue space, such that

$$W_t = \int_{-\frac{1}{2}}^{\frac{1}{2}} f_t(\lambda) dZ(\lambda), t \in T, \tag{44}$$

and the processes $\{X_t, t \in T\}$ and $\{W_t, t \in T\}$ have the same finite dimensional distributions, the integral in (44) being a Wiener (type) integral.

Here $f_t(\lambda)$ cannot be replaced by $e^{it\lambda}$ in general. Also, the argument is based on an analog of V-boundedness. Further analysis and extensions as well as sample path studies are detailed in Okazaki (1979), see also Marcus (1987), which should be of interest for related analysis and extensions of these works.

5.5 Vector-Valued Harmonizable Random Fields

To complete our discussion, it will be of interest to consider *briefly* the vector and (operator) valued harmonizable fields also, for the following studies.

If $\{X_g, g \in G\} \subset L_0^2(P)$ is a centered random field with r as its covariance, $r(g_1, g_2) = E(X_{g_1} \bar{X}_{g_2})$, then there are a right, left, and a two-sided stationarity concept and one needs to consider the dilation problem in each class. Thus X is left (or right) stationary if

$$r(gg_1, gg_2) = \tilde{r}(g_2^{-1} g_1), [r(g_1 g, g_2 g) = \tilde{r}(g_1 g_2^{-1})], \tag{45}$$

and it is *two-sided stationary* if it is both left and right stationary. Then for dilation problems, each class has to be studied separately. Now X is termed *hemi-homogeneous*, as defined by Ylinen (1996), if its covariance $r(\cdot, \cdot)$ can be expressed as:

$$r(g_1, g_2) = \rho_1(g_2^{-1} g_1) + \rho_2(g_1 g_2^{-1}), \qquad g_1, g_2 \in G, \tag{46}$$

for a pair of positive definite covariances ρ_1, ρ_2 on G. Then Ylinen's result gives the following using the early terminology.

Theorem 5.5.1 *Let G be a separable unimodular locally compact group, and $X : G \rightarrow L_0^2(P)$, a continuous random field in $L_0^2(P)$. Then X is weakly harmonizable iff it has a hemi-homogeneous dilation Y, $Y : \mathbb{C} \rightarrow L_0^2(\tilde{P}) \supset L_0^2(P)$ so that $X(g) = (QX)(g), g \in G$, where Q is the orthogonal projection of $L_0^2(\tilde{P})$ onto $L_0^2(P)$.*

This result, due to Ylinen (1987), is actually proved for all locally compact groups using the Fourier transform method of Eymard's (1964) which employs results from C^*-algebras. The dilation problem has additional difficulties for vector random fields. The result is presented here, to indicate the general patterns in the noncommutative case.

5.6 Cramér and Karhunen Extensions of Harmonizability Compared

We now have two types of extensions of (weak) harmonizability due first to Karhunen (1947) and later to Cramér (1951). In both extensions, the mappings are more general than the group characters $t \mapsto e^{it(\cdot)}, t \in \mathbb{R}$ which are replaced by a class of smooth functions $t \mapsto g(t, \cdot)$. Here we show that these interesting extensions retain many useful properties. However, the key dilation property may be lost, and that the mapping $t \to g(t, \cdot)$, being a group character, is special. The extension itself utilizes a new technical result, on a (generalized) domination of measures, due to Lindenstrauss and Pelcyziński (1968). The desired Cramér-Karhunen general result is then given as follows:

Theorem 5.6.1 *Let $X : \mathbb{R} \to L_0^2(P)$ be a process and $\{g(t, \cdot), t \in \mathbb{R}\}$ be a set of Borel functions with the following properties: If X is a Karhunen process relative to this g-family, and a σ-finite measure F on $\mathcal{B}(\mathbb{R})$ and $T : L_0^2(P) \to L_0^2(P)$ is a continuous linear map, then the process $\{Y(t) = TX(t), t \in \mathbb{R}\}$ is a Cramér process relative to the same g-family and a (suitable) covariance bimeasure. In the opposite direction, if $\{g(t, \cdot), t \in \mathbb{R}\}$ is of bounded Borel class, $X : \mathbb{R} \to L_0^2(P)$ is a Cramér process relative to this g-family and a suitable covariance bimeasure β, then there is an extension space $L_0^2(\tilde{P})(\supset L_0^2(P))$, determined by the given process, namely a Karhunen process $Y : \mathbb{R} \to L_0^2(\tilde{P})$, relative to some g-family and an appropriate finite Borel measure on $\mathbb{R}$ such that $X(t) = QY(t), t \in \mathbb{R}$, with $Q : L_0^2(\tilde{P}) \to L_0^2(P)$ as a projection (contained as a subspace) onto it.*

Remarks. 1. This is a generalization of the harmonizable class, but is weaker in that $l^2(\tilde{P})$ spaces change with each given (Cramér) process! Indeed the construction of the spaces depends on the particular vector integrals and applications (see E. Thomas (1970), which is his thesis).

This implies the differences and weaknesses between the Fourier analysis and its generalizations. The problem appears in the fact that $L^2(Z)$ and $L^2(\tilde{Z})$ where $\tilde{Z} = QZ$ for orthogonal projection Q can have the spaces so that the former may not be including the latter, in contrast to the harmonizable case in which the vector measures $Z(\cdot)$ are more restricted. [An example of this effect was constructed by E. Thomas (unpublished).]

2. The preceding observation and result show that the set of Cramér classes is quite large, and some cannot be easily dilated to Karhunen processes. But it is seen that the Cramér class is closed under transformation $T : \nu \to T \circ \nu$, when appropriate integrability conditions are imposed on T. This will not be detailed here.

We end this chapter with some complements and in the next (and final) chapter some extensions of the subject, motivated by applications, will complete the work of the volume.

5.7 Complements and Exercises

1. Let $X = \{X_t, t \in T = [a, b]\}$ be a Karhunen process, $E(X_t) = 0, t \in T$. If $A : L_0^2(P) \to L_0^2(P)$ is a bounded linear operator, let $Y_t = AX_t, t \in T$. Show that the $\{Y_t, t \in T\}$ is a Cramér process if the spectral measure of X_t is finite. Verify that the converse holds when the representing measure of the X_t-process is finite and the corresponding class $g(t, \cdot)$ of the X_t-process is bounded.

2. Verify that a weakly harmonizable process $X = \{X_t, t \in \mathbb{R}\} \subset L_0^2(P)$ is automatically of Karhunen's class relative to a suitable Borel measure μ and some Borel family $\{f_t, t \in \mathbb{R}\} \subset L^2(\mathbb{R}, \mu)$.

3. Let $\{X_t, t \in T = [a, b]\} \subset L_0^2(P)$ be a process with continuous covariance r. Then the operator $R : L_0^2(P) \to L_0^2(P)$ defined by $(Rg)(t) = \int_T r(s, t)g(s)ds$ is positive definite and compact and if $M_\rho = \{E(X_t), t \in T\}$, then we have $M_\rho \subset R^{\frac{1}{2}}(L^2(dt))$ i.e., $f \in M_p$ implies $f = R^{\frac{1}{2}}h$ for some $h \in L^2(T, dt)$. [This result is essentially due to T. S. Pitcher (1963) and is useful in studying measures and the process X_t.]

4. Let G be an LCA group and $X : G \to L_0^2(P)$ be a random field which is weakly harmonizable. Verify that there is a finite Borel measure μ on the Borel σ-algebra of the dual $\hat{G}$ of G and a family $\{g_s, s \in G\} \subset L^2(\hat{G}, \mu)$ such that we have the representation of

r as:

$$r(s,t) = \int_{\hat{G}} g_s(\lambda)\overline{g_t(\lambda)}\mu(d\lambda), s, t \in G$$

so that $t \in G$, and $g \in \hat{G}$.

[Regarding this and some related results, one can find the details and extensions in Rao (1985).]

5. On an LCA group G, let $X : G \to L_0^2(P)$ be a weakly harmonizable random field. Then show that there is a finite Borel measure μ on $\mathcal{B}(\hat{G})$ and a family $\{g_s, s \in G\} \subset L^2(\hat{G}, \mu)$ for a σ-finite measure μ on the Borel sets of $\hat{G}$ such that

$$r(s,t) = \text{cov}\,(X_s, X_t) = \int_{\hat{G}} g_s(\lambda)\overline{g_t(\lambda)}d\mu(\lambda), s, t \in G.$$

Thus each harmonizable field $X : G \to L_0^2(\mathbb{R})$ belongs to a Karhunen class.

6. A centered second order random field $X : \mathbb{R}^n \to L_0^2(P)$ on a probability space with covariance $r(s,t) = \tilde{r}(s-t)$, (so it is stationary), is termed *isotopic* (concept due to Bochner) if $\tilde{r}(\cdot)$ is invariant under rotations and reflexions, in addition to translation, and he has characterized it as:

$$r(s,t) = \tilde{r}(s-t) = 2^n \Gamma\left(\frac{n}{2}\right) \int_{\mathbb{R}^n} \frac{J_\nu(\lambda|s-t|)}{(\lambda(|s-t|)^\nu)} dG(\lambda),$$

where $\nu = \frac{n-2}{2}, G(\cdot)$ is a unique bounded Borel measure on $\mathbb{R}^+ = [0, \infty)$ and $|s - t|^2 = (s-t) \cdot (s-t), n \geq 2$. Here $J(\cdot)$ is the Bessel function of the first kind of order ν. Using the MT-integration we can now present an extension (of isotropy) to harmonizable fields and outline its consequences giving a real opening here. [We need to consider the weakly harmonizable isotropic class as it appears in applications and restricting it to a weakly stationary class excludes many interesting practical problems as pointed out in the monograph by Yadrenko (1983)]. We have to employ the move complicated notation, to present a key extension of Cramér's class (C) for illustration. Recall (at least here) that a bimeasure $F : \mathcal{S} \times \mathcal{S} \to C$ ($\mathcal{S}$ a semi-ring of a set T) with $\mathcal{S}_0 = \{A \in \mathcal{S} : \|F\|(A, A) < \infty\}$, $\mathcal{S}(A) = \{A \cap B : B \in \mathcal{S}\}$, with $\|F\|(\cdot, \cdot)$ denoting the Fréchet variation introduced earlier. The covariance $r : T \times T \to \mathbb{C}$ is then of *weak class* (C) if it is representable as:

$$r(s,t) = \int_\Lambda \int_\Lambda g_s(\lambda)\overline{g}_t(\lambda')F(d\lambda, d\lambda'), (s,t) \in T \times T,$$

the family $\{g(\cdot), s \in T\}$ verifies the conditions of the above strict MT-integral. We then have the integral representation of $X(\cdot)$ as:

$$X(t) = \int_A g_t(\lambda)dZ(\lambda), \quad t \in T \quad \text{(Dunford-Schwartz type)}$$

and

$$E(Z(A)\overline{Z(B)}) = F(A, B), \quad A, B \in \mathcal{S} = \sigma(T).$$

The converse of the above statement is also true. [More details are in Rao (1991), and extensions and application of these are useful.]

7. This problem indicates how an integral representation of a second order general Cramér type random field looks. Thus let $(A, \mathcal{S})$ be a measurable space, $F : \mathcal{S} \times \mathcal{S} \to \mathbb{C}$ a positive definite bimeasure, and let $X : T \to L_0^2(P)$ be a random field with covariance r admitting a representation

$$r(s,t) = \int_A \int_A g_s(\lambda)\overline{g}_t(\lambda')F(d\lambda, d\lambda'), s, t \in T,$$

relative to a class $g_s(\cdot)$ so that the above is an MT-integral. If $X : T \to L_0^2(P)$ with covariance r given by the above integral, then verify that there is a stochastic measure $Z : \mathcal{S} \to L_0^2(P)$ such that, as a Dunford-Schwartz integral, we have

$$X(t) = \int_A g_t(t, \lambda)Z(d\lambda), t \in T,$$

$$E(Z(A)\overline{Z(B)}) = F(A, B), \quad A, B \in \mathcal{S}.$$

The converse statement is also true.

As a consequence, we have an explicit representative of X

$$X(t,r) = \alpha \sum_{n=0}^\infty \sum_{l=1}^{h(m,n)} g_m^l(0) \int_{\mathbb{R}^+} \frac{J_{m+\nu}(\lambda r)}{(\lambda r)^\nu} Z_m^l(d\lambda)$$

with $Z_m^l(\cdot)$ satisfying the orthogonal relations. [See Rao (1991) for more.]

8. We now present an integral (series) representation of a weakly harmonizable isotropic random field $X(\cdot) : \mathbb{R}^n \to L_0^2(P)$, useful in some applications. Thus $X(\cdot)$ is representable as a series given by:

$$X(\underset{\sim}{t}) = X(r, \underset{\sim}{v}) = \alpha_n \sum_{m=0}^{\infty} \sum_{l=1}^{h(m,v)} S_m^l(v) \int_{\mathbb{R}^n} \frac{J_{m+\nu}(\lambda r)}{(\lambda r)^\nu} Z_m^l(d\lambda),$$

with $E(Z_m^l(B_1)\bar{Z}_{m'}^{l'}(B_2)) = \delta_{mm'}\delta_{ll'}F(B_1, B_2)$, $F(\cdot, \cdot)$ being of finite (bimeasures of) Fréchet variation. In the converse direction, the above representation gives $X(\cdot)$ to be an isotropic harmonizable random field. Moreover in the above representation $X(\cdot)$ is weakly stationary iff $Z(\cdot)$ is orthogonally valued. [More details on this representation and several applications can be found in this author's paper (Rao (1991)) along with some additional results on sampling of such random fields with some related extensions.]

9. If $\mathcal{H}$ is a Hilbert space let $\mathcal{T}(\mathcal{H})$ denote the collection of trace class operators on $\mathcal{H}$, so that with

$$\tau([f, g]) = \int_\Omega \langle f, g \rangle dP, \qquad \|f\|_{2,\tau}^2 = \tau([f, f]),$$

we see that $(L_0^2(P), \mathcal{H})$, becomes a Hilbert space with inner product $\tau([\cdot, \cdot])$ and $\langle f, g \rangle(\omega) \in \mathcal{H}$. Thus for $\mathcal{H}$, a Hilbert space, $\mathcal{T}(\mathcal{H}) \subset L(\mathcal{H})$, is the set of trace class operators, the mapping $[\cdot, \cdot] : \mathcal{X}_0 \times \mathcal{X}_0 \to \mathcal{T}(\mathcal{H})$ has the properties: (i) $[x, x] \geq 0$, $[x, x] = 0$ iff $x = 0$, (ii) $[x + y, z] = [x, z] + [y, z]$, (iii) $[Ax, y] = A[x, y]$ for $A \in L(\mathcal{H})$, (iv) $[x, y]^* = [y, x]$ with '*' for the adjoint operation on $L(\mathcal{H})$. The mapping $[\cdot, \cdot]$ is the Gramian on $\mathcal{X}_0$. If $\|x\|_\tau^2 = \tau([x, x]) = $ trace$([x, x])$ let $\mathcal{X}$ be the completion of $\mathcal{X}_0$ for the norm $\| \cdot \|_\tau$ and $\mathcal{X}$ is termed the *normal Hilbert $L(\mathcal{H})$ module* if $[x, y] = xy^*$ and $A \cdot x = Ax, x, y \in HS(K, \mathcal{H})$, and $A \in L(\mathcal{H})$.

Now let $X : G \to \mathcal{H}$ be a mapping on an LCA group G, into a normal $L(\mathcal{H})$-module $\mathcal{X}$. Then X is *weakly harmonizable* if we have

$$X(t) = \int_G \langle t, \lambda \rangle Z(d\lambda), \quad t \in G$$

the integral being in the Dunford-Schwartz sense, and $Z(\cdot) : \mathcal{B}(\hat{G}) \to \mathcal{H}$ being a vector measure of finite semi-variation, i.e., $A \in \mathcal{B}(\hat{G}) \Rightarrow$ there is a compact F and open O of $\hat{G}$, $F \subset A \subset O$ with

$\|Z\|(O - F) < \varepsilon$, for the semi-variation $\| \cdot \|$ of Z. With this elaboration, we have "A random field" $X : G \to \mathfrak{X}$, which is a normal $L(\mathcal{H})$-module, is weakly harmonizable iff it is V-bounded and continuous in the norm topology of $\mathfrak{X}$.

[Some technical machinery is needed to complete the proof and the reader will find the work of Kakihara (1986) useful here.]

10. We present a simple extension with its characterization, of class (C) random fields of interest in several applications. Thus let $(\Lambda, \mathcal{B})$ be a measurable space and $F : \mathcal{S} \times \mathcal{S} \to \mathbb{C}$ be a positive definite bimeasure. Suppose $X : T \to L_0^2(P)$ is a random function (centered) with covariance r of weak Cramér class hence representable as (for r):

$$r(s,t) = \int_\Lambda \int_\Lambda g_s(\lambda)\overline{g_t(\lambda')}F(d\lambda, d\lambda'), \quad s, t \in T,$$

where the bimeasure F has finite Fréchet variation on $\mathcal{S}$, to mean

$$\|F\|(A, B) = \sup\left\{ \left| \sum_{i,j=1}^n a_i b_j F(A_i, B_j) \right| : |a_i| \leq 1, |b_i| \leq 1, \{A_i\}, \right.$$

$$\left. \{B_j\} \text{ are disj. collections from } \mathcal{S}(A), \mathcal{S}(\mathcal{B}), n \geq 1 \right\}.$$

If $X : T \to L_0^2(P)$ whose covariance r is as above then it has a representation relative to a stochastic measure $Z : \mathcal{S} \to L_0^2(P)$:

(i) $X(t) = \int_\Lambda g(\lambda)Z(d\lambda), t \in T$ (Dunford-Schwartz integral).

(ii) $E(Z(A)\overline{Z(B)}) = F(A, B), \quad A, B \in \mathcal{S}$.

Conversely, if $X(\ )$ is given as above relative to a $Z(\cdot)$ and F, then X is a random field of weak class (C).

11. This problem gives a series representation of an *isotropic* weakly harmonizable random field and indicates the progress of work. Thus let $X : \mathbb{R}^n \to L_0^2(P)$ be a centered weakly harmonizable isotropic random field. Then it is representable as:

$$X(\underset{\sim}{t}) = \alpha_n \sum_{n=0}^\infty \sum_{l=1}^{h(m,n)} S_m^l(v) \int_{\mathbb{R}^+} \frac{J_{n+\nu}(\lambda r)}{(\lambda r)^\nu} Z_m^l(d\lambda),$$

where the $Z_m(\cdot)$ are orthogonal and satisfy $E(Z_m^l(B_1)\bar{Z}_{m'}^{l'}(B_2)) = \delta_{mm'}\delta_{ll'}F(B_1, B_2)$. In the opposite direction, $X(\cdot)$ given by the above series is a weakly harmonizable isotropic random field. Further, the above given $X(\cdot)$ is stationary iff $Z_m^l(\cdot)$ is just orthogonally valued so that the measure $F(\cdot, \cdot)$ concentrates on the diagonal of $\mathbb{R}^n \times \mathbb{R}^n$. [For further details and extensions of this problem, the reader may see the author's paper in the *Journal of Combinatorics, Information and System Sciences* **16** (1991), 207–220.]

5.8 Bibliographical Notes

The work here begins by outlining the necessary bimeasure integration, weaker than the standard one based on Lebesgue's ideas, the weakening is due to M. Morse and W. Transue who developed it around the 1950s, and these are utilized here. Then our main thrust is in obtaining extensions of the harmonizability on LCA groups. A reasonably detailed account of the analysis on locally compact groups following some classical methods and the work due to Mautmer (1955) are utilized and the corresponding extensions are detailed in our long Section 2. Actually, the paper by Mautmer is easily adapted for our cases, but this opens the way to utilize and extend the theory for the noncommutative cases leading the way to consider Harish-Chandra, A. Borel, A. Weil and other great harmonic analysts' works.

Also, K. Ylinen's method is briefly indicated to show the possibility of this analysis for more general random fields. Here Ylinen's work indicated how there is another direction and further treatment. A different procedure is to replace the LC groups as indexed with hyper groups. This is somewhat new to the subject but Heyer has indicated the possibility and had discussed directions in a recent survey, Heyer considers random fields on such objects as hyper groups which is of special interest for workers in other fields as well.

This leads to vector harmonizability and some new problems appear, as indicated in Section 5.5. We only sketched the work of Ylinen and the need to employ C^*-algebra technique, and this may be considered by the readers later on. An interesting aspect here is an extension of (weak) harmonizability by Karhunen (1942) followed by H. Cramér (1951). These are of special interest from both applicational point of view as well as theory. This usually goes beyond harmonic analysis, and the corresponding theory is not yet well developed, compared to the one using only harmonic (or Fourier) methods. There are some additional ideas of interest indicated in the complements section, but we end the abstract treatment here.

The problem of isotropy is important and we only indicated its analysis briefly. Much work is available in Swift's publications and his monograph on the subject to be completed soon. It would have the details to study more of this part of analysis, and we end this general survey here. On the hypergroup extension, we mention two theses that have some detailed applications of hypergroup structure in B. Englehardt (2002) and some another thesis from a student under Heyer at the University of Tubingen since the book appeared from Bloom and Heyer (1995).

6

Some Extensions of Harmonizable Random Fields

6.1 Introduction

This chapter concludes our treatment of harmonizable random fields whose extensions include the Karhunen and Cramér classes as well as some results on L^p-valued processes (or fields) when $1 < p < \infty$, with some applications. To understand the problem better, we restate the (equivalent) Naĭmark-Sz.-Nagy theorem connecting the harmonizable and stationary processes, the former being a projection of a suitably located (super) Hilbert space. The type or form that we use is as follows:

Theorem 6.1.1 *Let* $\{Y_t, t \in T\} \subset L_0^2(\Omega, \Sigma, P)$ *or* $L_0^2(P)$ *be a weakly (or strongly) harmonizable process* $(T \subset \mathbb{R})$. *Then there is an enlarged probability space* $(\tilde{\Omega}, \tilde{\Sigma}, \tilde{P})(\supset (\Omega, \Sigma, P))$ *so* $L_0^2(\tilde{P})$ *includes* $L_0^2(P)$ *with an orthogonal projection* $Q : L_0^2(\tilde{P}) \to L_0^2(P)$, *so that there is a stationary process* $\{X_t, t \in T\} \subset L_0^2(\tilde{P})$ *with* $Y_t = QX_t, t \in T$, *the* X_t-*process being termed a dilation of the* Y_t-*process. On the other hand, each continuous linear mapping* $A : L_0^2(P) \to L_0^2(P)$ *defines* $Y_t = AX_t, t \in T$ *to be a weakly harmonizable process in* $L_0^2(P)$.

This result is established without difficulty from the previous methodology and the proof is left to the reader. The following specializations (and modifications) will lead to several applications. To consider this in some detail and depth we introduce the key concept of *isotropy* for harmonizable fields and include their integral representations, with related discussion, along with some additional developments of our subject. Further, a few consequences (or adjuncts) will be indicated for some possible extensions and new applications, along with some multidimensional indications of these classes.

6.2 Harmonizability, Isotropy and Their Analyses

We recall that a centered second order random field $X = \{X_t, t \in T \subset \mathbb{R}^n\} \subset L_0^2(P)$, with covariance $r(s, t) = E(X_s \bar{X}_t)$ is *weakly stationary* if $r(s, t) = \tilde{r}(s - t)$ and

$$r(s, t) = \tilde{r}(s - t) = \int_{\hat{T}} e^{i(s-t)\lambda} F(d\lambda) \tag{1}$$

relative to a bounded increasing positive $F(\cdot)$ and is *weakly harmonizable* if $r(\cdot, \cdot)$ is just representable as:

$$r(s, t) = \int_{\hat{T}} \int_{\hat{T}}^* e^{is\lambda - it\lambda'} F(d\lambda, d\lambda') \tag{2}$$

where $F(\cdot, \cdot)$ is a (positive definite) function of *bounded Fréchet variation*. Hence

$$\|F\|(\hat{T} \times \hat{T}) = \sup \left\{ \sum_{i,j=1}^n a_i \bar{a}_j F(\lambda_i, \lambda_j) : \lambda_i \in \hat{T}, a_i \in \mathbb{C}, \right.$$

$$\left. |a_i| < 1, 1 \leq i \leq n, n \geq 1 \right\} < \infty, \tag{3}$$

the 'star' on the integral indicates that it is a Morse-Transue integral. A detailed exposition of this (weaker) integral is in Chang and Rao (1986), and it will be needed and used here. [*Strongly harmonizable* obtains if $\text{var}(F) < \infty$ and then we use Lebesgue integral in (2).]

A centered random field $\{X_t, t \in T\} \subset L_0^2(P)$ is *isotropic* if the covariance r verifies $r(gs, gt) = r(s, t), s, t \in T \subset \mathbb{R}^n$, and $g \in SO(n)$, the set of all orthogonal matrices on $\mathbb{R}^n$. The above classes of (centered) random fields have the (proper) inclusion relationships: stationary isotropic $X \subset$ strongly harmonizable isotropic $X \subset$ weakly harmonizable isotropic $X \subset$ isotropic X. Characterizations of the first (i.e. stationary isotropic) was given originally by S. Bochner, and the last one by M. I. Yadrenko, but the characterizations of the others were not done. All these cases are included here to understand the problem better.

We first present a basic characterization of each of the weak and strong harmonizable isotropic covariances:

Theorem 6.2.1 *Let X be a centered weakly harmonizable random field on $\mathbb{R}^n, n > 1$, with covariance $r : \mathbb{R}^n \times \mathbb{R}^n \to \mathbb{C}$. Then r is isotropic iff, using the strict Morse-Transue integral, we have:*

$$r(s,t) = 2^\nu \Gamma\left(\frac{\nu}{2}\right) \int_{\mathbb{R}^+} \int_{\mathbb{R}^+}^* \frac{J_\nu(|\lambda s - \lambda' t|)}{|\lambda s - \lambda' t|^\nu} \Phi(d\lambda, d\lambda'), \qquad (4)$$

where Φ is a positive definite function of bounded Fréchet variation, J_ν is a Bessel function (of the first kind) of index $\nu \left(= \frac{n-2}{2}\right)$, and the integral is in the strict Morse-Transue sense. The corresponding strongly harmonizable characterization obtains if Φ in (4) has finite Vitali variation and then the integral is in the standard Lebesgue sense.

Proof. We present the proof in a series of steps for convenience.

The direct (or the "if") part follows easily since $r(gs, gt) = r(s, t)$ for $g \in SO(n)$. This is because $|g(\lambda s - \lambda' t)| = |\lambda s - \lambda' t|$ since 'g' gives an isometry in (4) and only the converse is nontrivial which is detailed in the following four steps.

I. Let $X : \mathbb{R}^n \to L^2(P)$ be a harmonizable random field so that by the dilation theorem there is an enlarged probability space $(\tilde{\Omega}, \tilde{\Sigma}, \tilde{P})$, with $L_0^2(\tilde{P}) \supset L_0^2(P)$, a contractive (onto) projection $Q : L_0^2(\tilde{P}) \to L_0^2(P)$, and a stationary random field $Y : \mathbb{R}^n \to L_0^2(\tilde{P})$ satisfying $X = QY$ and that $X_t = QY_t$, $t \in \mathbb{R}^n$. Let ρ be the covariance of Y so that $\rho(s, t) = \tilde{\rho}(s - t)$. We assert that ρ and $\tilde{\rho}$ can also be taken isotropic when X is isotropic.

Now the random field Y in $L_0^2(\tilde{\rho}) = K = \mathcal{H} \oplus \mathcal{H}_0$ is representable as $Y = X + X_1$, where $X_1 \in \mathcal{H}_0$ is stationary, and the space $\mathcal{H}_0$ is obtained from $L^2(\mu)$, μ being a finite measure (due to Grothendieck). But now $\mathcal{H}_0$ may be realized as a subspace of some $L^2(\tilde{\mu})$ on a measure space $(\tilde{\Omega}, \tilde{\Sigma}, \tilde{\mu})$ (e.g. as a Gaussian probability space with a standard (Kolmogorov-Bochner) method whose construction gives even $\tilde{\mu}(gA) = \tilde{\mu}(A)$). This shows that the above type generated Hilbert space can be replaced by such an isometric space based on a Gaussian measure in this construction after the initial step. This modification makes the components of $X_1 = (X_{1t}, t \in \mathbb{R}^n)$ to have invariant distributions under g so that X_{1t} and X_{1gt} to be identically (Gaussian) distributed under g so X_{1gt} and X_{1t} to be identically distributed. Consequently, the random field X_1 is stationary, isotropic and orthogonal to X. Thus our $\tilde{P}$ is determined in the final construction by P and $\tilde{\mu}$. [The calculation here is the same in detail as in Rao (1982), the L'*Enseignment Math.,* 28, 295–351, paper and will not be reproduced here.] The invariance of the bimeasure F of X is verified simply as follows.

$$r(gs, gt) = \int_{\mathbb{R}^n} \int_{\mathbb{R}^n}^* e^{i(s\cdot(g\lambda)-t\cdot(g\lambda'))} dF(\lambda, \lambda')$$

$$= \int_{\mathbb{R}^n} \int_{\mathbb{R}^n}^* e^{i(s\tilde{\lambda}-t\tilde{\lambda}')} dF(g\tilde{\lambda}, g\tilde{\lambda}')$$

$$= r(s, t) = \int_{\mathbb{R}^n} \int_{\mathbb{R}^n}^* e^{i(s\cdot\lambda-t\cdot\lambda')} dF(\lambda, \lambda'). \tag{5}$$

Since $r(s, t) = r(gs, gt)$, $g \in SO(n)$, $s, t \in \mathbb{R}^n$ by hypothesis, it follows from the properties of the strict MT-integrals that $F(\lambda, \lambda') = F(g\lambda, g\lambda')$ implying the invariance. Thus the isometries of X and X_1 reflect in the same property of Y, and thus $E(Y_s \overline{Y_t}) = \rho(s, t) = \tilde{\rho}(s - t) = \rho(gs, gt) = \tilde{\rho}(g(s - t))$. This is the needed refinement of the dilation theorem, so $X = QY$, which is used here.

II. Next, the representation of stationary isotropic covariance. $\rho(\cdot)$ of Y constructed above is representable as (cf., e.g., Yaglom (1987), p. 353):

$$\rho(t) = C_n \int_{\mathbb{R}^+} \frac{J_\nu(\lambda\tau)}{(\lambda\tau)^2} d\Phi(\lambda), \tag{6}$$

where $t = (\tau, u)$, $|t| = \tau$, with u as the spherical polar of t, J_ν being the Bessel function of order $\nu = (n-2)/2$, with Φ as a bounded increasing function and $C_n = 2^\nu \Gamma\left(\frac{n}{2}\right)$. Now using the addition formula for Bessel functions, (6) above can be expressed as:

$$\rho(s, t) = C_n^2 \sum_{m=0}^{\infty} \sum_{l=1}^{h(m,n)} S_m^l(u) S_m^l(v) \times$$

$$\int_{\mathbb{R}^+} \frac{J_{m+\nu}(r_1\lambda) J_{m+\nu}(r_2\lambda)}{(r_1 r_2)^\nu \lambda^{2\nu}} d\Phi(\lambda), \tag{7}$$

where $s = (r_1, u), t = (r_2, v), h(m, n) = (2m + 2\nu) \frac{(m+2\nu-1)!}{\nu! m!}, m \geq 1$, and $= 1$ if $m = 0$, the $S_m^l(u)$ being the spherical harmonics which are orthogonal on the unit sphere S_m relative to the surface measure. For each m, there exist $h(m, n)$ of them in number. [This series representation originates from the classical Mercer theorem for positive definite kernels.] With (7) an integral representation of Y_t and then X_t is obtained. For this, it is convenient to express (7) as a triangular covariance so that we can invoke Karhunen's theorem.

III. Since only the second order properties of the process are considered, by the classical Kolmogorov theorem, we can find a centered stationary Gaussian random field with covariance function ρ. Here Y can be taken to be Gaussian with such a ρ for the ensuing computation.

Let $\tilde{N} = \{(m, l) \in \mathbb{N} \times \mathbb{N} : 1 \leq l \leq h(m, n), m \geq 0\}$, and $\mathcal{P}$ be its power set. Next, let $\tilde{F} : \mathcal{P} \times \mathcal{P} \times \mathcal{B}(\mathbb{R}^+) \to \mathbb{R}^+$ be defined as

$$\tilde{F}(A_1, A_2; B_1) = \zeta(A_1 \cap A_2)\Phi(B_1) \tag{8}$$

where $\Phi : A \mapsto \int_A d\Phi(\lambda)$ is the positive bounded measure given by the Φ of (7), $\zeta(\cdot)$ as the counting measure on $\mathcal{P}$, so that $\tilde{F}$ extends to a σ-finite measure. If $\tilde{g}$ is defined on $\tilde{N} \times \mathbb{R}^n \times \mathbb{R}^+ \to \mathbb{R}$ by:

$$\tilde{g}((m, l); t, \lambda) = S_m^l(u) \cdot \frac{J_{m+\nu}(\tau\lambda)}{(\lambda\tau)^\nu}, \tag{9}$$

where $t = (\tau, u)$ is the spherical polar representation, and $\nu, S_m^l(\cdot), J_\nu$ are defined earlier. Then $\tilde{g}$ is square integrable for $\tilde{F}$ and

$$r(s, t) = C_n \int_\Lambda \int_\Lambda \tilde{g}((m, l); s, \lambda)\bar{\tilde{g}}((m, l); t, \lambda)d\tilde{F}, \tag{10}$$

where $\Lambda = \tilde{N} \times \mathbb{R}^+$. But (10) implies that $r(\cdot, \cdot)$ is, on $\mathbb{R}^n \times \mathbb{R}^n$, a triangular covariance relative to $\tilde{F}$, (cf., e.g., Rao (1982) and Chang-Rao (1986), p. 53). From these facts we can conclude that there is a Gaussian measure $Z : \mathcal{P} \otimes \mathcal{B}(\mathbb{R}^+) \to L^2(P)$ that represents Y as follows:

$$Y_t = C_0 \int_\Lambda \tilde{g}((m, l); t, \lambda)dZ((m, l); \lambda)$$
$$= C_0 \int_{\tilde{N}} \int_{\mathbb{R}^+} \tilde{g}((m, l); t, \lambda)dZ((m, l); \lambda), \tag{11}$$

with $E(Z(A_1, B_1)\bar{Z}(A_2, B_2)) = \tilde{F}(A_1, A_2; B_1 \cap B_2) = \zeta(A_1 \cap A_2) \times \Phi(B_1 \cap B_2)$. If now we set $A_1 = \{(m, l)\}$, a singleton and similarly $A_2 = \{(m', l')\}$, then with $Z(A_1, B_1) = Z_m^l(B)$ and likewise for A_2, B_2, the last relation gives $E(Z_m^l(B_1)\bar{Z}_{m'}^{l'}(B_2)) = \delta_{mm'}\delta_{ll'}\Phi(B_1 \cap B_2)$. Now taking $B = (0, \lambda)$, interval, the associated processes $\{Z_m^l(\lambda), 0 \leq l \leq h(m, n), m \geq 0\}$ are orthogonal with orthogonal increments. Hence in the case that the Y-process is Gaussian, as it can be and so assumed, each of the $Z_m^l(\cdot)$-process becomes independent of the others, and are identically (Gaussian) distributed with independent increments. Thus (11) becomes:

$$Y_t = Y(\tau, u) = C_n \sum_{m=1}^{\infty} \sum_{l=1}^{h(m,n)} S_m^l(u) \int_{\mathbb{R}^+} \frac{J_{m+\nu}(\tau\lambda)}{(\tau\lambda)^\nu} Z_m^l(d\lambda), \quad (12)$$

which is the integral representation of the isotropic stationary random field Y, the series (12) converging in $L^2(\tilde{P})$-mean.

IV. Next let $X_t = QY_t$ with $Q : L^2(\tilde{P}) \to L^2(P)$ as the orthogonal projection in the (isotropic) dilation result of Step I above. Now with the mean convergence of the series in (12), it is possible to interchange Q and the sum in the integral to obtain:

$$X(\tau, u) = (QY)(\tau, u)$$

$$= C_n \sum_{m=0}^{\infty} \sum_{l=1}^{h(m,n)} S_m^l(u) \cdot \int_{\mathbb{R}^+} \frac{J_{m+\nu}(\tau\lambda)}{(\tau\lambda)^\nu} (QZ_m^l)(d\lambda), \quad (13)$$

the interchange of Q and the sum as well as the integral is permissible here based on a theorem due to Hille. Since the $\tilde{Z}_m^l(\lambda) = (QZ_m^l)(\lambda)$ are also Gaussian and Q linear, so that these are independently and identically distributed, we get

$$E(\tilde{Z}_m^l(A)\bar{\tilde{Z}}_m^{l'}(B)) = \delta_{ll'}\delta_{mm'}F(A, B), \quad (14)$$

where F is the bimeasure determined by the common distribution of $\tilde{Z}_m^l$, which now does not have orthogonal increments any longer. The key point is that F is independent of m, which is necessary for our work here. Also, note that F is just a bimeasure of finite Fréchet variation. Since our interest is only on the second order properties of the processes, the Gaussian assumption is convenient. Thus (13) is the integral representation of the isotropic harmonizable field X_t, $t = (\tau, u)$.

Let $s = (\tau_1, u), t = (\tau_2, v)$ be the spherical polar coordinates, so that (13) and (14) give the covariance r of X as:

$$r(s,t) = E(X_s \bar{X}_t) = C_n^2 \sum_{m=0}^{\infty} \sum_{l=1}^{h(m,n)} S_m^l(u) S_m^l(v) \times$$

$$\int_{\mathbb{R}^+} \int_{\mathbb{R}^+} J_{m+\nu}(\tau_1\lambda) J_{m+\nu}(\tau_2\lambda') F(d\lambda, d\lambda'), \quad (15)$$

the integrals being in the strict MT-sense.

V. The spherical harmonics $S_m^l(u) \cdot S_m^l(v)$ can be summed, and they satisfy the relation,

$$\sum_{l=1}^{h(m,n)} S_m^l(u) S_m^l(v) = \frac{h(m,n) C_m^\nu(\cos(u,v))}{\omega_n C_m^0(l)} \tag{16}$$

where ω_n is the surface area of the unit sphere $S_n \subset \mathbb{R}^*$ and $C_m^\nu(\cdot), m \geq 0$, are the Gagenbeuer or ultraspherical polynomials of order ν for each $m \geq 0$. Putting (16) into (15) and using the addition formula for Bessel functions one gets the conclusion of the theorem, since the converse is established by the above computation. (See also Swift (1994), Lemma 2.1.) $\square$

We next present an equivalent form of isotropy which is useful for applications.

Theorem 6.2.2 *Let $X = \{X_t, t \in \mathbb{R}^n\}$ be a centered (or mean zero) weakly harmonizable random field. Then the following four statements are equivalent:*

(i) X is isotropic, so the covariance satisfies $r(s,t) = r(gs, gt), g \in SO(n)$,

(ii) the covariance r of X admits the representation (14),

(iii) the covariance r of X is given by (15) relative to a positive definite bimeasure F of finite Fréchet variation,

(iv) X is representable as an $L^2(P)$-convergent series (12) where the stochastic integral is in the Dunford-Schwartz sense.

Sketch of Proof. (i)⇔(ii) is shown by the above theorem, and (iii)⇒(iv) follows on using Karhunen's theorem suitably. That (13)⇒(15) is established above. The rest is obvious and completes the proof. $\square$

Remark. A weaker form of the representations of $X(\tau, u)$ can be established without invoking the dilation theorem. This will be indicated below which may be of interest in the same applications. [A form of this result was established by A. Grothendieck, and so is also referred to as Grothendieck's theorem.] The following slightly weaker version is also useful in applications. So we include it with a sketch here for reference as well as for some applications.

Proposition 6.2.3 *Let $X : \mathbb{R}^n \to L_0^2(P)$ be a weakly harmonizable isotropic random field. Then it is representable as $t = (\tau, u)$:*

$$X(\tau, u) = \Gamma(\nu)\omega_n 2^\nu \nu \sum_{m=0}^{\infty} i^m \sum_{l=1}^{h(m,n)} S_m^l(u) \int_{\mathbb{R}^+} \int_{S_n} \frac{J_{m+\nu}(\tau\lambda)}{(\tau\lambda)^\nu} \times$$
$$S_m^l(v) Z(d\lambda, dv) \qquad (17)$$

where ω_n is the surface area of the unit sphere S_n of $\mathbb{R}^n$.

Proof. Since X is weakly harmonizable, it is representable as:

$$X(t) = X(\tau, u) = \int_{\mathbb{R}^n} e^{it \cdot s} Z(ds)$$
$$= \int_{\mathbb{R}^+} \int_{S_n} e^{i\tau\lambda \cos(u,v)} Z(d\lambda, dv) \qquad (18)$$

with $s = (\lambda, v)$ and $t = (\tau, u)$ as polar coordinates. Using the orthogonal ultraspherical polynomials $C_m^\nu(\cdot)$ on S_m expand e^{itx} in a series (cf. Vilenkin (1968), p. 557), we get for $|x| \leq 1$,

$$e^{itx} = \Gamma(\nu) \sum_{m=0}^{\infty} i^m (m + \nu) \frac{J_{m+\nu}(\tau)}{(\tau/2)^\nu} C_m^\nu(x). \qquad (19)$$

Putting this into (18) with $x = \cos(u, v)$ and the relation

$$C_m^\nu(\cos(u, v)) = \sum_{l=1}^{h(m,n)} S_m^l(u) S_m^l(v) \frac{\omega_m C_m^\nu(1)}{h(m, n)}$$

we get

$$X(\tau, u)$$
$$= \Gamma(\nu)\omega_n 2^\nu \nu \sum_{m=0}^{\infty} \sum_{l=1}^{h(m,n)} S_m^l(u) \int_{\mathbb{R}^+} \int_{S_n} \frac{J_{m+\nu}(\tau\lambda)}{(\tau\lambda)^\nu} S_m^l(v) Z(d\lambda, dv).$$

Since the series in (19) converges in $L^2(S_n, \mu_n)$ the simplification interchange of integral here is valid. This gives (17). $\square$

We include an adjunct to the above, with a Hilbert space parameter.

Proposition 6.2.4 *On a separable Hilbert space $\mathcal{H}$, let X be an isotropic weakly harmonizable random field $X : \mathcal{H} \to L_0^2(P)$. Then its continuous covariance $r : (x, y) \mapsto E(X_x \bar{X}_y)$ admits the representation for $x, y \in \mathcal{H}$.*

$$r(x, y) = \int_{\mathbb{R}^+} \int_{\mathbb{R}^+}^* \exp\{-(\lambda x - \lambda' y, \lambda x - \lambda' y)\} F(d\lambda, d\lambda'), \quad (20)$$

with $F : \mathbb{R}^+ \times \mathbb{R}^+ \to \mathbb{C}$ as a positive definite bimeasure of finite Fréchet variation, $(\cdot, \cdot)$ being the scalar product in $\mathcal{H}$ and the above integral is in the strict MT-sense.

Proof. If the random field is stationary, so $r(s, t) = \tilde{r}(s - t)$ and by isometry $\tilde{r}$ is invariant under the rotation group, so that $\tilde{r}(x) = \tilde{r}(\|x\|)$ which depends just on the length of x. Under these conditions Schoenberg's result (cf. Schoenberg (1935), Thm. 2) gives the representation:

$$E(X_x \bar{X}_y) = \tilde{r}(\|x - y\|) = \int_{\mathbb{R}^+} \exp[-\lambda(x - y, x - y)] d\Phi(\lambda), \quad (21)$$

for a bounded increasing left continuous function $\Phi : \mathbb{R}^+ \to \mathbb{R}^+$.

We now assert that $\tilde{r}$ is also a triangular covariance, so that the Karhunen integral representation argument of Theorem 6.2.1 can be used. Now employing an idea from Yadrenko (1983), one can consider an isometric image of the (separable) space $\mathcal{H}$ by l^2, the sequence Hilbert space. The desired isometry $\tau : \mathcal{H} \to l^2$ is obtained by the Fourier series expansion of each element relative to an orthonormal basis of $\mathcal{H}$, and then $r \circ \tau$ and r have the same property and thus they can be identified in our analysis. Then with $x \in \mathcal{H}$, let $t = t_x = \tau(x) \in l^2$, and consider

$$\begin{aligned} \exp\{-\lambda(s - t, s - t)\} &= \exp\{-\lambda[(s, s) - (t, t) - 2(s, t)]\} \\ &= \exp\{-\lambda\|s\|^2\} \cdot \exp\{-\lambda\|t\|^2\} \times \\ &\sum_{m=0}^{m} \sum_{j \in l_m} \frac{(2\lambda)^m}{k_1! \cdots k_m!} (s_1, t_1)^{k_1} \cdots (s_m, t_m)^{k_m} \end{aligned}$$

where $l_m = \{j = [(i_1, k_1), \cdots (i_m, k_m)] : i_m \geq 1, k_m \geq 0, k_1 + \cdots + k_m = m\}$. Next consider $\psi_j(\cdot)$ defined by

$$\psi_j(s, \lambda, m) = e^{-\|x\|^2} \frac{(2\lambda)^{m/2}}{\sqrt{(k_1! \cdots k_m!)}} s_{i_1}^{k_1} \cdots s_{i_m}^{k_m}$$

where $s = (s_{i_1} \ldots s_{i_m})$ is the vector of reals associated with j, so that (21) becomes

$$r(s,t) = \tilde{r}(s-t) = \sum_{m=0}^{\infty} \sum_{j \in l_m} \int_{\mathbb{R}^+} \psi_j(s,\lambda,m)\psi_j(t,\lambda,m)d\Phi(\lambda). \quad (22)$$

This is a triangular covariance relative to ψ if we set

$$\psi(x,\lambda) = \{\psi_j(x,\lambda,m) : j \in l_m, m \geq 0\}$$

so that

$$r(s,t) = \int_{\mathbb{R}^+} \psi(s,\lambda)\psi^*(t,\lambda)d\Phi(\lambda),$$

$\psi^*(\cdot,\lambda)$ being the transpose of the vector $\psi(\cdot,\lambda)$ which is similar to (10). Hence there exists a stochastic measure such that the following representation obtains:

$$X_t = \int_{\mathbb{R}^+} \psi(t,\lambda)dZ(\lambda) = \sum_{m=0}^{\infty} \sum_{j \in l_m} \int_{\mathbb{R}^+} \psi_j(t,\lambda,m)dZ_m^j(\lambda), \quad (23)$$

where $Z_m^j(\lambda) \in L_0^2(P)$, and $E(Z_m^j(A)\bar{Z}_{m'}^{j'}(B)) = \delta_{jj'}\delta_{mm'}\Phi(A \cap B)$. The remaining analysis is similar to the earlier case. Thus to proceed to the harmonizable case, we use the dilation theorem which is also valid in this case. Thus there is a Hilbert space $L_0^2(\tilde{P}) \supset L_0^2(P)$ and a stationary isotropic random field $Y : \mathcal{H} \to L_0^2(\tilde{P})$, such that $X_t = QY_t, t \in \mathcal{H}$ ($\cong l^2$), where Q is the orthogonal projection onto $L_0^2(P)$. Replacing X_t of (23) by Y_t and since the series converges in $L_0^2(P)$, one obtains that $X_t = QY_t$, so that

$$X_t = \sum_{m=0}^{\infty} \sum_{j \in l_m} \int_{\mathbb{R}^+} \psi_j(t,\lambda,m)(QZ_m^j)(d\lambda). \quad (24)$$

Here $\tilde{Z}_m^j = QZ_m^j$ satisfies the conditions: $E(\tilde{Z}_m(A)) = 0$, and $E(\tilde{Z}_m^j(A)\bar{\tilde{Z}}_m^{j'}(B)) = \delta_{jj'}\delta_{mm'}F(A,B)$, where $F(\cdot,\cdot)$ is the same for all processes $\tilde{Z}_m^j$ which are i.i.d. in the Gaussian case. Now one can conclude that for the harmonizable covariance $r(s,t) = r(gs,gt)$ for all rotations g and then

$$r(s,t) = \sum_{m=0}^{\infty} \sum_{j \in l_m} \int_{\mathbb{R}^+} \int_{\mathbb{R}^+}^* \psi_j(s\lambda,m)\psi_j(t\lambda',m)F(d\lambda,d\lambda'). \quad (25)$$

Interchanging the sum and the strict MT-integral, which is valid (cf. Chang and Rao (1986)), one obtains ($\mathcal{H} \cong l^2$)

$$r(s,t) = \int_{\mathbb{R}^+} \int_{\mathbb{R}^+}^* e^{-\lambda(s,t)-\lambda'(s,t)+2\lambda\lambda'(s,t)} F(d\lambda, d\lambda')$$

$$= \int_{\mathbb{R}^+} \int_{\mathbb{R}^+}^* e^{-(\lambda s - \lambda' t, \lambda s - \lambda' t)} F(d\lambda, d\lambda')$$

which establishes (20) and the proposition follows. $\square$

A consequence of the above representation is stated separately for a reference.

Corollary 6.2.5 *If* $X : l^2 \to L_0^2(P)$ *is an isotropic harmonizable random field, then it is representable as:*

$$X_t = \sum_{m=0}^{\infty} \sum_{j \in l_m} \int_{\mathbb{R}^+} \int_{\mathbb{R}^+} \psi(t, \lambda, m) \tilde{Z}_m^j(d\lambda), \tag{26}$$

where $E(\tilde{Z}_m^j)$ *verifies the same conditions as in (24) above and the series (26) converges in* $L_0^2(P)$-*mean.*

Thus there is an analogous spectral representation of isotropic weakly harmonizable random fields using the (strict) MT-integration and the dilation analysis. Extension of Yaglom's (1961) analysis may now be treated.

6.3 Some Moving Averages and Sampling of Harmonizable Classes

Let us introduce precisely a moving average representation of a process that is (strongly or weakly) harmonizable and give some results on their structural representations to be used for sampling and related problems on this class.

Definition 6.3.1 *Let* $\{X_t, t \in T\} \subset L_0^2(P)$ *be a harmonizable process where* $T = \mathbb{Z}$ *or* $\mathbb{R}$. *It is termed a (strongly or weakly) moving average represented, if it has the form:*

$$X_t = \begin{cases} \sum_{t \in \mathbb{Z}} e^{i(j-t)} \xi_t, & T = \mathbb{Z}, \\ \int_{\mathbb{R}} e^{i(\lambda-t)} \xi_t dt, & T = \mathbb{R}. \end{cases} \tag{27}$$

The integral in the continuous case is in Bochner's sense. Here $\hat{C}(\cdot)$, *as usual, is the Fourier transform of some* $C(\cdot) \in L^2(d\lambda)$, *and* $\text{cov}(\xi_s, \xi_t)$ *is the covariance* $\rho(s, t)$, *with* ρ *as a scalar covariance. In case the* ξ_t-*process is stationary or (strongly/weakly) harmonizable, then the moving average is termed stationary or harmonizable accordingly.*

Since the moving average property is important in many applications, we point out a useful specialization for some key applications, termed a *virile moving average*, defined now.

Definition 6.3.2 *A weakly (or strongly) harmonizable moving average process $\{X_t, t \in T\}$ introduced above in (27) is a virile (moving average) process if the (matrix valued) function $C(\cdot)$ falls off "to zero fast at infinity" in the sense that if*

$$C_N(\lambda) = \int_{|\theta|>N} \hat{C}(\theta)e^{i\theta\lambda}d\theta \left[or = \sum_{|j|>N} \hat{C}(j)e^{ij\lambda} \right],$$

then

$$\lim_{N\to\infty} \int_{(\lambda,\lambda')\in\hat{D}\times\hat{D}} C_N(\lambda)C_N^*(\lambda')\mu_\xi(d\lambda, d\lambda') = 0. \qquad (28)$$

This somewhat involved concept is useful in understanding the moving average processes. This property is detailed for n-dimensional strongly harmonizable processes as follows.

Theorem 6.3.3 *Let $\{X_t, t \in T\} \subset L_0^2(P)$ be an n-dimensional strongly harmonizable process. Then it has a virile moving average representation with full rank say m, so that*

$$X_t = \begin{cases} \sum_{t\in\mathbb{Z}} \hat{C}(i-t)\xi_t, \ t \in \mathbb{Z}, \ (discrete) \\ \int_{\mathbb{R}} \hat{C}(\lambda-t)\xi_t dt, \ t \in \mathbb{R}, \ (continuous) \end{cases} \qquad (29)$$

where $r_\xi(s,t) = \int\int_{\hat{D}\times\hat{D}} e^{it\lambda-it\lambda'}\mu_\xi(d\lambda, d\lambda')$, iff it is a strongly harmonizable process. Moreover, the X_t-process has spectral characteristic $C(\cdot)$ so that $X_t = \int_{\hat{D}} e^{it\lambda}\hat{C}(\lambda)Z_\xi(d\lambda)$.

This interesting representation is detailed in the paper by Mehlman (1991) and we refer the reader for the detailed proof and related results (e.g., one-sided moving averages and specializations for the stationary case and factorizable spectral measure where $r(s,t)$ is expressible as:

$$r(s,t) = \int\int_{\hat{D}\times\hat{D}} e^{i(\lambda s-\lambda' t)}C(\lambda)C^*(\lambda')\mu(d\lambda, d\lambda') \qquad (30)$$

with $C(\cdot)$ as an $n \times p$-matrix valued function (for both weak and strong harmonizable processes). Also, a centered process is called *splitting* if its covariance $r(s,t) = g(s)g^*(t)$ for suitable matrices $g(\cdot)$ which are detailed. A study of Mehlman's (1992) paper is helpful here.

In this connection it is useful to include a general nonstationary class of $L^2(P)$-valued processes that extend the main Cramér class (hence also of Karhunen's) for further study and applications, calling it a *weak class* (C) as it automatically includes the weakly harmonizable class:

Definition 6.3.4 *A process* $\{X_t, t \in \mathbb{R}\} \subset L_0^2(P)$ *with covariance* r *is termed of* weak class (C) *[C for Cramér], if for a covariance bimeasure* $F : \mathcal{B}(\mathbb{R}) \times \mathcal{B}(\mathbb{R}) \to \mathbb{C}$ *of class* (C), *there is a covariance bimeasure* $F : \mathcal{B}(\mathbb{R}) \times \mathcal{B}(\mathbb{R}) \to \mathbb{C}$ *of locally bounded semi-variation to mean that (i)* $F(A, B) = \bar{F}(B, A)$, *(ii)* $\sum_{i,j=1}^n a_i \bar{a}_i F(A_i, A_j) \geq 0, a_i \in \mathbb{C}$, *and (iii)* $\|F\|(A \times A) < \infty, A, A_i, B_j \in \mathcal{B}(\mathbb{R})$, *bounded, where* $\infty > \|F\|(A \times A) = \sup \left\{ \left| \sum_{i,j=1}^n a_i \bar{a}_j F(A_i, B_j) \right| : |a_i| \leq 1; A_i, B_j \right.$ *disjoint* $\}$, *then* $r(\cdot, \cdot)$ *is representable as an MT-integral written as* $r(s,t)(= I(g_s, \bar{g}_t))$:

$$r(s,t) = \int_{\mathbb{R}} \int_{\mathbb{R}} g_s(\lambda) \bar{g}_t(\lambda') F(d\lambda, d\lambda'), s, t \in \mathbb{R}. \tag{31}$$

We now present a general characterization of the classical weak *Cramér class*:

Theorem 6.3.5 *Let* $X : \mathbb{R} \to L_0^2(P)$ *be a weakly class* (C) *process relative to a bimeasure* F *and a family* $\{g_s, s \in \mathbb{R}\}$ *of MT-integrable class for* F, *as in the above definition. Then there exists a stochastic measure* $Z : \mathcal{B}_0 \to L_0^2(\tilde{P})$, *where* $\mathcal{B}_0$ *is the* δ-ring of bounded Borel sets of $\mathbb{R}$ and $(\tilde{\Omega}, \tilde{\Sigma}, \tilde{P})$ *is some enlargement of* (Ω, Σ, P) *so that* $L_0^2(\tilde{P}) \supset L_0^2(P)$, *such that*

$$(i) \qquad E(Z(A)\overline{Z(B)}) = (Z(A), Z(B)) = F(A, B), A, B \in \mathcal{B}_0$$

$$(ii) \qquad X(t) = \int_{\mathbb{R}} g(t, \lambda) Z(d\lambda), t \in \mathbb{R} \tag{32}$$

where the integral is in the Dunford-Schwartz sense for $\mathcal{B}_0$.

Conversely, if $\{X(t), t \in \mathbb{R}\}$ *is defined by (32) relative to* $Z : \mathcal{B}_0 \to L_0^2(P)$ *with* $\{g_t, t \in \mathbb{R}\}$ *a D-S integrable class, and where* $F : (A, B) \to E(Z(A)\bar{Z}(B))$, *with the* g_s-class MT-integrable, and $\mathcal{H}_x = \bar{sp}\{X(t), t \in \mathbb{R}\}$ *with* $\mathcal{H}_z = \bar{sp}\{Z(A), A \in \mathcal{B}_0\} \subset L_0^2(P)$, *then* $\mathcal{H}_x = \mathcal{H}_z$ *when and only when, the functions* $\{g_t, t \in \mathbb{R}\}$ *satisfy the following equation:*

$$\int_{\mathbb{R}}\int_{\mathbb{R}} f(\lambda)\bar{g}_t(\lambda')F(d\lambda, d\lambda') = 0 \iff \int_{\mathbb{R}}\int_{\mathbb{R}} f(\lambda)\bar{f}(\lambda')F(d\lambda, d\lambda') = 0$$

(33)

both being MT-integrals.

Proof. We keep the basic format of Cramér's, but now the integrals are in the Morse-Transue sense, and the details are spelled out for appreciating this general case.

First let the (centered) process be of class (C) so that, using the MT-integrals to represent $r(\cdot, \cdot)$, we have

$$r(s,t) = E(X_s \bar{X}_t) = \int_{\mathbb{R}}\int_{\mathbb{R}} g_s(\lambda)\bar{g}_t(\lambda')F(d\lambda, d\lambda').$$

(34)

Consider the L_F^2-space for the bimeasure F, defined by

$$L_F^2 = \left\{ f : \int_{\mathbb{R}}\int_{\mathbb{R}} f(\lambda)\bar{f}(\lambda')F(d\lambda, d\lambda') = (f, f)_F \right.$$

$$\left. \text{with } 0 \le (f, f)_F < \infty \text{ and } f \text{ is MT-square integrable for } F \right\}.$$

Then L_F^2 is a semi-inner product space. Define $T : L_F^2 \to \mathcal{H}_F$ by $(Tg_s) = X(s)$ and extend it linearly so that $(Tg_s, Tg_t) = (g_s, g_t)$, $s, t, \in \mathbb{R}$, and that T is an isometric mapping of $\Lambda_F^2 = \text{sp}\{g_t, t \in \mathbb{R}\} \subset L_F^2$ and is onto $\mathcal{H}_F$ (the space as in the theorem).

Now the MT-integration implies that each bounded Borel set A is integrable and this gives for $T : g \to X_g$

$$(Tg_s, Tg_t)_{\mathcal{H}_x} = (g_s, g_t)_F, s, t \in \mathbb{R},$$

(35)

so that T is an isometric mapping of $\Lambda_F^2 = \bar{\text{sp}}\{g_t, t \in \mathbb{R}\}$ onto $\mathcal{H}_x$, the space defined in the theorem. This implies if $T\chi_A = Z_A$ then for bounded Borel sets A, B, we have

$$E(Z_A \bar{Z}_B) = (T\chi_A, T\chi_B) = (\chi_A, \chi_B) = F(A, B),$$

and for disjoint A, B (Borel), we get

$$E(|Z_{A\cup B} - Z_A - Z_B|^2) = (\chi_{A\cup B} - \chi_A - \chi_B, \chi_{A\cup B} - \chi_A - \chi_B) = 0$$

since F is additive in these components. So $Z_{(\cdot)} : \mathcal{B}_0 \to L_0^2(P)$ is again additive. Its σ-additivity follows from continuity, so $A_i \in \mathcal{B}_0$, disjoint, $A_0 = \bigcup_{i=1}^\infty A_i \in \mathcal{B}_0$, implies

$$E((Z_A - \sum_{i=1}^{n} Z_{A_i})^2) = E(|Z_{\cup_{i\geq 1} A_i} - \sum_{i=1}^{n} Z_{A_i}|^2)$$

$$= E(|Z_{\cup_{i>n} A_i}|^2) = F(\cup_{i>n} A_i, \cup_{i>n} A_i) \to 0 \tag{36}$$

as $n \to \infty$ since F is suitably continuous. So $Z(\cdot)$ is also σ-additive.

Thus $Z(\cdot)$ is σ-additive on $\mathcal{B}_0$, and is a stochastic measure of finite semi-variation on compact sets, and $\mathcal{K}_Z \subset \mathcal{K}_X$.

From this a standard approximation argument shows that, since $\{g_t, t \in R\} \subset L_F^2$ is dense, we get with $\tilde{g}_n = \sum_{i=1}^{n} a_i g(t_i) \to \chi_A$ in $L^2 \Rightarrow E[|\sum_{i=1}^{n} a_i X(t_i) - Z(A)|^2] \to 0$ as $n \to \infty$ and hence $E\left[|\sum_{i=1}^{n} a_i X(t_i) - Z(A)|^2\right] \to 0$ as $n \to \infty$.

It follows that $\{Z_A, A \in \mathcal{B}_n\}$ is a dense subspace in $\mathcal{K}_X$ and each element in $\mathcal{K}_X$ corresponds uniquely to one in $\bar{L}_F^2$, the completion of L_F^2. If now $Y(t)$ is defined that

$$Y(t) = \int_{\mathbb{R}} g_t(\lambda) Z(d\lambda), (\in \mathcal{K}_Z = \mathcal{K}_X), \tag{37}$$

then it can be verified as

$$(Y(s), Y(t)) = \int_{\mathbb{R}} \int_{\mathbb{R}} g_s(\lambda) g_t(\lambda') F(d\lambda, d\lambda')$$

which holds if g_t is $\mathcal{B}_0$-step function and then generally by the MT-analysis. It also can be shown from this that

$$Z(A) = T(\chi_A) = l \cdot i \cdot m \cdot (T\tilde{g}_n) = l \cdot i \cdot m \cdot \sum_{i=1}^{n} a_i T(g_{t_i})$$

$$= l \cdot i \cdot m \cdot \sum_{i=1}^{n} a_i X(t_i) = l \cdot i \cdot m \cdot \tilde{X}_m \text{ (say)}.$$

Then the $L^2(P)$-limits imply

$$E(X(s)\bar{Z}(A)) = \lim_n E(X(t)\bar{\tilde{X}}_m) \tag{38}$$

$$= \lim_n \sum_{i=1}^n a_i E(X(s)\overline{X(t_i)}) \tag{39}$$

$$= \lim_n \sum_{i=1}^n a_i \int_{\mathbb{R}} \int_{\mathbb{R}} g_s(\lambda)\bar{g}_t(\lambda')F(d\lambda, d\lambda') \tag{40}$$

$$= \int_{\mathbb{R}} \int_{\mathbb{R}} g_t(\lambda)\chi_A(\lambda')F(d\lambda, d\lambda'). \tag{41}$$

By isometry, if $\tilde{\zeta}_n = \sum_{j=1}^n b_j Z(A_j)$, one has when $\tilde{h}_n = \sum_{j=1}^n b_j X_{A_j} \in L_F^2$,

$$E(X(s)\bar{\tilde{\zeta}}_n) = \int_{\mathbb{R}} \int_{\mathbb{R}} g_s(\lambda)\bar{\tilde{g}}_n(\lambda')F(d\lambda, d\lambda').$$

Since $g_s(\cdot)$ is MT-integrable, this gives on letting $n \to \infty$,

$$E(X(s)\bar{Y}(t)) = \int_{\mathbb{R}} \int_{\mathbb{R}} g_s(\lambda)g_t(\lambda')F(d\lambda, d\lambda').$$

From this, it follows that

$$E(|X(t) - Y(t)|^2) = 0.$$

Hence $X(t) = Y(t)$, a.e., $t \in \mathbb{R}$. So if Λ_F is dense in L_F^2, this shows that (31) is true in the case that $Y(t)$ is given by (37).

The general case follows from this result easily as shown below.

In the general case let $L_T^2 = \tilde{\Lambda}_F^2$, and $\{h_t, t \in \tilde{F}\}$ be a basis of the space $\tilde{\Lambda}_F^l$. If $\tilde{R} = R \dotplus \tilde{R}$, (a disjoint sum) as a new index, $\{h_i, i \in \tilde{R}\}$ forms a basis of $\tilde{\Lambda}_F$. By the preceding case, on extending T from $L_F^2(P)$ to $L_F^2(\tilde{P})$ where $(\tilde{\Omega}, \tilde{\Sigma}, \tilde{P})$ is an enlargement of (Ω, Σ, P) by an adjunction procedure with $T\chi_A = Z_A \in L^2(\tilde{P})$, one finds

$$\tilde{Y}(s) = \int_{\mathbb{R}} \tilde{g}_s(\lambda)Z(d\lambda), (\in L_0^2(\tilde{P})). \tag{42}$$

Hence all $\tilde{g}$ are Borel and MT-integrable. So as before, $\tilde{T}(s) = X(s), s \in \mathbb{R}$, and (31) holds again and so the inclusion $\mathcal{H}_Z \supset \mathcal{H}_X$ holds.

For the converse, let $X(t)$ be given by (31) and $F(A, B) = (Z(A), Z(B))$. If $g_n = \sum_{i=1}^n a_i \chi_{A_i} (A_i, A, B \in \mathcal{B}_0, A_i$ disjoint), then

$$F(A, B) = \sup\left\{\sum_{i,j=1}^{n} a_i \bar{a}_j F(A_i, A_j) : A_i \in \mathcal{B}(A), |a_i| \leq 1\right\}$$

$$\leq \sup\left\{\left\|\sum_{i=1}^{n} a_i Z(A_i)\right\| : |a_i| \leq 1, A_i \in \mathcal{B}(A)\right\} \leq \|Z\|^2(A) < \infty.$$

Thus if $X_{g_n} = \int_{\mathbb{R}} g_n(\lambda) dZ(\lambda)$, then by standard arguments

$$E(X_{g_n} \bar{X}_{h_n}) = \int_{\mathbb{R}} \int_{\mathbb{R}} g_n(\lambda) \bar{h}_m(\lambda') F(d\lambda, d\lambda'). \tag{43}$$

Next using the obvious properties of the MT-integrals, if $\{g_n\}_1^\infty \subset L_F^2$ with $g_n \to g_s$ boundedly a.e., one also gets similarly with $\tilde{g}_n \to g_t$,

$$I(g_n, \tilde{g}_n) \to I(g_s, g_t), I(|g_s|, |g_t|) < \infty.$$

Hence,

$$\int_{\mathbb{R}} \int_{\mathbb{R}} g(\lambda) \bar{g}(\tilde{\lambda}) F(d\lambda, d\bar{\lambda}) = \lim_n \int_{\mathbb{R}} \int_{\mathbb{R}} g_n(\lambda) \bar{\tilde{g}}_n(\lambda') F(d\lambda, d\lambda')$$

$$= \lim_n (X_{g_n}, X_{\tilde{g}_n})$$

$$= \lim_n \left(\int_{\mathbb{R}} g_n(\lambda) Z(d\lambda), \int_{\mathbb{R}} \tilde{g}_n(\lambda) Z(d\lambda)\right)$$

$$= \left(\int_{\mathbb{R}} g_s(\lambda) Z(d\lambda), \int_{\mathbb{R}} g_t(\lambda) Z(d\lambda)\right),$$

since the dominated convergence holds for the D-S integrals

$$= (X_s, X_t) = r(s, t).$$

Thus $\{X_t, t \in \mathbb{R}\}$ is of weak class (C).

The last assertion is easy since $\{g_s, s \in \mathbb{R}\}$ is a basis for L_F^2 iff $I(f, g_t) = 0, t \in \mathbb{R} \Rightarrow I(f, f) = 0$ so that $\mathcal{K}_Z = \mathcal{K}_X$. This implies all the assertions, and the result holds on stated. $\square$

If $\|F\|(\mathbb{R} \times \mathbb{R}) < \infty$, then each bounded measurable function is MT-integrable for F. So taking $g_t(\lambda) = e^{it\lambda}$, the above theorem gives the following result due to Rozanov (1959), and another proof is in Niemi (1975). We state it for comparison.

Theorem 6.3.6 *Let* $X : \mathbb{R} \to L_0^2(P)$ *be* $\|X(t)\|_2 \leq M < \infty, t \in \mathbb{R}$, *and weakly continuous. Then it is weakly harmonizable relative to some covariance bimeasure* F *of finite semi variation iff there is a measure* $Z : \mathcal{B} \to L_0^2(P)$, *(*$\mathcal{B}$ *is Borel field) with* $F(A, B) = (Z(A), Z(B))$, *and*

$$X_t = \int_{\mathbb{R}} e^{it\lambda} Z(d\lambda), t \in \mathbb{R}, \tag{44}$$

the integral in D-S sense and $\|Z\|(\mathbb{R}) < \infty$. *Further,* X *is strongly harmonizable iff the covariance bimeasure* F *of* Z *is of bounded variation, in* $\mathbb{R}^2$. *In either case the harmonizable process* X *on* $\mathbb{R}^2$ *is uniformly continuous, and is represented as the integral* (20). *If* $Z(\cdot)$ *is orthogonally valued so that*

$$F(A, B) = (Z(A), Z(B)) = \tilde{F}(A \cap B), r(s, t) = \int_{\mathbb{R}} g(\lambda)\bar{g}(\lambda)d\tilde{F}(\lambda),$$
$$\tag{45}$$

this reduces to a Karhunen *process.*

The following result gives a kind of summary statement on the class of weakly harmonizable processes on an LCA group and will be useful as a reference of this account.

Theorem 6.3.7 *Let* G *be an LCA group and* $X : G \to \mathcal{X}$ *be a weakly continuous mapping where* $\mathcal{X} = L_0^2(P)$, *separable. Then we have the following equivalent assertions:*

(i) X *is weakly harmonizable.*

(ii) X *is* V-*bounded (in Bochner's sense).*

(iii) X *is the Fourier transform of a regular vector measure on* $\hat{G}$ *into* $\mathcal{X}$.

(iv) *For each* $p \in \hat{L}^1(\hat{G})$, *the process* $Y_p = pX : G \to L_0^2(P)$ *is weakly harmonizable and bounded.*

Further the following statement implies all the four assertions above:

(v) *If* $\mathcal{K} = \bar{\mathrm{sp}}\{X(g) : g \in G\} \subset \mathcal{X}$, *then there is a weakly contractive positive type of operators* $\{T(g), g \in G\} \subset B(\mathcal{X})$, *such that* $T(0) = id$, *and* $X(g) = T(g)X(0), g \in G$.

In view of the desirable properties of harmonizable processes, let us consider some general classes of processes covered by it that are not necessarily stationary. One such class was introduced and extensively studied by the well known French probabilists J. Kampé de Fèrier and F. N. Frankiel (also called *class* (KF)) which includes the weak stationary as a subclass. It is related to a class called *asymptotically stationary* by E. Parzen (1953), but is somewhat more general. We discuss this class as an adjunct of weakly harmonizable family that is being studied above. [We actually consider it easily and recall.]

Definition 6.3.8 *If $X : \mathbb{R} \to L_0^2(P)$ is a (centered) random process with covariance $r : (s, t) \to E(X(s)\bar{X}(t))$, then it is of* class *(KF) (or of Kampé de Fériet and Frankiel (1995) class), if the limit*

$$r(h) = \lim_{T \to \infty} \frac{1}{T} \int_0^{T-|h|} K(s, s+h)ds = \lim_{T \to \infty} r_T(h), h \in \mathbb{R}, \quad (46)$$

exists. [Note $r_T(\cdot)$ as in (46): hence $r(\cdot)$ is positive definite.]

By the classical Bochner-Riesz theorem there is a unique increasing bounded function F, termed an *associated spectral* function of the process X. It can be verified that every strongly harmonizable process is in class (KF). However, all weakly harmonizable processes are not necessarily in class (KF). The following example due to H. Niemi verifies the negative condition, in order to present a simple (sufficient) condition for a weakly harmonizable X to be in class (KF) and obtain a positive result.

Example 1. Let $\{a_n, n \in \mathbb{Z}\}$ be a sequence such that $a_0 = 1, a_n = a_{-n}$, and for $k < \infty$ let a_k be given by

$$a_k = \sum_{n=0}^{\infty} (\chi_{C_n} + 2\chi_{D_n})(k), \quad (47)$$

where we take $C_n = [2^{2n}, 2^{2n+1})$ and $D_n = [2^{2n+1}, 2^{2n+2})$. So $C_n \cap D_n = \emptyset$, and for each $k \geq 0, 1 \leq a_k \leq 2$, let $\{\varepsilon_k, -\infty < k < \infty\} \subset l^2$ be a complete orthonormal sequence, and let $A : \varepsilon_n \mapsto a_n\delta_n, |a_n| \leq g$.

Let $a_n = a_{-n}, a_0 = 1$. We define for $k > 0, a_k$ as above, so $1 \leq a_k \leq 2, k \geq 0$. The covariance $r(k, l) = 0$ for $k \neq l$. Then

$$r_n(h) = \frac{1}{n} \sum_{k=0}^{n-k-1} r(k, k+h) = \begin{cases} 0 & \text{if } h \neq 0 \\ \frac{1}{n} \sum_{k=0}^{n-1} a_k^2, & \text{if } h = 0. \end{cases}$$

Thus $\lim_{n\to\infty} r_n(h) = 0$ for $h \neq 0$ and

$$r_n(0) = \begin{cases} \frac{5}{3} - \frac{1}{3 \cdot 2^{n-1}}, & \text{if } n = 2^{2m} - 1 \\ \frac{4}{3} - \frac{1}{3 \cdot 2^{2m}}, & \text{if } n = 2^{2m+1} - 1. \end{cases}$$

Hence $\lim_{m\to\infty} r_{2^m-1}(0) = \frac{5}{3}$, and $\lim_{m\to\infty} r_{2^{2m+1}-1}(0) = \frac{4}{3}$. So $\lim_{n\to\infty} r_n(0)$ does not exist, and $\{X_n, n \in \mathbb{Z}\}$ is not in class (KF). Thus $n \to \infty$ the $\{X_n = A_{\varepsilon_n}, n \in \mathbb{Z}\}$ is a weakly harmonizable process, and is not in class (KF) so not all harmonizable processes are in class (KF).

In view of this example, one may ask under what (further) conditions does a (weakly) harmonizable process belong to class (KF) that generalizes the (weakly) stationary classes. We include a family that is more general than (weakly) stationary but a somewhat restricted class of (weakly) harmonizable processes which will be shown now to belong to class (KF). Thus this class is also large.

Definition 6.3.9 *A stochastic measure* $Z : \mathcal{B}(\mathbb{R}) \to L_0^2(P)$ *is said to be of* second order *if* $\tilde{Z} = Z \otimes Z : \mathcal{B} \otimes \mathcal{B} \to L_0^2(P)$ *is also σ-additive in* $L_0^2(P)$ *and* $\tilde{Z}$ *is of the same type, so that* $\|Z \otimes Z\|(\mathbb{R} \times \mathbb{R}) \leq [\|Z\|(\mathbb{R})]^2 < \infty$. *The corresponding harmonizable process* X *is termed second order weakly harmonizable.*

This (strengthened) property of $Z(\cdot)$ for the stationary case, implies that $Z(\cdot)$ has finite Vitali variation so that $Z \otimes Z$ and even higher products are included.

The positive result, for applications, is as follows.

Theorem 6.3.10 *Let* $X : \mathbb{R} \to L_0^2(P)$ *be a second order weakly harmonizable process in the sense of Def. 6.3.9. Then* $X \in$ *class (KF) so that it has a well-defined associated spectral function.*

Proof. The hypothesis implies that we have

$$X(t) = \int_{\mathbb{R}} e^{it\lambda} Z(d\lambda), t \in \mathbb{R}, \tag{48}$$

for a stochastic measure Z of the type given in Definition 6.3.9

$$F(A, B) = (Z(A), Z(B)), \tag{49}$$

so that $F : \mathcal{B} \times \mathcal{B} \to \mathbb{C}$ is a bounded bimeasure. Also with $K(s, t) = E(X(s)\bar{X}(t))$, let

$$r_T(h) = \frac{T-h}{T} \cdot \frac{1}{T-h} \int_0^{T-h} k(s, s+h)ds. \tag{50}$$

To see that $\lim_{T \to \infty} r_T(h)$ exists, consider

$$\frac{1}{T} \int_0^t k(s, s+h)ds = E\left(\frac{1}{T} \int_0^T ds \int_{\mathbb{R}} e^{it\lambda} Z(d\lambda) \int_{\mathbb{R}} e^{-i(s+h)\lambda'} Z(d\lambda')\right). \tag{51}$$

We assert that the right side has a limit as $T \to \infty$. Now let $\mathcal{X} = \mathcal{Y} = \mathcal{L}_0^2(P)$, and $\mathcal{Z} = L^1(P)$. So $Z, \tilde{Z} : \mathcal{B} \to \mathcal{Y}$ are stochastic measures, using the bilinear mapping $(x, y) \to x \cdot y$ of $\mathcal{X} \times \mathcal{Y} \to \mathcal{Z}$, the product measure $Z \otimes \tilde{Z} : \mathcal{B} \times \mathcal{B} \to \mathcal{Z}$ is defined and satisfies the D-S integrals:

$$\int \int_{\mathbb{R} \times \mathbb{R}} f(s, t)(Z \otimes \tilde{Z})(ds, dt) = \int_{\mathbb{R}} Z(ds) \int_{\mathbb{R}} f(s, t)\tilde{Z}(dt)$$

$$= \int_{\mathbb{R}} \tilde{Z}(dt) \int_{\mathbb{R}} f(s, t)Z(ds) \tag{52}$$

for $f \in C_0(\mathbb{R} \times \mathbb{R})$, (cf. Dinculeanu ((1974), p. 388), and also see (Duchoñ and Kluvánek (1967)) for some relaxations that we use to get $\|Z \otimes \tilde{Z}\|(\mathbb{R} \times \mathbb{R}) \leq \|Z(\mathbb{R})\|^2 < \infty$, so the measure function $Z \otimes \tilde{Z}$ is stochastic.

This gives for $f_{s,t}(\lambda, \lambda') = e^{is\lambda} \cdot e^{-i(s+h)\lambda'}, f \in C_1(\mathbb{R} \times \mathbb{R})$,

$$\int_{\mathbb{R} \times \mathbb{R}} f(s, t)(Z \otimes \tilde{Z})(d\lambda, dt) = \int_{\mathbb{R}} e^{is\lambda} Z(d\lambda) \int_{\mathbb{R}} e^{-i(s+h)\lambda'} Z(d\lambda')$$

$$= \int_{\mathbb{R} \times \mathbb{R}} e^{is(\lambda-\lambda')-ih\lambda'} Z \otimes Z(d\lambda, d\lambda') \tag{53}$$

and the right side is in $L^1(P)$. With the same type of calculations and argument applied to $Z \otimes Z : \mathcal{B}(\mathbb{R} \times \mathbb{R}) \to \mathcal{X}$ and $\mu : \mathcal{B}([0, T]) \to \mathbb{R}^+$ with μ as Lebesgue measure, and $(x, a) \to ax$ on $\mathcal{Z} \times \mathbb{R} \to \mathcal{X}$, we can define

$$\mu \otimes (Z \otimes Z) : \mathcal{B}(0, T) \times \mathcal{B}(\mathbb{R} \times \mathbb{R}) \to \mathcal{X}.$$

Now writing $\underline{\lambda}$ for (λ, λ'), we get

$$\int_0^T \mu(dt) \int_{\mathbb{R} \times \mathbb{R}} f(t, \underline{\lambda})Z \otimes Z(d\underline{\lambda}) = \int_{\mathbb{R} \times \mathbb{R}} Z \otimes Z(d\underline{\lambda}) \int_0^T f(t, \underline{\lambda})\mu(dt).$$

This gives easily

$$
E\left(\frac{1}{T}\int_0^T ds\int_{\mathbb{R}\times\mathbb{R}} e^{is(\lambda-\lambda')-ih\lambda'}Z\otimes Z(d\lambda,d\lambda')\right)
$$
$$
= E\left(\int_{\mathbb{R}\times\mathbb{R}} e^{-ih\lambda'}\frac{e^{iT(\lambda-\lambda')}-1}{iT(\lambda-\lambda'}X_{\lambda\pm\lambda')}+\delta_{\lambda\lambda'}Z\otimes Z(d\lambda d\lambda')\right). \quad (54)
$$

But the quantity inside the integral on the right is bounded for all T and we can let $T\to\infty$ in (54) to get $=\int_{[\lambda=\lambda']}e^{ih\lambda}F(d\lambda,d\lambda')$ where F is the bimeasure of Z. Hence $\lim_{T\to\infty}r_T(h)=r(h)$ exists.

This implies that $\lim_{T\to\infty}r_T(h)=r(h)$ exists and $=\int_{\mathbb{R}}e^{ih\lambda}dG(\lambda)$ where $G : A\mapsto\int_{\pi^{-1}(A)}\delta_{\lambda\lambda'}F(d\lambda,d\lambda'), A\in\mathcal{B}$, is positive definite and thus is an associated spectral measure of X in class (KF). This finishes the proof of the theorem. $\square$

As an application of the preceding result, we can give a slight extension of a useful application obtained originally by Yu. A. Rozanov (1959) that has some engineering as well as Fourier analytic interesting consequences.

Theorem 6.3.11 *Let $X : \mathbb{R}\to L_0^2(P)$ be a weakly harmonizable process with $Z : \mathcal{B}\to L_0^2(P)$ as its representing stochastic measure. Then for $(\lambda_1,\lambda_2)\subset\mathbb{R}$ with $Z(\cdot)$ as a stochastic measure where $Z(A)$ is $Z((-\infty,\lambda))$, we have*

$$
l\cdot i\cdot m.\int_{-T}^T\frac{e^{-it\lambda_2}-e^{it\lambda_1}}{-it}X(t)dt
$$
$$
{}_{0<T\to\infty}
$$
$$
=\frac{Z(\lambda_2+)-Z(\lambda_1-)}{2}-\frac{Z(\lambda_1+)+Z(\lambda_1-)}{2} \quad (55)
$$

where, as usual, $l\cdot i\cdot m\cdot$ stands for limit in mean. Moreover, the covariance bimeasure F of Z can be obtained for $A=(\lambda_1,\lambda_2)$ and $B=(\lambda_1',\lambda_2')$ as:

$$
F(A,B) \quad (56)
$$
$$
=\lim_{0\leq T_1,T_2\to\infty}\int_{-T_1}^{T_1}\int_{-T_2}^{T_2}\frac{e^{-i\lambda_2 s}-e^{-i\lambda_1 s}}{-is}\cdot\frac{e^{-i\lambda_2't}-e^{-i\lambda_1't}}{it}r(s,t)dsdt
$$

for A,B as continuity intervals of F in the usual sense, where $r(\cdot,\cdot)$ is the covariance function of the X-process. If the mapping $S : \mathbb{R}\to\mathbb{C}$ is continuous, $\frac{1}{T}\int_0^T S(t)dt\to a_0$ exists as $T\to\infty$ and $r(s,t)\to 0$ as

$|s|, |t| \to \infty$, then for the observed process $\tilde{Y}(t) = S(t) + X(t)$, $S(\cdot)$ is signal (nonstochastic) and $X(\cdot)$ is harmonizable noise, then $\hat{S}_T = \frac{1}{T} \int_0^T Y(t)dt \to a_0$ in $L_0^2(P)$ as $T \to \infty$. Thus S_T estimates consistently the constant a_0 and so both X- and $\tilde{Y}$-processes obey the usual law of large numbers.

Proof. The method employed is to reduce the result to the classical stationary case through an application of the dilation theorem. Thus an enlarged $(\tilde{\Omega}, \tilde{\Sigma}, \tilde{P})$ with $\tilde{L}_0^2(P) \supset L_0^2(P), Y : \mathbb{R} \to L_0^2(\tilde{P}), X(t) = QY(t)$ where Q is the orthogonal projection on $L_0^2(\tilde{P})$ onto $L_0^2(P)$. Then the stationary process $Y : \mathbb{R} \to L_0^2(P)$, satisfying $X(t) = QY(t), t \in \mathbb{R}$ with Q as the orthogonal projection onto. Also there is an orthogonally valued measure $Z : \mathcal{B} \to L_0^2(\tilde{P})$ satisfying the relation

$$Y(t) = \int_{\mathbb{R}} e^{it\lambda} \tilde{Z}(d\lambda), t \in \mathbb{R},$$

and $Z(A) = Q\tilde{Z}(A)$, $A \in \mathcal{B}$, with $Z : \mathcal{B} \to L_0^2(P)$ representing the X-process. Now Q commutes with the integral, giving the representation (55) using the classical computations, as in Doob's classic (1953, p. 527). Thus our statement follows on applying Q to both sides and obtaining (55).

Now consider (56). With (55) we get

LHS

$$= \lim_{T_1, T_2 \to \infty} F\left(\int_{-T_1}^{T_1} \int_{-T_2}^{T_2} \left[\frac{e^{-is\lambda_2} - e^{-is\lambda_1}}{-is} X(s) \right] \right.$$

$$\left. \times \left[\frac{e^{-it\lambda_2'} - e^{-it\lambda_1'}}{-it} X(t) \right] \right) ds\, dt$$

$$= E\left[\left(\int_{-T_1}^{T_2} \frac{e^{-is\lambda_2} - e^{-is\lambda_1}}{-is} X(s)ds \right) \left(\int_{T_2}^{T_2} \frac{e^{-it\lambda_2'} - e^{-it\lambda_2'}}{-it} X(t)dt \right) \right]$$

$$= E\left[\left(\frac{Z(\lambda_2+) + (Z\lambda_2-)}{2} - \frac{Z(\lambda_1+) + Z(\lambda_1-)}{2} \right) \right.$$

$$\left. \cdot \left(\frac{Z(\lambda_2'+) + Z(\lambda_1'-)}{2} - \frac{Z(\lambda_1'+) + Z(\lambda_1'-)}{2} \right) \right.$$

$$= E(A, B),$$

by the continuity conditions on F, and taking expectations. This gives (56).

Finally, if $\tilde{Y}(t) = S(t) + X(t), t \in \mathbb{R}$, let

$$a_T = E(\hat{S}_T) = \frac{1}{T} \int_0^T S(t)dt.$$

Since $\tilde{Y} \in$ class (KF), because X is, and $a_T \to a_0$ and letting $T \to \infty$ we get

$$E\left(|\hat{S}_T - a_0|^2\right) = \frac{1}{2T} \int_{-T}^T r_T(h)dh - 2|a_T - a_0|^2,$$

where $r(h) = \lim_{T\to\infty} \frac{1}{T} \int_0^{T-|k|} k(s, s + h_1)ds = \lim_{T\to\infty} r_T(h)$.

It follows that $|r(s,t)|^2 \leq r(s,s)r(t,t) \leq M^2 < \infty$, when $\|X(t)\|_2 \leq M < \infty$ since $X(\cdot)$ is V-bounded. So $r_T(h) \to r(h) = 0$ as $T \to \infty$ for each $h \in \mathbb{R}$ by the Cesaro summability properties. We may now conclude that $E(|\hat{S}_T - a_t|^2) \to 0$, so that $\hat{S}_T$ is a consistent estimator of a_t. This means (stated differently) that both X_t- and $\tilde{Y}_t$-processes obey the law of large numbers. $\square$

6.4 Multivariate Harmonizable Random Fields

The multivariate or vector harmonizable random fields are needed in some applications as indicated in Bochner's (1956) basic analysis of the subject which we shall discuss it now. Thus if $f = (f_1, \ldots, f_n)$: $\Omega \to \mathbb{C}^n$ is a (standard) measurable n-vector on (Ω, Σ, P) such that $|f|^2 = (\sum_{i=1}^n |f_i|^2)$ is P-integrable, $E(f) = \int_\Omega f dP = 0$ (vector), and $\|f\|^2 = (f, f)$ also exists, or f is termed square integrable relative to P, let $L^2(P, \mathbb{C}^n)$ on (Ω, Σ, P) be the square integrable (for P) space of the f's with inner product

$$(f, g) = \sum_{i=1}^n \int_\Omega (f_i \bar{g}_i)(\omega)dP(\omega) \tag{57}$$

existing so that $\mathfrak{X} = L_0^2(P, \mathbb{C}^n)$, with the above inner product, is well-defined and is an n-vector (complex) Hilbert space of members with vanishing means. Also $X : G \to \mathfrak{X}$, with $Y_a = a \cdot X = \sum_{i=1}^n a_i X_i$ is *weakly* or *strongly harmonizable* if $Y_a : G \to \mathfrak{X}$ is (a scalar) harmonizable field of the same type for each $a \in \mathbb{C}^n$.

An integral representation of such a random field can be obtained easily with our earlier analysis. It is given as follows.

Theorem 6.4.1 *On an LCA group G let $X : G \rightarrow L^2(P, \mathbb{C}^k)$ be a bounded weakly continuous mapping. It is weakly harmonizable iff there is a stochastic vector measure $Z : \mathcal{B}(\hat{G}) \rightarrow \mathcal{X}^n$ (i.e. if $\tilde{Z}(A) = (Z_1(A), \ldots, Z_k(A))$, $A \subset \hat{G}$, is a Borel set), then each Z_i is a stochastic measure on $\mathcal{B}(\hat{G}) \rightarrow \mathcal{X}$, $j = 1, \ldots, n$, such that the following representation holds:*

$$X(G) = \int_{\hat{G}} \langle g, s \rangle Z(ds), g \in G, \tag{58}$$

where $\hat{G}$ is the dual group of G. The field $X(\cdot)$ is strongly harmonizable on $\hat{G}$ if $F = (F_{j,l}; j, l = 1, \ldots, k)$, with each $F_{j,l}(A, B) = (Z_j(A), Z_l(B))$ of bounded variations on $\hat{G}$. The covariance matrix R is then representable as:

$$R(g, h) = \int_{\hat{G}} \int_{\hat{G}} \langle g, s \rangle \overline{\langle h, t \rangle} F(ds, dt), g, h \in G, \tag{59}$$

the integral being in the MT-sense, defined componentwise. The integrals in (59) are taken in the Lebesgue-Stieltjes sense componentwise, in the strongly harmonizable case. The converse statements hold in the analogous way as in the usual one variable case.

Proof outline. We add a sketch of proof so that extensions of this work for Cramér and Karhunen classes of interest in applications can be included. Consider $Y_a = a \cdot X, a \in \mathbb{C}^k$. If X is weakly harmonizable then Y_a has also the same property, and by our earlier work, with the index $\mathbb{R}$ replaced by G, there is a stochastic measure Z_a on the dual $\hat{G}$ of G into $\mathcal{X}$ such that

$$Y_a(g) = \int_{\hat{G}} \langle g, s \rangle Z_a(ds), \quad g \in G. \tag{60}$$

It follows that $Z_{(\cdot)}(A) : \mathbb{C}^k \rightarrow \mathcal{H}$ is linear and continuous. Then there is a $\tilde{Z} : G \rightarrow \mathcal{X}^{**} (= \mathcal{X}$ by reflexivity) and that $Z_a(A) = a \cdot \tilde{Z}(A), \tilde{Z} : \mathcal{B}(G) \rightarrow \mathcal{X}$ is σ-additive and we get

$$Y_a(g) = a \cdot X(g) = \int_{\hat{G}} \langle g, s \rangle a \cdot \tilde{Z}(ds) = a \cdot \int_{\hat{G}} \langle g, s \rangle \tilde{Z}(ds). \tag{61}$$

Here the integral on the extreme right defines a member of $\mathcal{X}$. This gives the direct part, and the converse is as in the earlier case. The other statements are similarly obtained from the earlier result. $\square$

The above type of problem arises in filtering studies and an abstract formulation was given by Bochner (1956) as follows. If G is an l.c.a. group as above, and $X : G \to \mathcal{X}$ is a random field, where $\mathcal{X}$ is a Hilbert space, thus a linear operator $\Lambda : \mathcal{X} \to \mathcal{X}$ is termed a *filter* of X, if Λ commutes with the translation operator, in that if $(\tau_a X)(g) = X(a, g)$, for all $g \in G$ (an lca group), to mean that $\tau_a(\Lambda X) = \Lambda(\tau_a X)$, and dom $(\Lambda) \supset \{\tau_h X(g), g \in G, h \in G\}$. Now it is desired to find the process X of the filter equation:

$$\Lambda X = Y \tag{62}$$

when the output process Y is a weakly or strongly harmonizable field so must X be. Special cases (e.g. if Y is stationary or i.i.d.) were considered before, and now we discuss this general problem.

If G is an LCA group and $\mathcal{X}$, a Banach space, a mapping $X : G \to \mathcal{X}$ and $\nu : \mathcal{B}(G) \to \mathcal{X}$, a vector measure such that for each $g \in G$, $X(g) = \int_{\hat{G}} g(s)\nu(ds)$, where $\hat{G}$ is the dual group of the LCA group G, then X is called the (generalized) Fourier transforms of ν. This concept as an extension of Bochner's definition for $\mathcal{X} = \mathbb{C}$ was introduced by R. S. Phillips (1950) which includes V-*boundedness* and he also established some of its key properties. Then I. Kluvánek (1967) extended the result further and obtained the following one of interest for our study. We state Kluvanek's result in order to use it in an extension of (the vector) harmonizability.

It is useful first to recall the V-boundedness for Banach space valued functions.

Definition 6.4.2 *Let $f : \mathbb{R} \to \mathcal{X}$, a Banach space, be a function. Then f is V-bounded if (i) $f(\mathbb{R}) \subset \mathcal{X}$ is bounded, (ii) f is strongly measurable, and (iii) $W = \left\{ \int_{\mathbb{R}} f(t)g(t)dt : \|g\|_\infty \le 1, g \in L^1(\mathbb{R}) \right\}$ is weakly compact in $\mathcal{X}$, the integral being in Bochner's sense.*

With this concept, the vector V-boundedness of $f : \mathbb{R} \to \mathcal{X}$ can be characterized, and is now given in Kluvanek's form:

Theorem 6.4.3 *Let $f : \mathbb{R} \to \mathcal{X}$, a Banach space, be a mapping. It is the Fourier transform of a vector measure ν on the Borel σ-algebra $\mathcal{B}$ into $\mathcal{X}$ iff $\nu : \mathcal{B}(\mathbb{R}) \to \mathcal{X}$ is V-bounded and weakly continuous.*

A detailed proof of this result and an extension of it can be found in the author's graduate text, entitled "Measure Theory and Integration" (2nd expanded edition (2004), p. 550) and need not be reproduced here.

There are also extensions to the Wiener integral as well as related discussion on its extension to the Bochner-Minlos theorem. Since the work is easily available to readers, the details will be omitted here and we refer the reader to the source noted above.

Our earlier analysis and characterization of Cramér's class of processes give the following result on Karhunen processes which is useful in applications, and will be presented.

Proposition 6.4.4 *Let S be a locally compact set, and $X : S \to L_0^2(P)$ be a Karhunen random field relative to a locally finite regular measure F, on the Borel field $\mathcal{B}$ of S and a family $\{g_t, t \in S\} \subset L^2(F)$ on $(S, \mathcal{B}, F)$. Then there exists a locally bounded regular stochastic measure $Z : \mathcal{B}_0 \to L_0^2(P)$ where $\mathcal{B}_0 \subset \mathcal{B}$ is the δ-ring of bounded sets such that $Z(\cdot)$ is orthogonally valued, or $E(Z(A)\bar{Z}(B)) = F(A \cap B), A \in \mathcal{B}_0$, and $X(t)$ is representable as*

$$X(t) = \int_S g_t(\lambda)Z(d\lambda), t \in B, \tag{63}$$

the integral being in the Dunford-Schwartz sense.

Conversely, if $X : S \to L_0^2(P)$ is given by (63) relative to an orthogonally valued Z on $\mathcal{B}(S)$, and $\{g_t, t \in S\}$ and F are given for (63), and orthogonally valued Z on $\mathcal{B}(S)$; the $\{g_t, t \in S\}$ as above, satisfy these conditions, then it defines a Karhunen process relative to this g_t-family and $F(A \cap B) = (Z(A), Z(B))$. Moreover, we have

$$\mathcal{K}_X = \bar{sp}\{X(t), t \in S\} \subset \mathcal{K}_Z = \bar{sp}\{Z(A), A \in \mathcal{B}_0\} \subset L_0^2(P)$$

and $\mathcal{K}_X = \mathcal{K}_Z$ iff $\{g_t, t \in G\}$ is dense in $L^2(F)$.

Proof Sketch. It is well-known that in the Dunford-Schwartz integration, a bounded linear operator and the integral of vector measure commute, and as was shown by E. Thomas, in his thesis, published in 1970, that the result also holds for measures of locally finite semi-variation, with an easy extended argument. This applies to the Karhunen process so that we have

$$TX(t) = \int_S g_t(\lambda)(T \circ Z)(d\lambda), \tag{64}$$

and then $\tilde{Z} = T \circ Z$ is a stochastic measure of locally finite semi-variation. By our earlier work, this implies that the process is of weak class (C).

In the opposite direction, we need to restrict the F's somewhat. Suppose if $g(\cdot)$ is a bounded Borel function and $M(S)$ is the uniformly closed algebra determined by $\{g_t, t \in S\}$, then $M(S) \subset L^2(F_\infty)$. Now if Tg is given by

$$Tg_t = X(t) = \int_S g_t(\lambda)Z(d\lambda),$$

let T be extended linearly to $M(S)$. So $T \in B(M(S), \mathcal{H})$ where $M(S)$ is given the uniform norm, and if at least one g_t has noncompact support, then F_X will be of finite semi-variations so that T is 2-absolutely summing. Hence one gets

$$\|Tf\| \le \|f\|_{2,\mu}, f \in M(S),$$

for a finite μ on S. The further discussion can be omitted. $\square$

Analogous result for class (C) may also be given. We state it here leaving the details to the reader. [Projection Q is bounded linear, $Q^2 = Q$.]

Theorem 6.4.5 *Let S be locally compact and $X : S \to L_0^2(P) = \mathcal{H}$ be a Karhunen process relative to a Radon measure F, and a set,*

$$\{g_t, t \in S\} \subset L^2(F).$$

If $Q : \mathcal{H} \to \mathcal{H}$ is a bounded projection, then $\tilde{X}(t) = QX(t)$, $t \in S$, is a process which is in the weak class (C). On the other hand, if $\{X(t), t \in S\}$ is of weakly class (C) then it is representable in the form:

$$X(t) = \int_S g(\lambda)Z(d\lambda), t \in S \tag{65}$$

for a family $\{g_t, t \in S\} \subset L^2(F_x)$, defined as above, then there is an extension Hilbert space $\mathcal{K} \supset \mathcal{H}$ with $\mathcal{K} = L_0^2(\tilde{\Omega}, \tilde{\Sigma}, \tilde{P})$ on some probability space and a Karhunen process $Y : S \to \mathcal{K}$ such that $X(t) = QY(t), t \in S$ where Q is an orthogonal projection onto $\mathcal{K}$.

A proof of this result will be left to the reader with the comment that it is essentially analogous to the one-dimensional case already detailed in our representation of the Cramér class. It is a useful exercise, but the reader can also look up in the author's (1982) paper on the structural analysis of harmonizable and related classes.

The approximations on L^2-spaces and the Hilbert space geometry are important in much of the above work. If we replace $L^2(P)$ by $L^p(P), 1 < p < \infty$, difficulties arise, and we present one general result here to amplify the situation and to open up a new class of problems for readers to study and extend the subject. We consider $L^p(P, \mathfrak{X})$ (or $L^p(\Omega, \Sigma, P, \mathfrak{X})$ in full) as a Lebesgue space of $\mathfrak{X}$-valued (strongly) measurable $f : \Omega \to \mathfrak{X}$, so that $f^{-1}(A) \in \Sigma$, for each Borel set $A \subset \mathfrak{X}(A$, closed or open or a limit of such sets). We take $\mathfrak{X}$ to be separable for simplicity. Then the space is complete and is reflexive if $1 < p < \infty$ for reflexive $\mathfrak{X}$, with norm $(f \in L^p_{\mathfrak{X}}(P) \Rightarrow)$:

$$\|f\|_p = \left[\left(\int_\Omega |f|^p dP \right)^{\frac{1}{p}} \right], \quad (L^p_{\mathfrak{X}}(P), 1 < p < \infty). \tag{66}$$

It can be shown that $L^p_{\mathfrak{X}}(P)$ is reflexive if $\mathfrak{X}$ is also reflexive and uniformly convex if $\mathfrak{X}$ is such and $1 < p < \infty$. We now establish the following key general result useful for most applications.

Theorem 6.4.6 *Let $\{X_t, t \in T\} \subset L^p_{\mathfrak{X}}(P)$, and $X_{t_0}, t_0 \notin T$, be in $L^p_{\mathfrak{X}}(P), 1 < p < \infty$, where $\mathfrak{X}$ is a separable uniformly convex B-space and (for ease) (Ω, Σ, P) along with $\mathfrak{X}$ be separable. [For $\mathfrak{X} = \mathbb{R}^n$ or $\mathbb{C}^n$, we have the usual n-variate process and the familiar case.] If we know the joint distributions of X_{t_0} and $\{X_t, t \in T\}$ as given, for each finite set of t-points in T, then there exists a unique $Y_0 \in L^p_{\mathfrak{X}}(P)$, such that, if $\mathfrak{B}_0$ is the σ-algebra generated by $\{X_t, t \in (T - t_0)\}$, and Y_0 is then given by*

$$Y_0 = P^{X_0}_{\mathfrak{B}_0} X_0, \tag{67}$$

where if $p = 2$, $P^{X_0}_{\mathfrak{B}_0}$ will be the usual conditional expectation, and if $1 < p \neq 2 < \infty$, then $P^{X_0}_{\mathfrak{B}_0}$ exists as a closed linear idempotent operator (with several properties of the conditional expectation not necessarily bounded but, is closed). [If $p = 2$, it is, of course, the usual conditional operator.] Also if $\mathfrak{B}_t = \sigma(X_t)$, then $\|Y_0 - Y_n\|_2 \to 0$ as $n \to \infty$, and $Y_n = P^{X_n}_{\mathfrak{B}_n}(X_n)$, if the X_t-process has no pathologies, such as fixed discontinuity points.

Proof. In order to explain the structure of the problem, we include the details of arguments so that similar work can be shortened later on. Also we consider the case $1 < p < \infty$. If $T' \subset T$ then $L^p(\mathfrak{B}_{T'}) \subset L^p(\mathfrak{B}_T)$

is separable and is a closed subspace and there is a one-to-one correspondence between the subfields $\mathcal{B}_t$ of Σ and $L^p(\mathcal{B}_t, \mathfrak{X}), 1 \leq p < \infty$. If $\mathcal{M}_t = L^p(\mathcal{B}_t, \mathfrak{X})$ and $d = \inf_{X \in \mathcal{M}} \|X_t - X\|_p$ with $\mathcal{M} = L^p(\mathcal{B}_t, \mathfrak{X})$, then $d > 0$. By the uniform convexity of $\mathcal{M}$, there exists a unique $Y \in \mathcal{M}$, with $d = \|X_t - Y\|_Y$. It is asserted that Y satisfies the requirements. Consider the set $N = \{h : \|h\| \leq \|g + h\|, g \in \mathcal{M}\}$ which is not necessarily linear. But, as F. J. Murray (1943) showed each $f \in L^p$ can be expressed as $f = g + h$, $g \in \mathcal{M}$ and $h \in N$, and since $\|h\| = \|f - g\|$, where $g \in \mathcal{M}$ is closest to f in $\mathcal{M}$. Thus $X_{t_0} = Y + Z, (Y \in \mathcal{M}, Z \in N)$ in such a decomposition, $(\mathcal{M} \cap N = \{0\})$.

[The argument depends on the work of F. J. Murray (1945).] But $\mathcal{M}$ is a closed subspace of the separable L^p, and so there is a (at least one) quasi-complement N' of $\mathcal{M}$, which is a closed subspace such that $\mathcal{M} \cap N' = \{0\}$ and $\mathcal{M} \oplus N'(= \mathcal{D})$ is dense in L^p. Here N' may be compressed or enlarged having the same property. So, by enlargement if necessary, we can assume that $Z \in N$. After this is done, so $X_{t_0} = Y + Z$, we get $P^{\mathcal{B}_t}_{X_{t_0}} : \mathcal{D} \to \mathcal{M}$ with $P^{\mathcal{B}_t}_{X_{t_0}} X_{t_0} = Y$, and if $p \neq 2$ this may be an unbounded operator. If $p = 2$ this operator becomes $E^{\mathcal{B}}(\cdot)$ and it is unique and the operator does not depend on X_{t_0}. This establishes the first part of the theorem.

For the second part, let $\mathcal{B}_n \subset \mathcal{B}_{n+1}$ be defined as before, and $\mathcal{M}_n = L^p(\mathcal{B}_n, \mathfrak{X})$ and $P^{\mathcal{B}_n}_{\mathfrak{X}}$ be the corresponding (closed) projection, and since the range space $\mathfrak{X}$ is separable, we can assume the process to be separable for this proof, and it has no fixed discontinuities, and as $n \to \infty$, $\mathcal{M}_n \uparrow$ and determine $\mathcal{M}$. If $Y_n = P^{X_0}_{\mathcal{B}_n} X_0$, then $\{Y_n, n \geq 1\}$ becomes a martingale with these closed operators, i.e., $P^{X_0}_{\mathcal{B}_n} Y_{n+1} = Y_n$ a.e. It will be shown that $Y_n \to Y$ in L^p as $n \to \infty$. Since the $\{Y_n, n \geq 1\}$ is not the usual martingale, the following modification of the classical argument is needed.

Let $\|Z_n\| = d_n = \|X_{t_n} - Y_n\| = \inf_{X \in M_n} \|X_n - X\|, \mathcal{M}_n \subset \mathcal{M}_{n+1}$. Then $d_n \geq d_{n+1}$ so $d_n \to d_0$, as $n \to \infty$. But $d_0 = d = \|X_{t_0} - Y\|$, since $\bigcup_{n \geq 1} \mathcal{M}_n$ is dense in $\mathcal{M}$, and we easily see $d_n \to d$ from this. Note that $\{Z_n, n \geq 1\}$ is norm bounded. Since $L^p, 1 < p < \infty$ is uniformly convex, the above implies that $Y_n \to Y_0$ weakly and $\|Z_n\| \to \|Z_0\|$ as well as $Z_n \to Z$ weakly. Now the uniform convexity of $L^p, 1 < p < \infty$, implies that $Z_n \to Z$ strongly, and then $Y_n \to Y_0$ in norm, [more details in Rao (1967), Z. W. paper]. This completes the proof. $\square$

We finally present a result on the construction of the best element Y_n of the above theorem, because of its use in some applications, and possible extensions.

Proposition 6.4.7 *Let* $L^p(\Sigma, \mathfrak{X}), 1 < p < \infty, \mathfrak{X}$ *being also a uniformly convex Banach space, on a probability space* (Ω, Σ, P). *If the functional* $Y \mapsto \|X_0 + nY\|$ *is (strongly) differentiable at* X_0, *for each* Y *then the element* $Y_n \in \mathfrak{M}_n$ *of the above theorem is given as a unique solution of the integral equation*

$$\int_\Omega |X_{t_0} - Y_n|^{p-1} \left(\frac{d}{du} |X_{t_0} - Y_n + nY| \right)_{t=0} du = 0, Y \in \mathfrak{M}_n,$$

and the solution does not depend on Y. *In general (without the differentiability hypothesis), there is a functional* $l \in (L^p(\Sigma, \mathfrak{X}))^n$ *such that we always have* $\|l\| = 1, l(X_0) = \|X_0\|$ *and*

$$D_-[\phi_Y(u)]\,|_{u=0} \geq \max(0, u_y), D_+[\phi_Y(u)])_{u=0} \leq \min(0, l(X))$$

where D_+, D_- *are the right and left derivatives, and* $Y \in \mathfrak{M}_n$. *[For details of proof, see Rao (1967), pp. 57–59.]*

6.5 Optimum Harmonizable Filtering with Squared Loss

If $\{X_t, t \in T \subset \mathbb{R}\}$ is a centered second order process, assumed expressible as an additive signal plus noise:

$$X_t = Y_t + Z_t, \quad -\infty < t < \infty \tag{68}$$

where the processes $\{Y_t, t \in \mathbb{R}\}$ and (noise) $\{Z_t, t \in \mathbb{R}\}$ are usually assumed uncorrelated and it is desired to obtain an estimator of $Y_t, \infty < t < \infty$, based on a set of observations, that is optimal for squared loss. If the covariance r is of bounded variation in finite domains of $\mathbb{R}^2$, then H. Cramér (1951) has given a solution for estimating the Y_t-process (the signal) if the X_t-process is assumed to have its covariance to be locally of bounded variation, i.e., in every relatively compact domain in $\mathbb{R}^2$. His result was extended by Grenander (1951). We state the basic result by Cramér for reference, and present a version that includes both the cases and some others. First, we recall b.v. in finite domains.

Definition 6.5.1 *Let $r(\cdot, \cdot)$ be the covariance function of $\{X_t, t \in \mathbb{R}\} \subset L_0^2(P)$. It is of bounded variation in a finite domain D if we have:*

$$\sum_{k=1}^{N} |\Delta^2_{i_k j_k} r(s,t)| \leq a < \infty, N > 1, \tag{69}$$

where $\Delta^2_{i_k j_k} r(s,t)$ denotes a standard two-dimensional increment of $r(s,t)$.

With this concept H. Cramér (1951) has derived an integral representation of centered processes with covariances satisfying condition (69) and it was studied in a general form by Grenander (1950) in his thesis. We now present here a general second order case that includes both the above noted works and may be used in more applications.

Theorem 6.5.2 *Let $\{X(t), t \in \mathbb{R}\} \subset L^2(P)$ be a process representable as:*

$$X(t) = Y(t) + E(t), \quad t \in \mathbb{R} \tag{70}$$

where the $Y(t)$ is termed, the "signal" and $E(t)$ the "noise" process, uncorrelated and centered with covariances r_Y and r_E respectively which have the representations as:

$$r_Y(s,t) = \int\!\!\int_{\mathbb{R}^2} f_Y(s,x)\overline{f_Y(t,y)} d^2 \rho_1(x,y), \tag{71}$$

and

$$r_E(s,t) = \int\!\!\int_{\mathbb{R}^2} f_E(s,x)\overline{f_E(t,y)} d^2 \rho_2(x,y), \tag{72}$$

where both ρ_1, ρ_2 are covariance bimeasures of bounded variation on each bounded domain of $\mathbb{R}^2$, each satisfying (69) so that the f_i are of locally bounded variation $(i = 1, 2)$. Then the linear estimator $\hat{Y}$ of Y in $\mathbb{R}^2$ that minimizes the mean squared error is given by

$$\hat{Y}(a) = \int_{-\infty}^{\infty} K_a(\lambda) Z_X(d\lambda), -\infty < a < \infty, \tag{73}$$

where $K_a(\cdot)$ is a solution of the integral equation

$$\int_{-\infty}^{\infty}\!\!\int_{-\infty}^{\infty} K(x) \left[d^2 \rho_1(x,y) + d^2 \rho_2(x,y) \right] = \int_{-\infty}^{\infty}\!\!\int_{-\infty}^{\infty} f_Y(a,x) d^2 \rho(x,y), \tag{74}$$

and $Z(\cdot)$ is the random measure representing the X-process. Moreover, the optimal filter satisfying the equation (74) is unique.

Remark 30. A detailed proof of the theorem is in the author's paper (Rao (1967), pp. 64–67) and will be referred to it, as the journal is available easily. The following special case due to Grenander (1950) is of interest in some applications.

Corollary 6.5.3 *Let $Y(t)$ and $E(t), t \in \mathbb{R}$, be stationary processes with spectral densities f_1 and f_2 (relative to the Lebesgue measure). Then the optimal filter (for squared error as loss) at 'a' is given by the integral equation.*

$$\hat{Y}(a) = \int_{\mathbb{R}} \frac{f_1(\lambda)}{f_1(\lambda) + f_2(\lambda)} e^{ia\lambda} Z_x(d\lambda), \tag{75}$$

where the stochastic measure has orthogonal increments.

Proof. By stationarity, the spectral measure concentrates on the diagonal $x = y$ and $g(t, x) = e^{itx}$ in the theorem. Since the F_i have densities $f_i, i = 1, 2$, we have

$$r_Y(t) = \int_{\mathbb{R}} e^{itx} f_1(x) dx; \quad r_E(t) = \int_{\mathbb{R}} e^{it\lambda} f_2(\lambda) d\lambda.$$

Hence $K_a(x) = \frac{f_1(x)e^{iax}}{f_1(x)+f_2(x)}$ in (74), since $Z(\cdot)$ has orthogonal increments. Then (75) follows from (73). $\square$

The above result can be specialized for the weakly (strongly) harmonizable processes where ρ_i will be of bounded variation for the strongly (or Loéve) harmonizable case.

Starting in the 1930s, the multidimensional prediction problem was investigated in detail by Norbert Wiener, and also popularized by him and some of his followers. Here we consider a brief analysis of this useful extension. Thus let $Y, X_1, X_2, \ldots$ be a sequence of vector-valued random variables on (Ω, Σ, P) valued in a normed linear space $(\mathfrak{X}, \|\cdot\|)$. If the joint distribution of Y and $(X_1, \ldots, X_n)$ is known then it is desired to find a function g_0 such that $E[W(\|Y - g_0(X_1, \ldots, X_m)\|)]$ is a minimum for a suitable function $g : \mathbb{R}^n \to \mathfrak{X}$. We present a solution here for a reasonable "loss function" $W(\cdot) \geq 0$. The background ideas and some familiar contexts of the problems were discussed by M. H. DeGroot and the author (1966) where its relation to some key statistical problems is studied particularly if $\mathfrak{X}$ is a finite dimensional vector space. We stretch it to an infinite dimensional version here at the same time.

Let $W : \mathbb{R} \to \mathbb{R}^+$ be a convex function, symmetric and increasing, for instance $W(x) = |x|^p, p \geq 1$, is included. Consider a sequence $Y, X_1, X_2, \ldots$ of r.v.'s strongly measurable such that $E(W(|Y|))$ and $E(W(|X_i|)), i \geq 1$, exist and it is also assumed that the joint finite-dimensional distributions of Y and $X_1, X_2, \ldots$ are known. We now describe a multivariate prediction question with a general solution to indicate an interesting class of prediction problems with answers.

Theorem 6.5.4 *Let $W(\cdot)$ be a symmetric convex increasing function on $\mathbb{R}$ such that (i) $W(0) = 0$, (ii) $W(\alpha x + \beta y) \leq \alpha W(x) + \beta W(y)$, $0 \leq \alpha \leq 1$, $\beta = 1 - \alpha$, (iii) $W(2x) \leq cW(x)$, $c > 0$, $x \geq 0$, (iv) if $W'(\cdot)$ is the derivative of W (which always exists), $W'(ax) > k_a W'(x)$ for $a > 1$ and some $k_a > 1$ for all $x \geq 0$. Then for a sequence $\{Y, X_1, X_2, \ldots\}$ of random variables on Ω for which $E(W(\|Y\|)) < \infty$, $E(W(\|X_n\|)) < \infty$, $n \geq 1$, and $\mathcal{B}_n = \sigma(X_1, \ldots, X_n)$, there exists a unique $Y_n \in L^W(X_1, \ldots, X_n)$ which depends only on $X_1, \ldots, X_n$ such that $E(W(Y - Y_n)) \to 0$ as $n \to \infty$.*

Remark. This result says that, to predict Y based on observations $X_1, \ldots, X_n$ all with values in $L^W(P)$, a normed complete vector space determined by the sequence $X_1, X_2, \ldots$, of vector-valued random variables, there exists a unique function $Y_n(= Y_n(X_1, \ldots, X_n))$ such that $E(W(Y_n)) < \infty$, and $E(W(Y - Y_n))$ is a minimum. The space of W-integrable Y's valued in $\mathcal{X}$ denoted $L^W(P : \mathcal{X})$ is called the $\mathcal{X}$-valued Orlicz space which is a Banach function space that includes all $L^p(P)$-spaces, with $W(x) = |x|^p$ above. We include a quick sketch of proof, as in DeGroot and Rao (1965). Unfortunately we need to recall again some abstract results here.

Sketch of Proof: Consider the convex function U defined by

$$U(Z) = \int_\Omega W(|Y - Z|)dP, \quad Z \in L^W(\Sigma, P) = L^W(\Sigma). \quad (76)$$

Since $L^W(\mathcal{B}_0) \subset L^W(\mathcal{B})$ and $Y \notin L(Z)$, (otherwise there is nothing to prove), we can conclude from earlier works, that there is a unique $Y_0 \in L^W(\mathcal{B}), [Y_n \in L^W(\mathcal{B}_n)]$, such that $U(Y_n)$ is a minimum on $L^W(\mathcal{B})$. So we only need to prove that $Y_n \to Y_0$ as asserted.

Since the integrals $U(Y_n)$ and $U(Y_0)$ are finite and satisfy $0 \leq U(Y_0) \leq U(Y_n) \leq U(Y_i)$ by definition of the Y_n-sequence, the sequence is bounded in the $L^W(\Sigma)$-norm (the Orlicz space norm). This can be seen to be equivalent to $0 \leq \|Y - Y_0\| \leq \|Y - Y_n\| \leq \|Y - Y_i\|$, and $\|Y - Y_n\|$ converges monotonely to a limit α_0 (say). In fact $\alpha_0 = \|Y - Y_i\|$. This argument is similar to that in Rao (*Indag. Math.* **27**, 100–112), and using the uniform convexity of $\bar{L}(\Sigma)$, we may conclude that $\|Y_n - Y_0\| \to 0$. (The argument of the above-stated result is for scalar processes, but it extends easily to the present vector-valued case.) Thus if $\varepsilon_n = \|Y_n - Y_0\|$, then for large n, $0 < \varepsilon_n < 1$, and

$$E[(W(Y_n - Y))] \leq \int_\Omega W\left(\frac{\varepsilon_n|Y_n - Y|}{\varepsilon_n}\right) d\mu \leq \varepsilon_n < 1. \qquad (77)$$

This is a consequence of simple Orlicz space analysis. Since $\varepsilon \to 0$ as $n \to \infty$, the outline of the proof that $Y_n \to Y_0$, is norm is concluded. $\square$

We also include a result as an adjunct to the last part of the above theorem that will be of interest for applications, as it shows how solutions on finite dimensional spaces approximate the global solution.

Theorem 6.5.5 *Let the hypothesis of the above theorem hold and the norm of $\mathcal{Y}$, the range space which is Banach space, is differentiable and the conditional distribution of Y given $(Y_1, \ldots, Y_n)$ or $\mathcal{B}_n$ exist. Then the predictor Y_n is obtained as a (unique) solution in $L(\mathcal{B}_n)$ of the integral equation:*

$$\int_{\mathcal{Y}} W(|y - Y_n(x)|)L(Z)(dy|x_1, \ldots, x_n) = 0 \qquad (78)$$

for a.a. $(x_1, \ldots, x_n)$ where $L(Z) = \frac{d}{dt}(|Y - Y_n + tZ|)\,|_{y=0}$ for all $Z \in L(\mathcal{B}_n)$, and $W(0) = 0, W(\cdot) \geq 0$ is a symmetric convex function. Moreover, if Y_n, Y_0 are the predictors of Y in $L(\mathcal{B}_n)$ and $L(\mathcal{B})$, then $Y_n \to Y_0$, as $n \to \infty$ with probability one.

The standard method of proof will be omitted, but the details are also seen in the paper by DeGroot and Rao (1966). The following summary statement will be of interest and is given for a ready reference.

Theorem 6.5.6 *Let $\{X_t, t \in T = (a, b)\}$ be a strongly measurable (Banach space) $\mathcal{Y}$-valued process on (Ω, Σ, P) without fixed discontinuities, where $\mathcal{Y}$ is a uniformly convex Banach space. If $X_{t_0}, t_0 \notin T$, is a $\mathcal{Y}$-valued r.v. whose joint distribution with each $X_{t_1}, \ldots, X_{t_n}$ is given, and $E(W(|X|))$ is finite, then the following conclusions hold:*

(a) there is a best predictor Y_0 relative to $W(\cdot)$ based on a sample path of $\{X_t, t \in T\}$, $[E(W(|Y_0|)) < \infty]$,

(b) if $\tilde{T}$ is a finite or countable set and Y_n is a predictor as in (a), then $E(W(|Y_n - Y_0|)) \to 0$ as $n \to \infty$, (as $t_n \to t_0$),

(c) if the norm of $\mathcal{Y}$ is differentiable, then Y_0 is the solution of the integral equation

$$\int_{\Omega} W(|X_{t_0} - Y_0|) \frac{d}{du} [|X_{t_0} - Y_0| + uZ]_{y=0} dP = 0$$

for all $Z \in L(\mathcal{B}_T)$, the σ-field $\mathcal{B}_T$ is determined by $\{X_t, t \in T\}$. Replacing Y_0 and t_0 by Y_n and t_n in the above, the limit as $n \to \infty$ may easily be calculated. If $W(x) = |x|^p, 1 < p < \infty$, and $\mathcal{Y}$ is the line then the special cases have been studied earlier by various authors, including the present one. [We leave the details to the reader.]

6.6 Applications and Extensions of Harmonizable Fields

A class of bounded linear mappings on $L^p(\Omega, \Sigma, \mu)$, $p \geq 1$, denoted $L^p(\mu)$ is considered and let $\mathcal{D}$ be a subset of linear operators T on $L^p(\mu)$, which are contractions along with their adjoints T^*, on $L^q(\mu)$, $q^{-1} + p^{-1} = 1$. Then a classical theorem of F. and M. Riesz brothers implies that T defined on $L^1(\mu)$ and $L^\infty(\mu)$ into the same space and is a contraction there is also defined on all $L^p(\mu)$, $1 \leq p \leq \infty$ and is again a contraction on each $L^p(\mu)$. These operators T have interesting structural properties as well as many applications studied by I. Schur in 1923, and later by Hardy-Littlewood-Pólya in 1929 and by G. Birkhoff in 1946 as well as many later researchers. Here we consider a few results for our analysis. Recall that a matrix $A = (a_{ij})$, $n \times n$ is called *doubly stochastic* if $a_{ij} \geq 0, \sum_{i=1}^n a_{ij} = 1 = \sum_{j=1}^n a_{ij}$ these constraints playing a key role here.

Definition 6.6.1 *Let (Ω, Σ, μ) be a finite measure space and for a real measurable f on Ω, the right continuous decreasing function of f as $d_f(\cdot)$ is given by: if $d_f(t) = \mu\{w : f(w) \geq t\}$ and if $\mu(\Omega) = 1$, then $d_f(t) = 1 - F_f(t)$, where $F_f(t) = \mu\{w : f(w) \leq t\}$, the "distribution" of f which is the usual one for $\mu(\Omega) = 1$. Then the right inverse of $d_f(\cdot)$, denoted f^*, is given by, for $0 \leq u \leq \mu(\Omega) = a$, as:*

$$f^*(u) = \inf\{t : d_f(t) \leq a\} = \sup\{t : d_f(t) > a\}.$$

[Observe that for $0 \leq t \leq \mu(\Omega) < \infty$, we have

$$f^*(a) = \inf\{t : d_f(t) \leq a\}, \quad f^* \in L^l((0,a),\mu),$$

if $\mu(\cdot)$ is the Lebesgue measure on $\mathbb{R}$. So $\int_\Omega f d\mu = \int_0^a f^*(t)dt.]$

We consider a partial ordering "$\prec$" on $L^1(\mu)$ and use it to characterize the well-known bistochastic class of operators on the real $L^1(P)$-space.

Definition 6.6.2 *(a) If $f, g \in L^1(\mu)$, then $f \propto g$ iff $\int_0^t f^*(u)du \leq \int_0^t g^*(u)du$ where f^*, g^* are defined in Definition 6.6.1, with equality for $t = a = \mu(a) < \infty$.*

(b) $T : L^1(\mu) \to L^1(\mu)$ is bistochastic (or b.s.) iff for each $f \in L^1(\mu)$ we have $Tf \propto f$.

We now present a characterization of b.s. operators for completeness of this analysis (and for information), as this knowledge will be of interest for some applications.

Theorem 6.6.3 *If (Ω, Σ, μ) is a complete finite measure space, and $f, g \in L^1(\mu)$ such that $g \propto f$ as defined above, then there exists a bistochastic operator $T : L^1(\mu) \to L^1(\mu)$ such that $g = Tf$, and conversely.*

The result has interest in understanding the structure of this class of operators. We omit its proof here and refer the reader for complete details in Rao (1978) and for another proof, see Day (1973) as well as for some applications in K. M. Chong (1976). The work of the author's in the above (done independently of Day) is motivated by the analysis of Ryff (1965), and it completes and extends an interesting representation of b.s. operators. We state the extension for readers to use it in applications.

Theorem 6.6.4 *Let (Ω, Σ, μ), $0 < \mu(\Omega) < \infty$, be a complete (finite) measure space and $T : L^1(\mu) \to L^1(\mu)$ be a bounded linear operator. Then the following statements are equivalent:*

(i) T is bistochastic.

(ii) $T \in \mathcal{D}$ and $T1 = 1$ where $\mathcal{D}$ is the class of linear operators on $L^1(P) \to L^1(P)$ and $L^\infty(P) \to L^\infty(P)$, which are contractions.

(iii) T admits an integral representation with a kernel K where $K : \Omega \times \Sigma \to \mathbb{R}$ satisfies the conditions (a)–(c) below:

(a) $K(w, \cdot)$ is σ-additive, vanishes on μ-null sets, and if $g(w) = K(w, \cdot), w \in \Omega$, then $\|g\|_\infty \leq 1$.

(b) $K(\cdot, A)$ is measurable on $\Omega, A \in \Sigma$, and $K(w, \Omega) = 1$ for almost all $w \in \Omega$.

(c) $\nu_f : A \to \int_\Omega K(w, A) d\mu(w), A \in \Sigma$, is a measure and $\nu_f(A) > 0$, for $f \geq 0$, and $\frac{d\nu_f}{du} = 1$ if $f = 1$, a.e.

(iv) T is on $L^1(\mu)$-continuous operator, $T1 = 1$ and $T^*1 = 1$.

Finally, the kernel K in (d) can be differentiated by μ iff T is weakly compact as well.

This class of the D-S operators is of interest in many applications. Consequently, we end this section with the types of operators that are in $\mathcal{D}$. The next result gives a description of this class.

Proposition 6.6.5 *The class $\mathcal{D}$ contains the following types of operators where for (Ω, Σ, μ), with $\mu(\Omega) < \infty$.*

(a) $T : L^p(\mu) \to L^p(\mu)$, a contraction for all $1 < p < \infty$ and $T(\mathcal{M}) \subset \mathcal{M}$, where $\mathcal{M}$ is the set of real bounded functions on Ω as usual.

(b) $T : L^p(\mu) \to L^p(\mu)$ is a contraction for $p = p_1$ or $p = p_1'$, $T1 = 1$ where $p_1^{-1} - (p_1')^{-1} = 1$ and $Tf \geq 0$ for $f \geq 0$ $(p_1 \geq 0)$.

(c) $T : L^p(\mu) \to L^p(\mu)$ is a contraction for $p = 1$ and for some $p = p_1 > 1$. Also T is positive satisfying $\sum_{n=1}^\infty T^n f = \infty$ a.e., $f > 0$ in $L^1(\mu)$.

In particular, if for some $0 < f_0 \in L^1(\mu)$, $Tf_0 = f_0$, T being a contraction then T must be a bistochastic operator.

Proof. (a) Let $\alpha = p^{-1}$ and $\psi(a) = \log \|T\|_\alpha$, so by M. Riesz's convexity theorem $\psi(\cdot)$ is convex on $(0, 1)$ and is continuous there.

Since $\|T\|_p = e^{\psi(\mathbb{R})} \leq 1$, it follows from the convexity of $\psi(\cdot)$ that the inequality holds as $\alpha \to 0$ and $\alpha \to 1$ so that $T \in \mathcal{D}$, giving (a).

(b) Again by the convexity theorem, T is a contraction on $L^p(\mu)$ for all $p \geq 1$, and in particular for $p = 2$. Then by a classical theorem of F. Riesz and B. Sz.-Nagy, the fixed points of T and T^* are the same on $L^2(\mu)$ and so $T1 = 1 = T^*1$. Since T is positive, if $g \in L^2(\mu)$, and $h = \text{sgn}(Tg) \in L^2(\mu)$, one has

$$\int_\Omega |Tg| d\mu = \int_\Omega Tg \cdot h d\mu \leq \int_\Omega |g|(T * h) d\mu \leq \int_\Omega |g| |T^*(\cdot)| d\mu$$

$$= \int_\Omega |g| d\mu.$$

It then follows that T extends to be a contraction on $L^p(\mu), 1 \leq p \leq 2$. Since $T1 = 1$ (so $T^k 1 = 1$ for $k \geq 1$), T is regular, and T has an extension to $L^\infty(\mu)$ renaming a contraction. So $T \in \mathcal{D}$ and hence T is bistochastic. Thus (b) holds.

(c) The first part is similar to (and simpler than) the above. The last part is similarly established, and then the positivity of T is as in (b) above, and completes the argument. $\square$

6.7 Complements and Exercises

1. Let $L^p(\mu), p \geq 1$, on a finite measure space (Ω, Σ, μ), and consider the class of (linear) operators $T : L^p(\mu) \to L^p(\mu), \|T\| \leq 1$, i.e., the T's are contractions). If this is true for $p = 1$ and $p = \infty$, then by the classical M. Riesz-Thorin theorem T is defined and is a contraction on all $L^p(\mu), 1 \leq p \leq \infty$. This class is termed the D-S (or Dunford-Schwartz) family on $L^p(\mu)$. Show that if for some $0 < f_0 \in L^p(\mu)$, $0 \leq \mu(\Omega) < \infty, \sum_{n=1}^{\infty} T^n f_0 = +\infty, (\text{a.e.}) \|T\| \leq 1$, then T is necessarily a bistochastic operator on both L^p and $L^{p'}$ $(p^{-1} + p'^{-1} = 1)$.

2. Let $T_a : L^p(\mu) \to L^p(\mu), i \geq 1$ be contractions on L^p and $L^{p'}$ where $p^{-1} + (p')^{-1} = 1, p \geq 1$ and $T_n^k = T_n T_{n-1} \cdot T_k$, so $V_n = T_n * T_n$ the T_n^* being a Dunford-Schwartz (or DS) operator on $L^\infty(\mu)$ so that

$$T_n^* f(\cdot) = \int_\Omega f(w) K_n(\cdot, dw), \quad f \in L^\infty(\mu).$$

Hence K_n is a bounded additive measure and then $T_n^* \chi_A = K_n(\cdot, A)$ is valued in $L^\infty(\mu)$. Thus $T_{1n} = T_n T_{n-1} \ldots T_1$ gives the representation:

$$T_{1n}^* f = \int_\Omega \tilde{K}_1(\cdot, dw_1) \int_\Omega K(w_1, dw_1) \ldots \int_\Omega f(w) \tilde{K}_{n-1}(w_{n-1}, dw).$$

[Using an extended form of Tulcea's (1949) theorem (1949/50) or the author's book (*Stochastic Process: General Theory* (1995), pp. 224–225) T_{1n}^* can be identified to obtain the above formula.]

With further analysis, N. Starr (1955) shows that $\{T_{1n} f, n \geq 1\}$ becomes a decreasing submartingale, leading to new developments.

3. Let $\{X_1, \ldots, X_n, 1 < n < \infty\}$ be a set of independent (real) random variables with a common distribution F which is strictly increasing and continuous. Let $\{Y_1, \ldots, Y_n\}$ be *order statistics* of the set so that

$Y_1 = \min\{X_1, \ldots, X_n\}$ and $Y_n = \max\{X_1, \ldots, X_n\}$, Y_i being the ith smallest, hence $\{Y_1 < Y_2 < \ldots < Y_n\}$ with probability one. Verify that $\{Y_i, 1 \le i \le n\}$ forms a (nonstationary) Markov process. Moreover, if $Z_i = -\log(Y^*_{n+1-i}) = \log \frac{1}{F(Y_{n+1-i})}$, then the process has independent increments in addition. (See the author's (1962) paper, and A. Renyí's (1953) paper for additional properties.)

4. Here is an extension and application of Exercise 1 above. Let $\{X_n\}_{n=1}^{\infty} \subset L^1(P)$ be a sequence of integrable (real) random variable on (Ω, Σ, P). If $\Sigma_m = \sigma(X_m) \subset \Sigma$ the σ-algebra generated by the r.v., X_m and $T_k f = E^{\Sigma_k}(f)$, for integrable f on (Ω, Σ, P), so that if $f = \chi_A, A \in \Sigma, T_k f = E^{\Sigma_k}(f)$, $\Sigma_k = \sigma(X_k)$, and so $T_k f = P^{\Sigma_k}(A) = P(A|\Sigma_k)$, the conditional probability, and $P^{\Sigma_k} : \Sigma \to L^1(\Sigma_k, P) \subset L^1(\Sigma, P)$ is defined, we then have the following assertion (related to the Starr-Rota result above):
Let $X_n \in L^1(P), n \ge 1$, and $T_{1n} = T_1 \ldots T_n$, where $T_k = E^{\sigma(X_k)}(\cdot)$ for the above, and it is a noncommutative product of conditional expectations. Show that the following assertions on (noncommutative) conditional operations hold:
(i) $(T_{1n}f)(\omega) = E^{\mathcal{F}_n}(f)(\omega), \mathcal{F}_n \supset \mathcal{F}_{n+1}$, where $\mathcal{F}_n$ is the σ-subalgebra of $\tilde{\Sigma} = \otimes_{i \ge 1} \Sigma^i (\Sigma^i = \Sigma)$, so that $\{T_{1n}f, n \ge 1\}$ is a decreasing martingale, whence $\lim_{n \to \infty} T^*_{1n} T_{1n} f = \tilde{f}$ holds with probability one, on the larger space, and also in $L^1(P)$-mean.
(ii) In the above, T^*_{1m} admits the integral representation, as in (i), forming a decreasing martingale, whence $\lim_{n \to \infty} T^*_n T_n f = \tilde{f}$ with probability one and in mean of $L^1(P)$. (See Rao (2007) for details.)

5. So far we considered the (spectral) analysis of weakly harmonizable class in $L^2(\mathbb{R}^k, P)$, and linear combinations of such processes, but a product of a pair of such processes takes us out of these spaces, and new problems arise. Here we consider such a problem and present some new results of this operation. This was originally considered by D. Dehay (1991) and gives a useful insight. Thus $\{X_n, n \in \mathbb{Z}\} \subset L^p(P), p \ge 1$, is $(L^p(P))$-harmonizable if there is a vector measure $\mu : \mathcal{B}(\mathbb{R}) \to L^p(P), p \ge 1$, such that

$$X_t = \int_{\mathbb{R}} e^{itu} d\mu(u), t \in \mathbb{Z},$$

and μ is again called the *spectral measure of the process*.

(i) If $\{X_t, t \in \mathbb{R}\}$ is L^p-harmonizable with spectral measure μ, as defined above, then the weighted L^p-mean limit exists, in the sense that

$$\operatorname*{plim}_{N \to \infty} \frac{1}{N} \sum_{k=0}^{N} e^{ika} X_n = \mu(\{a\}), \text{ for all } a \in \mathbb{R}.$$

(ii) Let $\{X_n^j, n \in \mathbb{Z}\}, j = 1, 2$ be a pair of L^p-harmonizable series with stochastic measures ν_j on $\mathcal{B}(I^2)$. Then $\{(X_1 \cdot X_2)_n = (X_n^1 \cdot X_n^2), n \geq 1\}$ is L^1-harmonizable with its stochastic measure existing $\mu : A \mapsto (\nu_1 \times \nu_2)(u_1 + u_2 \in A)$.

6. This is a companion of the preceding problem, extending it to a product of $p_i (i = 1, 2)$-harmonizable series. Thus denote by $X_j = \{X_{j_n}, n \in \mathbb{Z}\}, j = 1, 2$ be L^{p_j}-harmonizable, $\frac{1}{p_1} + \frac{1}{p_2} = \frac{1}{p} \leq 1$, where ν_j is the (spectral) representing measure of X_j, as in the preceding problem, $j = 1, 2$ and $\nu_1 \times \nu_2$ is the product representing the L^p-harmonizable process $\{X_j, j \in \mathbb{Z}\}$ where $\{X_j = X_{1j_n} \cdot X_{2j_n}, j_n \in \mathbb{Z}\}$ is well-defined, and the measure $\mu(A) = \nu_1 \times \nu_2(u_1 + u_2 \in A)$ so that the product process X_j is L^p-harmonizable defined on I. [Thus $(\mu(A) = \nu_1 \times \nu_2(u_1 + u_2 \in A))$ is the representing measure.] [The product harmonizable series to be harmonizable needs such (somewhat unnatural) conditions, rendering the resulting product somewhat less intuitive in applications, but for the theoretical analysis one has to complete the study for possible future formulations. This result is also due to Dehay (1991).]

7. The preceding one is an extended version of a [weakly/strongly] harmonizable mapping $X : G \to L_0^2(P)$ on an LCA group G, to be called a centered second order random field on G, with covariance $r : G \times G \to \mathbb{C}$. If X, is mean continuous for $L^2(P)$, which is true for harmonizable or stationary classes that we studied earlier, then it admits an integral representation relative to a (random) measure $Z : \mathcal{B}(G) \to \mathbb{C}$, with $(Z(A), Z(B)) = \beta(A, B)$, for a bimeasure $\beta(\cdot, \cdot)$, of (finite) Vitali or Fréchet type variation so that in either case, verify (from our earlier work) that

$$X(g) = \int_{\hat{G}} \langle g, \lambda \rangle dZ(\lambda), \lambda \in G,$$

$\hat{G}$ being the dual of G, is defined where the integral is in the D-S sense. Deduce that, if $T : L_0^2(P) \to L_0^2(P)$ is a bounded linear operator then we get for $Y(t) = TX(t)$, that

$$Y(g) = \int_{\hat{G}} \langle g, \lambda \rangle d(T \circ Z)(\lambda), g \in G$$

$\langle g, \cdot \rangle : \hat{G} \to \mathbb{C}$ is a continuous character. [See the author's book Rao (2013) (Sec. 9.1 for details and related remarks).]

8. The following result on convolution of bimeasures representing harmonizable fields (due to J. H. Park (2016)) is of interest in this analysis. Let $B_i : C_0(G \times G) \to L^2(P), i = 1, 2$ be bounded bilinear forms induced by the σ-additive random measures $Z_i : \mathcal{G}_i \to L_0^2(P), i = 1, 2$ (so is σ-additive) ($\mathcal{G}_i, i = 1, 2$, where $\mathcal{G}_i$ is Borel σ-fields of G). Then there exists a unique random measure $Z : \mathcal{G}(= \mathcal{G}_1 \otimes \mathcal{G}_2) \to L_0^2(P)$ that represents the event $B = B_1 * B_2$ in the sense that $(B_1 * B_2)^\wedge = \hat{B}_1 \cdot \hat{B}_2$ (the pointwise product). [Here we use the known calculus that for each bounded bilinear form $B_i : C_0(G_1) \times C_0(G_2) \to \mathbb{C}, i = 1, 2$ with Fourier transforms $\hat{B}_i : \hat{G}_1 \times \hat{G}_2 \to \mathbb{C}i$, there is a $\hat{B} \in B(\hat{G}_1, \hat{G}_2)$ such that $\hat{B} = \hat{B}_1 \cdot \hat{B}_2$. This is not a trivial assertion, and is detailed in (Rao (2012), p. 58, 178) wherein it was noted that the Morse-Transue type of integration, in place of the Lebesgue concept, need be used. The necessary details are in Park's paper. The use of the MT-integrals is avoidable in the (weakly) harmonizable analysis.]

9. This problem indicates a possible infinite dimensional form of weakly harmonizable random fields in $L^2(P)$, by discussing a case of (weakly) stationary fields indexed by an LCA group that is not a simple formulation itself, indicating a motivation for a good future study. [This class was originally studied by G. M. Molchan and Yu. A. Rozanov who characterized the wide sense Markov property and we give a second order extension here.] [We recall that X is wide sense Markov if $S, T \subset \mathbb{R}^d, d > 1$, without a convex boundary Γ so that $T = \mathbb{R}^d - (S \cup \Gamma), S \subset \mathbb{R}^d - (T \cup \Gamma)$, and $X(t) = P(\Gamma)X(t) \perp X(s) - P\Gamma X(s), s \in S, t \in T$ are mutually orthogonal. The characterization problem is nontrivial.] [See H. Heyer and M. M. Rao (2012), for details and extensions.] For instance, a second order process $X : \mathbb{R} \to \mathcal{H}$ has a wide sense. Markov property iff its correlation characteristic R defined by

$$R(s, t) = E(X_s \bar{X}_t)/E(|X_s|^2)^{-1} \text{ with } \{s \in \mathbb{R} : E(|X_s|^2) > 0\}$$

satisfies $R(r, t) = R(r, s)R(s, t)$ for $r < s < t$. (See Rao (1962) for details of this example.) The problem is thus not very simple.

6.8 Bibliographical Notes

This chapter completes the basics of harmonizable processes and applications as well as some indications of further work on the theme of the book especially furthering the harmonizability class. We focussed on Cramér's as well as the initial ideas of Karhunen's in extending the harmonizable classes to second order random fields and processes. This shows clearly the generality and limitations of extending weak stationarity, implying also the extent that is profitable to go beyond the use of the classical harmonic analysis and ideas. This study uses the Morse-Transue method of (multiple) integration that is slightly simplified and refined as needed for analysis of weak harmonizability and the consequent extensions. The work uses the corresponding integration detailed in the analysis of Chang and Rao (1986), and further analysis on it by Mehlman (1991) and Swift (1997) as well as some others.

It should be noted that a great amount of harmonizable analysis was done by D. Dehay in a series of papers, some with his associate (or mentor) R. Moché. Most of the published (and some related unpublished) work by both of them should be mentioned. In fact, D. Dehay has kindly given me a collection of the papers (both published and some not yet out in print) which show the interesting developments of the harmonizable class and its applications. The works of Dehay and Moché will be of interest. Also from our vantage point, generalizing the subject in the directions of I. M. Gelfand and N. Yu. Vilenkin and their followers, indicated and developed by R. J. Swift, M. Mehlman and J. H. Park on the extensions of harmonizability as well as its applications should be continued by the current and future workers on the subject. Indeed it has a good potential both for the stochastic and general functional analytic methods and several applications of interest in addition.

Bibliography

N. K. Artemiadis,

[1] *History of Mathematicians, From a Mathematics Vantage Point*, (English translation of a Greek text published in 2000 in Athens, and later in English by the AMS, Providence, R.I. in 2004).

N. Aronszajn,

[1] "Theory of Reproducing Kernels", *Trans. Amer. Math. Soc.*, (**68**), (1950), 337–404.

A. V. Balakrishnan,

[1] "On A Characterization of Covariances", *Ann. Math. Statist.*, **30**, (1959), 670–675.

D. R. Brillinger,

[1] "The Spectral Analysis of Stationary Interval Functions", *Proc. Sixth Berkeley Symposium on Math. Statistics and Prob.*, Univ. Calif. Press, Berkeley, CA, (1972), no. 453–513.

C. Berg, J. P. R. Christansen, and P. Ressel,

[1] *Harmonic Analysis on Semi-Groups*, Springer Verlag, New York, (1978).

C. S. K. Bhagavan,

[1] Nonstationary Processes, Spectra, and some Ergodic Theorems, Ph.D. Thesis. *Andhra. Univ. Press*, Walter Indic., (1974).

P. Billingsley,

[1] *Probability and Measure*, (3rd ed.) Wiley, New York, (1995).

R. H. Bing,

[1] "Metrization Problems", in *General Topology and Modern Analysis*, Proceedings of Conf. in Honor of F. B. Jones' Retirement, (L. F. McArley and M. M. Rao, eds.) Academic Press, New York, (1981), 3–16.

W. R. Bloom and H. Heyer,

[1] *Harmonic Analysis of Probability Measures on Hypergroups*, de Gruyter Publisher, Berlin and New York, (1995).

S. Bochner,

[1] "Monotone Functionen, Stieltjesche Integrale und Hermonische Analyse", *Math. Ann.* **108**, (1933), 378–410.
[2] "A theorem on Fourier-Stieltjes integrals", *Bull. Amer. Math. Soc.*, **40**, (1934), 271–276.
[3] *Harmonic Analysis and The Theory of Probability*, Univ. of Calif. Proc., Berkeley and Los Angeles, U.S.A., (1955).
[4] "Stationarity, Boundedness, Almost Periodicity of Random Valued Functions", *Proc. 3rd Berkeley Symp. on Math. Statist. and Prob.* **2**, (1956), 7–27.

J. M. Borwein,

[1] "Brower-Heyting Sequences Converge", *Mathematical Intelligencer*, **28**, (1998), 14–15.

J. Bram,

[1] "Subnormal Operators", *Duke Math. J.*, **22**, (1955), 75–94.

D. K. Chang,

[1] *Bimeasures, Harmonizable Processes, and Filtering*, Ph.D. Thesis, UCR Library, Riverside, CA, (1983).

D. K. Chang and M. M. Rao,

[1] "Bimeasures and Sampling Theorems for Weakly Harmonizable Processes" *Stochastic Anal. and Applic.*, (1983), 21–55.
[2] "Bimeasures and Nonstationary Processes", in *Real and Stochastic Analysis*, Wiley, New York, (1986), 7–115.

R. V. Checón and N. Friedman,

[1] "Additive Functionals", *Arch. Rational Math. Anal.*, **18**, (1965), 230–240.

G. Y.-H. Chi,

[1] "Nonlinear Prediction and Multiplicity Theory of Generalized Random Processes", Ph.D. thesis at *Carnegie Institute of Technology* (now Carnegie-Mellon University), Pittsburg, Pa., (1969).

J. A. Clarkson and C. R. Adams,

[1] "On Definitions of Bounded Variation of Two Variables", *Trans. Am. Math. Soc.* **35**, (1933), 824–854.

H. Cramér,

[1] "A Contribution to The Theory of Stochastic Processes", *Proc. Second. Berkeley Symp. Math. Statist. and Prob.*, (1951), 324–339.

[2] "On The Structure of Purely Nondeterministic Stochastic Processes", *Arkiv. f. Mat.*, **4**, (1951), 249–266.

[3] "Structural and Statistical Properties for A Class of Stochastic Processes", *Princeton Univ. Press.*, Princeton, New Jersey, (1971).

M. M. Crum,

[1] "On Positive Definite Functions", *Proc. London Math. Soc.*, (3) **6**, (1956), 548–560.

A. Davinatz,

[1] "Integral Representations of Positive Definite Functions-II", *Trans. Amer. Math. Soc.*, **77**, (1954), 455–480.

P. W. Day,

[1] "Decreasing Rearrangements and Doubly Stochastic Operators", *Trans. Amer. Math. Soc.*, **178**, (1973), 383–392.

M. H. DeGroot and M. M. Rao,

[1] "Multidimensional Information Inequalities and Prediction", *Proc. Internat. Symp. on Multi-variate Analysis*, Dayton, Ohio, (1965), pp. 287–313.

D. Dehay,

[1] "On The Product of Two Harmonizable Time Series", *Stochastic Processes and Appl*, (1991), **37**, 347–358.

A. Denjoy,

[1] "L'Hypothése de Riemann sur le Distribution des Transue di $\zeta(s)$, Meliée À La Theorie de Probabilities", *C. R. Acad. Sci. (Paris)*, **193**, (1931), 656–658.

J. Derbyshire,

[1] *Prime Obsession*, J. Henry Press, Washington D.C., (2003), 906–915.

N. Dinculeanu,

[1] *Vector Measures*, Pergamon Press, London, (1966).
[2] *Integration on Locally Compact Spaces*, Nordhoff International Publisher, Leyden, The Netherlands (1974).
[3] *Vector Integration and Stochastic Integration in Banach Spaces*, John Wiley and Sons, New York, (2000).

I. Dobrokov,

[1] "Integration in Banach spaces VII–IX", *Čzeck Math. J.* **37–40**, 1987–'89, 487–506 and later.

E. L. Dolph and M. A. Woodbery,

[1] "On The Relation Between Green Functional Covariation of Certain Stochastic Processes and its Application to Unbiased Linear Prediction", *Trans. AMS.*, **72**, (1953), 519–550.

M. Duchoň and I. Kluvánek,

[1] "Injective Tensor Product of Operator Valued Measures", *Mat. Casopes*, **17**, (1967), 108–112.

N. Dunford and J. T. Schwartz,

[1] *Linear Operator*, Parts I, II and III, Wiley-Interscience, New York, 1958, 1963, 1971.

H. Dym,

[1] "Stationary Measures for The Flow of a Linear Differential Equation Driven by White Noise", *Trans. Amer. Math. Soc.*, **123**, (1966), 130–164.

H. M. Edwards,

[1] *Riemann Zeta Function*, Academic Press (1974) A New York Wowr. Publication (2006), New York, (1974).

B. Engelhardt,

[1] *Representation and Statistical Inference of Random Sequences on Convolution Structures*, Ph.D. Thesis, Techinsche Universität München, Munich, Germany, (2002).

A. Friedman,

[1] *Generalized Functions and Partial Differential Equations*, Prentice-Hall, Englewood Cliffs, New Jersey, (1963).

N. Friedman and M. Katz,

[1] "A Representation Theorem for Additive Functionals", *Arch. Rational Math. Anal.*, **21**, (1965), 49–57.

[2] "On Additive Functionals", *Proc. Amer. Math. Soc.*, **21**, (1969) 557–561.

R. Gangolli,

[1] "Positive Definite Kernels on Homogenous Spaces and Certain Stochastic Processes Related to Lévy's Brownian Motion of Several Parameters", *Ann. Inst. Henri Poincaré Series B*, **3**, (1967), 121–225.

B. R. Gelbaum,

[1] "von Neuman's Theorem on Abelian Families of Operators", *Proc. Amer. Math. Soc.*, **15**, (1964), 391–392.

I. M. Gel'fand and N. Ya. Vilenkin,

[1] *Generalized Functions* Vol. 4, (Applications of Harmonic Analysis), (Translation from Russian, by A. Feinstein), Academic Press, New York, (1964).

R. K. Getoor,

[1] "The Shift Operator for Nonstationary Stochastic Processes", *Duke Math. J.*, **23**, (1956), 175–183.

[2] "On Semi-Groups of Unbounded Normal Operators," *Proc. Amer. Math. Soc.*, **6**, (1957), 387–391.

I. I. Gikhman and A. V. Skorokhod,

[1] *Introduction to the Theory of Random Processes (English Translation)*, W. B. Sanders Co., Philadelphia, (1969).

[2] *The Theory of Stochastic Processes-I*, Springer-Verlag, New York, (1974).

E. G. Gladyšev,

[1] "Periodically Correlated Random Sequences", *Soviet Math.*, **2**, (1962), 385–388.

J. P. Gram,

[1] "Note sur des Zeros de la Fonction $\zeta(s)$ de Riemann" *Acta Math.* **27**, (1903), 289–304.

U. Grenander,

[1] *Abstract Inference*, J-Wiley and Sons, New York, (1981).

H. Helson,

[1] "Isomorphism of Abelian Group Algebras", *Ark. Mat.* **2**, (1953), 475–487.

E. Hewitt and K. A. Ross,

[1] *Abstract Harmonic Analysis I*, Springer-verlag, New York, (1963).

H. Heyer,

[1] *Structural Aspects in the Theory of Probability* (Second Enlarged Edition), World Scientific Publishing Co., Singapore, (2010).

[2] *Random Fields and Hypergroups, (Real and Stochastic Analysis)*, (Ed. by M. M. Rao) World Scientific Publishing Co., Singapore, (2014).

H. Heyer and M. M. Rao,

[1] "Infinite Dimensional Stationary Random Processes over a Locally Compact Abelian Group", *International J. of Math.*, **23**, (2012), (23 pages).

T. Hida and M. Hitsuda,

[1] *Gaussian Processes* (English translation from Japanese), *Amer. Math. Soc.*, Providence, RI., (1976).

T. Hida and N. Ikada,

[1] "Analysis on Hilbert Space with Reproducing Kernel Arising from Multiple Wiener Integral", *Proc. Fifth Berkeley symp. on Prob. and Math. Statistics*, **2**, Part 1, (1967), (1967), 117–143.

Y. Hosoya,

[1] "Harmonizable Stable Processes", *Z. Watrsh* **60**, (1982), 517–533.

H. L. Hurd,

[1] "Representation of Harmonizable Periodically Correlated Processes and Their Covariances", *Technical Representation Centre for Stochastic Processes*, Univ. of North Carolina, Chapel-Hall, (1987).

[2] "Correlation Theory of Almost Periodically Correlated Processes", *J. Multivariate Anal.*, **37**, (1991), 24–45.

[3] "Almost Periodically and Unitary Stochastic Processes", *Stoc. Processes and Appl.*, **43**, (1992), 99–113.

C. Ionescu Tulcea,

[1] "Measures Dans les Espaces Products", *Att. Acad. Naz. Linces Rendt. et. Sci. Fis. Mat. Nat.*, **7(8)** (1989/90) 208–211.

R. Joyeux,

[1] "Slowly Changing Processes and Harmonizability", *J. Time Series. Anal.*, **8**, (1987), 425–431.

M. Kaç,

[1] *Statistical Independence in Probability, Analysis and Number Theory*, Wiley, New York, (1959).

T. T. Kadota and L. A. Shapp,

[1] "Conditions for Absolute Continuity Between A Certain Pair of Probability Measures", *Z. Wahrs.* **16**, (1970), 250–260.

G. Kallianpur and V. Mandrekar,

[1] "Multiplicity and Representation Theory of Purely Nondeterministic Stochastic Processes", *Theo. Veroyat. iee. Prememin* **10**, 514–544.

Y. Kakihara,

[1] *Multi-Dimensional Second Order Stochastic Processes*, World Scientific, Singapore, (1997).

J. F. Kelsh,

[1] *"Linear Analysis of Harmonizable Time Series"* Ph.D. thesis, UCR Library, Riv., CA, (1978).

J. Kampé de Feriet, and F. N. Frankiel,

[1] "Contribution for Truncated Samples of A Random Function". *Proc. International Congress of Math.*, Amsterdam, *Noordhoff*, (1954), 291–292.

[2] *"Correlation and Spectra of Nonstationary Random Functions"*, *Math. Comput.* **16**, (1962), 1–21.

K. Karhunen,

[1] "Uber Linear Methoden in der Wahrcheinlichkeitsrechnung", *Ann. Acad. Sci. Kenn. SesA Math-Phys,* **37**, (1947), 1–79.

T. Kawata,

[1] "Fourier Analysis of Nonstationary Stochastic Process", *Trans. Amer. Math. Soc.,* **118**, (1997), 276–302.

A. Ya. Khintchine,

[1] "Korrelations Theorie der Stateonarin Stochastic Processe", *Math. Ann.* **109**, (1934), 604–615.

I. Kluvánek,

[1] "Characterization of Fourier-Stieltjes Transformations of Vector and Operator Valued Measures", *Čzech. Math. J,* **17** (92), (1967), 361–377.

[2] "Remarks on Bimeasures", *Proc. Amer. Math. Soc.,* **8**, (1981), 233–239.

J. Kuelbs,

[1] "A Representation Theorem for Symmetric Stable Processes and Stable Measures", *Z. Wahrs.*, **26**, (1973), 259–271.

R. Lasser,

[1] "Orthogonal Polynomials and Hypergroups", *Rend. Math. Ser.* **73**, (1983), 185–208.
[2] "Convolution Semi-Groups on Hypergroups" *Pacific. J. Math.*, **127**, (1983), 358–371.

R. Lasser and M. Leitner,

[1] "Stochastic Processes Indexed by Hypergroups", *J. Theor. Prob.*, **2**, (1989), 301–311.

A. Laurinčikas,

[1] *Limit Theorems for Riemann's Zeta Function*, Kluwer Acaademic Pub. Dordrecht, The Netherlands, (1996).

N. Levinson,

[1] "Almost All Roots of $\zeta(s) = 0$ Are Arbitrarily Close to $\sigma = \frac{1}{2}$", *Proc. Nat. Acad. Sci of U.S.A.*, **72**, (1975), 1322–1324.

P. Lévy,

[1] *Processes Stochastique et Movement Brownian*, Gantherer Villars, Paris, (1948).

J. Lindenstrauss and A. Pelegzylski,

[1] "Absolutely Summing Operators in $\mathcal{L}_p$-spaces and Their Applications", *Studio. Math.*, **29**, (1968), 275–326.

R. S. Lipster and A. N. Shiryayer,

[1] *Statistics of Random Processes-I, II*, Springer, New York, (1977).

J. E. Littlewood,

[1] 'Euclques Conse'quences L'hypothése Que La Fonction $\zeta(s)$, n' A Pas de Zeros dan Le Denom Plan $Re(s) > \frac{1}{2}$', *C. R. Acad. Sci. (Paris)*, **154**, (1912), 263–266.

M. Loéve,

[1] *Probability Theory*, D. Van Nostraud Co., Princeton, N.J. (4th Edition), (1955, 61, 63, 74).

[2] "Fonctions Aléotoines du Second Order", Supplement to Lévy's, *Processes Stochastiques et Movement Brownian*, Paris, (1981).

L. H. Loomis,

[1] *An Introduction to Abstract Harmonic Analysis*, P. Van Nostrand Co., New York, (1953).

D. R. Lewis,

[1] "Integration with Respect to Vector Measures", *Pacific J. Math.*, **33**, (1970), 157–165.

V. Mandrekar,

[1] "A Characterization of Oscillatory Processes and Their Predictions", *Pre. Amer. Math. Soc.*, **32**, (1972), 280–284.

H. von Mangoldt,

[1] "Beweis der Gleichung $\sum_{k=1}^{\infty} U(t)/k = 0$", *S.-B. Kgl. Press. Acad. Wies., Berlin*, (1897).

M. B. Marcus,

[1] "ζ-radial Processes and Random Fourier Series", *Mem. Amer. Math. Soc.*, **368**, (1987), 181 pp.

P. Masani,

[1] "Orthogonally Scattered Measures", *Adv. in Math.* **2**, (1968), 61–117.

F. I. Mautner,

[1] "Unitary Representations of Locally Compact Groups, II", *Ann. Math.*, **52**, (1950), 528–556.

[2] "Note on Fourier Inversion Formula for Groups", *Trans. Amer. Math. Soc.*, **78**, (1985), 371–384.

M. H. Mehlman,

[1] "Structure and Moving Average Representation for Multidimensional Strongly Harmonizable Processes", *Stochastic Anal. and Appl.*, **9**, (1991), 322–361.

[2] "Prediction and Fundamental Moving Averages for Discrete Multidimensional Harmonizable Processes", *J. Multivar. Anal.*, **43**, (1992), 147–170.

[3] "Moving Average Representation and Prediction for Multidimensional Harmonizable Processes", *Dekker series of Lecture Notes in Pure and Applied Math.*, **238**, (2004), 265–276.

A. G. Miamee and H. Salehi,

[1] "Harmonizability, V-Boundedness and Stationary Dilations of Stochastic Processes", *Indiana Math. J.*, **27**, (1978), 37–50.

V. J. Mizel,

[1] "Characterization of Nonlinear Transformation Processing Kernels", *Cand. J. Math.*, **22**, (1970), 449–471.

V. J. Mizel and M. M. Rao,

[1] "Quadratic Equations in Hilbertian Operators and Applications", *Int. J. Math.*, **20**, (2009), 1431–1454.

M. Morse and W. Transue,

[1] "$\mathbb{C}$-Bimeasures and Their Integral Extensions", *Ann. Math.*, **64**, (1956), 480–504.

C. Müller,

[1] *Spherical Harmonics*, Springer Lect. Notes in Math. Vol. **17**, New York, (1966).

F. J. Murray,

[1] "Quasi-Complements and Closed Projections in Reflexive Banach Spaces", *Trans. Amer. Math. Soc.*, **58**, (1945), 35–62.

K. Nagabhushanam,

[1] "The Primary Process of a Smoothing Relation", *Arkin für Math.* **1**, (1951), 421–488.

B. Sz.-Nagy,

[1] "Transformations de L'space de Hilbert, Fonctions de Type Positif sur un Groupe", *Acta. Sci. Math. Seged* **15**, (1954), 104–114.

[2] "Extensions of Linear Transformations in Hilbert Space which Extend Beyond This Space", Appendix to the book *Functional Analysis*, (by F. Riesz and B. Sz.-Nagy) Unger, New York, (1955).

M. A. Naĭmark,

[1] *Normed Rings*, Nordhoff, Groningen (2nd ed.), (1964).

F. I. Nelson,

[1] "A Class of Orthogonal Series Related to Martingales", *Ann. Math. Statist.*, **41**, (1970), 1684–1694.

J. Neveu,

[1] *Processus Alatoires Gaussiens*, Presses de l'Universite de Montreal, (1968).

H. Niemi,

[1] "Stochastic Processes as Fourier Transforms of Stochastic Measures", *Ann. Acad. Sci. Fenn. AI Math.*, **581**, (1975), 1–47.

Y. Okazaki,

[1] "Wiener Integral by Stable Random Measure", *Math. Fac. Sci. Kyushu Univ, Ser A, Math.*, **33**, (1979), 1–70.

J. H. Park,

[1] "A Random Measure Algebra under Convolution", *J. Statist. Theory and Practice*, **10**, (2016), 1559–1571.

[2] *Random Measure Algebras under Convolution*, Ph.D. Thesis, UCR Library, Riverside, CA. (2015).

E. Parzen,

[1] "Spectral Analysis of Asymptotically Stationary Time Series", *Bull. Inst. Internat. Statist.*, **39** (Li version), (1960), 87–103.

R. S. Phillips,

[1] "On Fourier-Stieltjes Integrals", *Trans. Amer. Math. Soc.*, **69**, (1950), 312–323.

T. S. Pitcher,

[1] "Likelihood Ratios of Gaussian Processes", *Ark. Mat.*, **4**, (1959), 35–44.

[2] "The Admissible Mean Values of A Stochastic Process", *Trans. Amer. Math. Soc.*, **108**, (1963), 538–546.

M. B. Priestly,

[1] "Evolutionary Spectra and Nonstationaries Processes", *J. Ray. Statist. Soc., Ser B*, (1965), 204–237.

M. M. Rao,

[1] "Theory of Lower Bounds for Risk Functions in Estimations", *Math. Ann.*, **143**, (1961), 379–398.
[2] "Theory of Order Statistics", *Math. Ann.*, **142**, (1962), 398–415.
[3] "Characterization and Extension of Generalized Harmonizable Random Fields", *Proc. National. Acad. Sci*, **58**, (1967), 1213–1219.
[4] "Inference in Stochastic Processes-III", *Zeitz Wahrenshein.*, **8**, (1967), pp. 49–72.
[5] "Representation Theory of Multidimensional Generalized Random Fields", *Multiv. Anal.*, Acad. Press, New York. Vol. 2 (P. R. Krishnaiah, Editor), (1969), 411–436.
[6] "Local Functionals and Generalized Random Fields with Independent Values, *Theory Prob. and its Appl.*' **16**, (1971), 466–483.
[7] "Inference in Stochastic Processes V: Admissible Means", *Sankhyā Ser A*, **37**, (1975), 538–549.
[8] "Bistochastic Operators", *Comment. Math.*, **21**, (1978), 301–313.
[9] "Local Functionals on $C_\infty(G)$ and Probability", *Functional Anal.* **39**, (1980), 23–41.
[10] "Harmonizable Processes: Structure Theory", *L'Enseignement mathematique*, **28**, (1982), 295–351.
[11] "Domination Problem for Vector Measures and Applications to Nonstationary Processes", *Springer Lect. Notes in Math.*, **945**, (1982), 296–313.
[12] "The Spectral Domain of Multivariate Harmonizable Processes", *Proc. Nat. Acad. Sci. U.S.A.*, **81**, (1984), 4611–4612.

[13] "Harmonizable, Cramér, and Karhunen Classes of Processes", *Handbook of Statistics*, Vol. 5, North-Holland, Amsterdam, (1985), pp. 279–310.

[14] "Bimeasures and Harmonizable Processes", *Springer Lect. Notes in Math.*, **1379**, (1989), 251–298.

[15] "Harmonizable Signal Extraction, Filtering and Sampling", in *Topics in Non-Gaussian Signal Processing*, Edited by E. J. Wegman, S. C. Schwartz and J. B. Thomas, Springer-Verlag, New York, Berlin, (1989), 98–117.

[16] "Sampling and Prediction for Harmonizable Isotropic Random Fields", *J. Combinatorics, Information and System sci.*, **16**, (1991), 207–220.

[17] "Stochastic Integration: A Unified Approach", *C. R. Acad. Sci. Paris* t.**314**, série 1, (1992), 629–633.

[18] "Harmonizable Processes and Inference: Unbiased Prediction for Stochastic Flows", *J. Stochast. Planning and Inference*, **30**, (1994), 187–209.

[19] "Characterization of Isotropic Harmonizable Covariances and Related Representatives", (privately circulated), (1995).

[20] "Characterizing Covariances and Means of Harmonizable Processes", in *Trends in Contemporary Infinite Dimensional Analysis and Quantum Probability*, Kyoto, (2000), (T. Hide Feedshift Volume) 363–381.

[21] "Representations of Conditional Means", *Georgia Math. J.*, **8**, (2001), 363–376.

[22] *Measure Theory and Integration*, (Wiley-Interscience, New York, 1987. (2nd ed.), Marcel Dekker, Inc, New York.), (2004).

[23] "Exploring Ramifications of The Equation $E(Y|X) = X$", *Statist. Theory and Practice*, **1**, (2007), 73–82.

[24] "Integral Representation of Second Order Process", *Nonlinear Anal.*, **69**, (2008), 979–986.

[25] "Application and Aspects of Random Measures", *Nonlinear Anal.*, **71**, (2009), 1513–1518.

[26] "Harmonic and Probabilistic Approaches to Zeros of Rieman's Zeta Function", *Stochastic Anal. Appl.*, **30**, (2012), 906–915.

[27] *Random and Vector Measures*, (World Scientific, Singapore, London, Hackensack, New Jersey.), (2012).

M. M. Rao and Z. D. Ren,

[1] *Theory of Orlicz Spaces*, Marcel Dekker Inc., New York, (1991).

M. M. Rao and V. Sazonov,

[1] "A Projective Limit Theorem for Probability Spaces and Applications", *Theory Probab. Appl.*, (1993), 307–315.

A. Renyi,

[1] "On The Theory of Order Statistics", *Acta. Math. Hong.* **19**, (1953), (91–231).

F. Riesz and B. Sz.-Nagy,

[1] *Functional Analysis*, F. Unger Publishing Co., New York, (1953).

M. Rosenberg,

[1] "The Square-Integrability of Matrix-Valued Functions with Respect to A Nonnegative Hermitian Measure", *Duke Matrix. J.*, **31**, (1964), 291–298.

H. L. Royden,

[1] *Real Analysis, (3rd ed.)*, Prentice Hall, Upper Saddle River, N.J., (1988).

Yu. A. Rozanov,

[1] "Spectral Analysis of Abstract Functions", *Theor. Prob. Appl.*, **4** (1959), 271–287.

V. Sazonov,

[1] "On the Glivenko-Cantelli Theorem", *Theory Probab. Appl.*, **8**:3 (1963), 282–286.

M. Schilder,

[1] "Some Structure Theorems for Symmetric Stable Laws", *Ann. Math. Statist.*, **41**, (1970), 412–421.

I. J. Schoenberg,

[1] "Metric Spaces and Completely Monotone Functions", *Ann. of Math.*, **39**, (1938), 811–841.

[2] "Metric Spaces and Positive Definite Functions", *Trans. Amer. Math. Sre*, **44**, (1938), 522–536.

B. M. Schriber,

[1] "Asymptotically Stationary and Related Processes", *Stochastic Processes and Functional Analysis, Dekker Lect. Notes in Pure and Appl. Math.*, **238**, (2004), 363–397.

L. Schwartz,

[1] *Théorie des Distributions*, Vols. 1 and 2 (2nd ed.) Hermann and Cie, Paris, (1957).

I. E. Segal,

[1] "An Extension of Plancherel's Formula to Separable Unimodular Groups", *Ann. Math.*, **52**, (1950), 272–292.

A. V. Skorokhod,

[1] "On Admissible Translates of Measures in Hilbert Space", *Theor. Veroyatnost, Premium*, **15**, (1970), 557–580.

R. Spector,

[1] "Measures Invariantes sur Les Hypergroups", *Trans. Amer. Math. Soc.*, **239**, (1978), 147–165.

N. Starr,

[1] "Operator Limit Theorems", *Trans. Amer. Math. Soc.*, **121**, (1966), 90–115.

R. J. Swift,

[1] "The Structure of Harmonizable Isotropic Random Fields", *Stoch. Anal. Appl.*, **12**, (1994), 583–616.

[2] "An Operator Characterization of Oscillatory Harmonizable Process", Marcel Dekker, *Lect. Notes in Pure and Applied Math.*, Vol. **186**, (1997), 235–244.

[3] "Some Aspects of Harmonizable Processes and Fields", in *Real and Stochastic Analysis (Recent Advances)* CRC Press, New York, (1997).

[4] "Isotropic Random Measures", *Amer. J. Math. and Management Sci.*, **21**, (2001), 283–293.

[5] "Covariance Analysis and Associated Spectra for Classes of Nonstationary Processes", *J. Statist. Planning and Inference*, **100**, (2002), 145–157.

[6] "A Spectral Representation of A Class of Nonstationary Processes", *J. Statist. Theory and Practice*, **5**, (2011), 515–523.

N. Tatsuma,

[1] "A Duality Theorem for Locally Compact Groups", *J. Math. Kyoto Univ.*, **6**, (1967), 187–293.

E. Thomas,

[1] "L'Integration par Rapport a une Mesure de Radon Vectorille", *Ann. Inst. Fourier Grenoble*, **20**, (1970), 35–191.

M. L. Thornett,

[1] "A Class of Second Order Stationary Random Measures", *Stochastic Processes Appl.*, **8**, (1979), 323–334.

H. F. Trotter,

[1] "Approximation of Semi-Groups of Operators", *Pacific J. Math.*, **8**, (1958), 887–919.

R. C. Verm,

[1] "Harmonic Analysis on Compact Hypergroups", *Pacific J. Math.*, **85**, (1979), 239–251.

[2] "Hypergroup Joins and Their Dual Objects", *Pacific J. Math.*, **111**, (1984), 483–495.

N. Ya. Vilenkin,

[1] *Spectral Functions and the Theory of Group Representations, (English Translation) Amer. Math. Soc. Providence*, RI, (1968).

J. Wermer,

[1] "Commuting Normal Operators on Hilbert Space", *Pacific J. Math.*, **4**, (1954), 355–361.

N. Wiener,

[1] 'Differential Space,' *J. Math. and Physics*, **2**, (1923), 132–174.

M. I. Yadrenko,

[1] *Spectral Theory of Random Fields (English Translation)* Optimization, Software, Inc. New York, (1983).

A. M. Yaglom,

[1] "Second Order Homogeneous Random Fields", *Proc. 4th Berkely Symp. Math. Statist. and Prob.*, Vol. 2, (1961), 593–622.

[2] *Introduction to the Theory of Stationary Random Process*, (English Translation) Prentice-Hall, Englewood Chel. N.J., (1962).

[3] *Correlation Theory of Stationary and Related Random Functions, I, II*, Springer, New York, (1987).

K. Ylinen,

[1] "On Vector Bimeasures", *Ann. Math. Pure. and Appl.*, **117**, (1978), 115–138.

[2] "Random Fields and Noncommutative Locally Compact Groups", *Prob. Measures on Groups-VIII, Lect. Notes in Math.*, No. **1210**, Springer Verlag, Berlin, (1986), pp. 365–386.

N. D. Ylvisaker,

[1] "A Generalization of A Theorem of Balakrishnan", *Ann. Math. Statist.*, **32** (1961), 1337–1339.

A. Zygmund,

[1] *Trigonometrical Series, Vol. I.*, Cambridge Univ. Press, Cambridge, UK, (1959).

Notation Index

Author Index

Subject Index

Printed in the United States
by Baker & Taylor Publisher Services